ADVANCES IN
GROUP PROCESSES

Volume 10 • 1993

ADVANCES IN GROUP PROCESSES

A Research Annual

Editors: EDWARD J. LAWLER
BARRY MARKOVSKY
KAREN HEIMER
JODI O'BRIEN

VOLUME 10 • 1993

 JAI PRESS INC.

Greenwich, Connecticut   *London, England*

Copyright © 1993 JAI PRESS INC.
55 Old Post Road, No. 2
Greenwich, Connecticut 06836

JAI PRESS LTD.
The Courtyard
28 High Street
Hampton Hill
Middlesex TW12 1PD
England

ISBN: 1-55938-280-5

Manufactured in the United States of America

CONTENTS

v

LIST OF CONTRIBUTORS

Julianne F. Brand

Department of Business Administration
DePaul University

Stephen Brand

Department of Psylchology
University of Illinois at Urbana-
 Champaign

Carol A. Burton

Institute of Labor and Industrial
 Relations
University of Illinois at Urbana-
 Champaign

Raymond A. Eve

Department of Sociology and
 Anthropology
University of Texas at Arlington

Gerald R. Ferris

Institute of Labor and Industrial
 Relations
University of Illinois at Urbana-
 Champaign

Noah E. Friedkin

Graduate School of Education
University of California, Santa Barbara

David C. Gilmore

Department of Psychology
University of North Carolina at Charlotte

Francis B. Harrold

Department of Sociology and
 Anthropology
University of Texas at Arlington

Karen A. Hegtvedt

Department of Sociology
Emory University

David R. Heise

Department of Sociology
Indiana University

Michael W. Macy   Department of Sociology
                                    Brandeis University

Per Månson   Department of Sociology
                                    University of Gothenburg, Sweden

Kendrith M. Rowland   Department of Business Administration
                                    University of Illinois at Urbana-
                                        Champaign

Steven J. Scher   Department of Psychology
                                    University College of the Cariboo,
                                        Canada

Richard T. Serpe   Department of Sociology
                                    California State University, San Marcos

Robert K. Shelly   Department of Sociology and
                                        Anthropology
                                    Ohio University

Sheldon Stryker   Department of Sociology
                                    Indiana University

Jacek Szmatka   Department of Sociology
                                    Uniwersytet Jagiellonksi, Poland

David Willer   Department of Sociology
                                    University of South Carolina

PREFACE

EDITORIAL POLICY

The purpose of this series is to publish theoretical analyses, reviews, and theory-based empirical papers on group phenomena. The series adopts a broad conception of "Group processes." In addition to topics such as status processes, group structure, and decision making, the series considers work on interpersonal behavior in dyads (i.e., the smallest group), individual-group relations, intergroup relations, and social networks. Contributors to the series include not only sociologists but also scholars from other disciplines such as psychology and organizational behavior.

The series is an outlet for papers that are longer, more theoretical, and more integrative than those published by the standard journals. The series places a premium on the development of testable theories and on theory-driven research programs. The editors are particularly receptive to work falling into the following categories:

1.   Conventional and unconventional theoretical work, from broad meta-theoretical and conceptual analyses to refinements of existing theories and hypotheses. One goal of this series is to advance the field of group processes by promoting theoretical work.

2.   Papers that review and integrate programs of research. The current structure of the field often leads to the piecemeal publication of different parts

of a program of research. This series offers those engaged in programmatic research on a given topic an opportunity to integrate their published and unpublished work into a single paper. Review articles that transcend the author's own work are also of considerable interest.

3. Papers that develop and apply social psychological theories and research to macrosociological processes. One premise underlying this series is that relationships between macro- and microsociological processes warrant more systematic and testable theorizing. The series encourages development of the macrosociological implications embedded in social psychological work on groups.

CONTENTS OF VOLUME 10

This volume contains ten chapters representing diverse interests and approaches in the study of group processes. Two chapters address issues of distributive justice and injustice. Hegtvedt examines the historical reasons for the development of distinct research traditions in the areas of distributive and procedural justice. She argues for an approach that brings together knowledge on procedural and distributive justice processes within the same domain of inquiry. Scher and Heise use Affect Control Theory to develop an original approach to the perception of distributive injustice. They assert that justice-relevant emotions arising in the course of interaction may trigger justice evaluations, a sequence of events opposite that which is generally assumed in other justice theories.

Three chapters employ ideas from social network theories in considering issues of power and collective action. Friedkin offers a method for predicting the location of power in social exchange networks. His analytic model holds promise for making predictions that are considerably more precise than those of others in its genre. Willer and Szmatka review their research on exchange networks conducted in cross-national experimental settings and find clear support for the generality of their theory. The authors also discuss issues concerning the logic of cross-cultural research methodology. Macy employs assumptions from social learning theory in his computer simulation models of collective action. Among other findings, a "weak tie" effect of social network linkages is shown to affect whether actors will successfully mobilize collective action. This paper was the 1993 winner of the Theory Prize awarded by the Theory Section of the American Sociological Association.

The contributions by Serpe and Stryker and by Eve and Harrold address certain effects of social interaction and identification in natural contexts. The former draw on identity theory to derive hypotheses about the effect of commitments to prior social groups on movement into relationships with new groups. They assess their expectations using data from college freshman and structural equation modeling. Eve and Harrold contend that pseudoscientific beliefs are the product of normal group processes. Through a series of case studies including Satanism and Creationism, the authors demonstrate that pseudoscientific beliefs and errors in reason result from and are reinforced by the groups with which one interacts.

Shelly contributes to the expectation states literature by presenting evidence that sentiments organize interactions, although possibly with lesser impact that those of status structures. Also interested in small group phenomena, Månson develops a multilevel analytical framework in discussing basic (ontological and methodological) problems in the study of small groups, and suggests some strategies to address them. Finally, the paper by Ferris and his collaborators proposes a model relating perceptions of organizational politics to a variety of work outcomes such a stress and satisfaction. Their research suggests that such outcomes are mediated by perceived control, which can cause political processes to be perceived as either threatening or as providing opportunities.

Edward J. Lawler
Barry Markovsky
Karen Heimer
Jodi O'Brien

Volume Co-Editors

SOCIAL LEARNING AND THE STRUCTURE OF COLLECTIVE ACTION

Michael W. Macy

ABSTRACT

Rational choice theory suggests that participation in collective action may be prompted by the marginal per capita return (MPCR) on individual investments in public goods. Members of many interest groups, however, lack the disposition, information, and cognitive skills to calculate MPCR. Social learning theory relaxes this constraint by replacing causality with propinquity as the link between action and outcome. Hence, the enthusiasm for participation may reflect the experience of *collective* rather than *personal* efficacy. Computer simulations based on a stochastic learning model reveal the systemic logic of interactive reinforcement in social dilemmas. Reward-seeking, penalty-aversive actors lead one another away from mutually beneficial outcomes and into social traps where everyone suffers. The simulations also identify the structural conditions for escape. Foremost are threshold effects in which each new participant triggers others. Successful mobilization then depends on the network of social ties that channel the chain reactions by linking socially distant actors, supporting Granovetter's case for the strength of weak ties.

Advances in Group Processes, Volume 10, pages 1-35.
ISBN: 1-55938-280-5

*[W]hen we attempt to explain behavior in more complicated social
contexts...then we shall add to the descriptive power of our models by
recognizing that man has not only a future, but also a past.*

—Dennis Mueller,
1986 Presidential Address
to the Public Choice Society

INTRODUCTION: THE ENIGMA OF COOPERATION

Cooperation allows for the more efficient allocation of resources, the spread
of innovations and skills, insurance against temporary setbacks, and economies
of scale. Isolated self-reliant individuals living in a hypothetical "state of nature"
would incur widespread redundancies of effort and invention. They would also
suffer the law of averages, which can be unmerciful when $N = 1$. Worse yet,
a Hobbesian world risks degeneration in a downward spiral of recrimination
and mutually exhausting conflict. That human cultures have evolved with a
predilection for group solidarity makes obvious sense.

Why, then, the continued fascination with the problem of collective action?
Since it is unthinkable that the social fabric would unravel into a state of nature,
why do scholars across the social sciences expend such effort to explain a
condition that everyone else takes for granted as unavoidable? Why are the
advantages (or perhaps one should say, the necessities) of combination not
a sufficient explanation?

A moment's reflection shows why. We need look no further than the
proliferation of nuclear arms, the depletion of the ozone layer, or the
greenhouse effect to see that *the necessity for collective rationality is no
guarantee that it will obtain.* That is perhaps the great paradox of the human
condition. We cannot live without one another, but we may not be able to
live with one another, either. Given what is at stake, the effort to unlock this
paradox becomes as compelling as any that science finds on its agenda.

The paradox centers on the social trap that arises when decisions that make
sense for each aggregate to produce outcomes that make no sense for all. The
trap has two jaws, the free-rider problem and the problem of efficacy. The
free-rider problem stems from the inability to restrict the advantages of
combination to the population of contributors, hence the temptation to let
others shoulder the burden. The efficacy problem is created by the pooling
of contributions such that "only a fraction of the benefits of one person's action
accrues to that person" (Coleman 1986, p. 59). The marginal impact of
individual contributions to a common cause may appear insignificant
compared to the effect of efforts we invest in private pursuits. The logic of
collective action thus leads each member of an interest group to the same
mutually ruinous conclusion: I may get little or no additional benefit from my

own efforts should I choose to contribute, yet I will enjoy the benefits of others' efforts even if I fail to contribute.

Olson illustrates the two problems. "The rational individual in the economic system does not curtail his spending to prevent inflation...because he knows, first, that his own efforts would not have a noticeable effect, and second, that he would get the benefits of any price stability that others achieved in any case. For the same two reasons, the rational individual in a sociopolitical context will not be willing to make any sacrifices to achieve the objectives he shares with others" (1965, p. 166). Hence, commuters are generally unwilling to pay the cost of converting their cars to burn propane or to suffer the inconvenience of public transportation, and those who do are not acting rationally, even as all choke on congested freeways.

Is Altruism the Answer?

How, then, can cooperation thrive despite the comparative advantage of shirking? The simplest and most obvious explanation is that cooperation provides a shared benefit, while exploitation favors only the feckless individual. Hence, free-riding will occur only to the extent that people place their own welfare above that of their group. As Fireman and Gamson have noted, "calculations of what may be gained or lost through collective action are very important to actors' decisions to join in." However, this instrumental logic, they continue, need not imply egoistic preferences. On the contrary, "we think that actors assess what their *group* may gain or lose as well as what they may gain or lose as individuals" (1979, p. 15). The rational altruist thus resists the temptation to free-ride so long as the enhancement of the general welfare exceeds the cost to the contributor.

This argument addresses only one side of the social trap. Since the motive remains instrumental, cooperation may be stymied by the efficacy problem even if the temptation to free-ride is tempered by a utilitarian interest in the greater good. Normative altruism, on the other hand, avoids not only the free-rider problem but the efficacy problem as well: cooperation is prompted by internalized rules and obligations, not by an instrumental interest in the collective welfare. For example, Elster's (1989) "everyday Kantian" responds to the hypothetical musing, "what if everyone acted that way." Similarly, norms of fairness impose an obligation to others that is independent of the ability to make a real difference in group welfare. The obligation depends instead on the level of contribution by others. Hence, group mobilization requires the right mix of Kantian norms that instigate the process and fairness norms that cause it rapidly to escalate.

The problem with the normative argument, Elster argues, is the need to explain how the norms arise. The impracticality of monitoring universal compliance shows the *need* for internalized norms, but does not explain the reproduction of the mechanisms by which unthinking conformity is secured. The argument introduces a troublesome circularity: internalized norms are needed for social order and cohesion, and a functioning society is needed to internalize norms. Any weakening of civic virtue is thus likely to precipitate a downward spiral, the relentless "rust of civilization" that de Tocqueville (1969) forewarned.

Reciprocity and Selective Incentives

The willingness to set aside considerations of personal welfare, out of obligation to or concern for the welfare of the group, thus strikes some scholars as a rather thin reed on which to hang the fate of civilization. Rational choice theory aims instead to harness collective action not to civic virtue or altruistic concerns, but to shared interests in the benefits of cooperation. The project resonates with the moral philosophy of Adam Smith: *can social order be secured without having to effect a moral transformation that demands the suspension of self-interest?*

One of the promising breakthroughs in this line of research is the game-theoretic demonstration that reciprocity (or "tit for tat") is a highly robust strategy, one that can thrive even in a world dominated by aggressive predators (Axelrod 1984). The strategy never seeks to exploit others, but also never turns the other cheek. Axelrod has used computer simulations of an iterated Prisoner's Dilemma game to show how reciprocity might evolve, even though it can never beat more aggressive strategies in head-to-head encounters, and even though it does not do as well as aggressors when interacting with unconditionally cooperative strategies. Yet, reciprocity is able to gain a foothold in an asocial world, to thrive while interacting with a wide range of other strategies, and to resist invasion by mutant predators once fully established.

Reciprocity compensates for its refusal to exploit by securing cooperation with like-minded players who know better than to cooperate with overly aggressive contestants. The more chances it has to interact with similar strategies, the better it performs. More aggressive strategies may do well initially by victimizing overly cooperative strategies, but in doing so, they dig their own graves (Axelrod 1984, p. 52).

Unfortunately, reciprocity turns out to have very restricted applicability. The strategy "unravels rather quickly as the number of players in the game increases. With large numbers, one player's defection has an imperceptible impact on the outcome of the game, and should not induce defections by other players. Thus, all rational players should defect" (Mueller 1986, p. 4). Moreover, as the number of contestants increases, free-riders can more easily become "lost in a sea of anonymous others" where they "anticipate getting away without retaliation" (Coleman 1986, pp. 66-77; see also Axelrod 1984, p. 49).

Following Olson, most rational choice theorists have concluded that collective action in all but very small groups requires "selective incentives" that detour the free-rider problem by providing supplemental benefits that are conditional upon individual contributions. Public goods, in other words, emerge only as byproducts of conventional market behavior—investments in private, excludable goods. Analysts are thus led to search for the hidden agenda that motivates rational actors to contribute to a common cause. For example, unions may obtain higher wages for free-riders, but they provide pension plans, health benefits, and grievance procedures only to dues-paying members (Olson 1965). In short, the collective benefit from public goods (e.g., strength in numbers, a cleaner environment, or a more peaceful neighborhood) is merely a byproduct of the pursuit of an ulterior objective, such as a union pension plan, social approbation, or the avoidance of arrest.

A well-known difficulty with this line of argument is the "second-order free-rider problem" (Oliver 1980; Heckathorn 1989). The selective allocation of benefits and penalties presumes the ability to monitor and sanction individual behavior (Hechter 1987). But, systems of social control are not costless. The problem is that rational actors know they will benefit from enforcement of collective obligations, even if they let others bear the cost of social control. How, then, can selective incentives explain contributions to collective action if they require a system of social control that is itself a public good? Olson thus seems to have it backwards: Collective action does not depend on the existence of selective incentives; rather, it is the other way around[1]

There is now substantial experimental evidence that human subjects do not free-ride nearly as much as the rational egoist model predicts (Marwell and Ames 1979, 1980). On the contrary, van de Kragt, Dawes, and Orbell (1983) find that subjects continue to contribute to public goods even after additional efforts are not needed. Yet, altruism does not appear to be the explanation; in a later study they show that increasing the benefit to others while holding constant the individual consequences does not increase the propensity to contribute to public goods (Van de Kragt, Dawes, and Orbell 1988). In short, members of interest groups seem willing to sacrifice for the common good, even in the absence of selective incentives (e.g., where contributions are anonymous). The level of effort, however, appears not to respond to increases in the benefit yielded to other persons. Neither selective incentives nor altruism can account for these findings.

Jointness of Supply and the Critical Mass

A plausible explanation is that actors are motivated by self-interest, but their interest centers primarily on the collective benefits, not the side payments. This of course brings us head up with the nonexcludability of the public goods. Marwell and Oliver (1993) propose an intriguing rational choice solution to the free-rider problem. They spring the social trap by solving the efficacy problem, which then makes free-riding innocuous. Public goods are not only nonexcludable, they also tend to have "jointness of supply," that is, their cost does not increase with the size of the group that consumes them. Public goods may be enjoyed by free-riders with little or no reduction in the benefits to those who paid for them. The classic illustration is the lighthouse that costs the same no matter how many benefit, and "one person's consumption of it does not reduce the amount available to anyone else" (Hardin 1982, p. 17).

This generates an important result. If the marginal per capita return (MPCR) exceeds the cost of investment, rational actors will invest in public goods no matter "how many others there are 'out there' in the interest group who might benefit" for free (Oliver and Marwell 1988, p. 6).

Moreover, the authors continue, the production function for public goods is often nonlinear, due to the costs of infrastructure that absorb initial investments. As these start-up costs are paid, the marginal impact of each additional contribution is ever greater. When the MPCR increases with the level of investment in the public goods, each new volunteer paves the way for others to follow. A "critical mass" of contributions can thus trigger a chain reaction that spreads throughout the group.

Unfortunately, this solution to the efficacy problem does not fully negate the free-rider difficulty. The problem is that nonlinear production functions rarely accelerate forever. More commonly, they are S-shaped, such that investments beyond the point of inflection suffer diminishing returns (Elster 1989, p. 32). If the actors expect the final contribution level to reach this "flat" portion of the benefit curve (the point where additional efforts are largely redundant), they may gamble and "let George do it."

In short, cost-effectiveness is a sufficient condition only for rational investment in private (excludable) goods. For goods that are not excludable, there is no bargain like getting something for free. Even when benefits to volunteers outweigh the costs, returns must be discounted by the probability that others will step forward and pay the costs instead.

Moreover, fully rational actors know that by volunteering, they alter everyone else's expectations of their behavior in future rounds, making it less likely that others will contribute next time. Hence, sophisticated players may be "motivated to appear less interested than they really are, and strategic gaming and misleading statements about one's interests may result" (Oliver, Marwell, and Teixeira 1985, p. 548).

Of course, not everyone has the instrumental disposition or cognitive skill of a seasoned strategist, as the authors emphasize, particularly "in the social and political sphere" (Oliver, Marwell, and Teixeira 1985, p. 548). However, if lay actors are unable to appreciate the "gaming strategies" posed by nonexcludable benefits, then it is equally doubtful that they are sufficiently tough-minded to cipher the changing marginal impact of each additional contributor and each additional consumer.

The problem with MPCR is that it cannot be readily observed. Actors cannot just alternate strategies and see what happens, since the results will be confounded by the decisions of others. Nor can volunteers simply compare their payoff with that of free-riders; they must compare it with the hypothetical payoff they would receive were they to make a different choice. Suppose Ego volunteers and Alter free-rides. It will always be the case that Alter will do better than Ego. Yet, Alter might have done better still by contributing, depending on the marginal return at the current level of contribution. The problem is that the MPCR is counterfactual and hence cannot easily be determined through experience. The estimate requires detailed knowledge of the "technology of collective action," knowledge that "is heavily shrouded in uncertainty" (Elster 1989, p. 34).

"As If" Rationality

Even for market actors, these calculations can place "extremely severe strain on information-gathering and computing abilities" (Arrow 1986, p. 213). Decision-makers must not only invest in the acquisition of costly information, they must know when to stop, so that they do not end up like Buridan's ass, who starves while trying to choose between two equidistant haystacks (Cook and Levi 1991, p. 7). The opposite error is to spend too little. Myopic actors, like Elster's (1979) hapless "gradient climbers," tend to get trapped in local maxima, like the "false peaks" that can fool the novice mountaineer.

Laboratory studies of economic decision-making bear out the difficulty. For example, Tversky and Kahneman (1990a) found that subjects were risk-averse about gambling to increase gains but risk-seeking in trying to reduce losses. Their reserach led them to conclude that the deviations of observed behavior from the rational choice model "are too widespread to be ignored, too systematic to be dismissed as random error, and too fundamental to be accommodated" (1990b, p. 60).

Other behavioral experiments have demonstrated that subjects tend to respond to the *average* rate of reinforcement (the "matching" principle), not the marginal rate (Herrnstein 1981, 1988; Baum 1979; for a critique, see Rachlin, Battalio, Kagel, and Green 1981). To illustrate the problem, suppose a hunter must allocate effort between two forests, one of which has an unlimited supply of game. The forest with limited supply has an initially higher yield, but it drops off with overforaging, while the other forest has a constant yield. If choices are based on marginal utility, the hunter will allocate effort so as to equalize the marginal yields, spending most of its time in the unlimited-supply forest, and returning to the other only long enough to earn the high yields that have built up in the interim. However, experimental evidence suggests that the hunter will instead learn to match effort with the average returns, which means it will spend too much time in the forest with limited supply (Herrnstein 1988, p. 21).

What Winter calls the "classic defense" of rational choice theory is that we need not assume that market actors are rationally animated, only that they eventually end up acting as if they were. Hence, most economists show a "willingness to concede that the rationality assumptions of economic theory are not descriptive of the process by which decisions are reached and, further, that most decisions actually emerge from response repertoires developed over a period of time by what may broadly be termed 'adaptive' or learning processes" (Winter 1986, p. 244). As long as the environment is reasonably stable, the invisible hand can usually be counted on to ruthlessly select and replicate optimal routines, as if the unwitting survivors had foreseen their success right from the beginning. As Mueller notes, "in a competitive environment the less profitable firms perish, and the surviving firms adopt decision rules *as if* they had been consciously trying to maximize profits, whatever the criteria actually employed to make decisions" (1986, p. 18).

This allows economists to calculate equilibria on the assumption of maximizing behavior without concern that the calculations do not describe the actual process by which adaptive actors converge with the analytic shortcut. The poor fit between the behavioral assumptions and the findings from experimental economics has thus not prevented rational choice models from generating reasonably accurate predictions about aggregate tendencies in markets for private goods.

Unfortunately, public goods are not so well behaved. Nonexcludability, nonrivalness, and the absence of competitive pressures in most nonmarket transactions prevent systemic mechanisms from weeding out suboptimizers. Hence, the "as if" principle, while useful for the private goods studied by economists, tends to be unreliable for the public goods that intrigue sociologists. Ironically, sociologists working with rational choice models have thus been forced to assume that actors in the social and political sphere really are tough-minded optimizers, which economists, focusing on market behavior, have not needed to do. Yet the ability and temperament to calculate MPCR, if scarce among market actors, seems even more remote for the typical volunteer in a collective action.

Learning to Cooperate

There are important applications where nonlinear cost-benefit calculations may be appropriate, such as party discipline among legislators, the formation of economic cartels and political conspiracies, and cooperation among institutional actors (Hechter 1987; Stinchcombe 1990). Some of the most interesting social dilemmas, however, arise in "everyday life" among actors who may lack the necessary information, cognitive skill, or instrumental disposition. In a series of papers, I have questioned whether lay actors can realistically be expected to appreciate the strategic implications of an iterated Prisoner's Dilemma (Macy 1989, 1991a) or changing marginal returns on investments in public goods (Macy 1990, 1991b). These behavioral constraints can be greatly relaxed by assuming that actors learn through experience, adapting their decision rules in response to social feedback, as broadly suggested by models of "backward-looking" decision-making proposed by Cross (1983) for market behavior, by Scott (1971) for the internalization of social norms, by Rapoport and Chammah (1965) for cooperation in Prisoner's Dilemma, and by Homans (1961) for social exchange.

In his 1986 Presidential Address to the Public Choice Society, Dennis Mueller called for a new approach to the study of collective action, using learning theory in place of conventional rationality assumptions. The key difference is temporal. "To describe an individual's expectations at time t using the rational expectations assumption we look at what happens after t, to describe them under adaptive expectations we look at what happened before t" (Mueller 1986, p. 19). Based on a survey of the evidence from experimental studies, Mueller concluded that the assumption of backward-looking behavior is likely to "add realism and descriptive power to our modeling of human behavior" (1986, p. 14). Learning models avoid the need to assume that actors have the information, cognitive capacity, or disposition needed to calculate in advance the level of participation required to make their contributions cost-effective. More plausibly, each charts his course on the fly, feeling his way along in response to the social feedback associated with alternative choices, guided by the outcomes that those choices collectively generate. Simply put, satisfactory outcomes increase the likelihood that the associated action will be repeated, while unsatisfactory outcomes reduce it.

The associated action need not be causally related to the outcome. "Backward-looking" volunteers do not need to estimate how much of the outcome was due to their individual contribution. All they need to know is whether their efforts were wasted, given what the group was able to accomplish. For example, members of a village may regularly contribute time to a community garden that benefits all who stroll through the commons. If their collaboration produces striking results, volunteers may feel encouraged by their accomplishments and become even more committed, even though their individual contributions may be largely redundant. The problem, of course, is that this success may also encourage shirkers to take the garden for granted. They may feel that their efforts were not needed and conclude that "one person can't make a difference." In an anomic community, on the other hand, the village green may deteriorate sufficiently to induce rejuvenating efforts, but volunteers will eventually lose heart if enough others do not join in to make the project successful. In short, as the level of individual and community involvement changes, so too do the signals received by each member, which, in turn, modify each member's propensity to contribute.

This is a key implication of the social learning[2] model: it need not matter whether the individual's efforts can make a difference, as long as the group in which they participate does. Propinquity replaces causality as the link between effort and attainment. Successful mobilization reinforces the associated participation, whatever the actual (and largely unknowable) marginal impact of individual efforts. Hence, the enthusiasm for participation reflects the experience of *collective* rather than *personal* efficacy.

Consider the classic riddle of the rationality of voting (Olson 1965). The common belief that one's vote can make a difference, inexplicable in terms of the MPCR, is readily explained by social learning theory. While individual votes are largely meaningless, the aggregation is not, and it is this collective power of voting that individual respondents come to associate with their own behavior. Simply put, rational choice theory has it backwards. *It is not individual efficacy that explains participation in collective action, it is successful collective action that explains why participants tend to feel efficacious.*

A recent cross-national study of participation in political protest is suggestive (Opp 1989). Opp finds that volunteers tend to believe their efforts can make a difference, but this appears to be largely a byproduct of their "strong attachment to political protest" (1989, p. 236). Participants typically ignore the real marginal impact of their individual effort and instead act as they would have others act to attain common goals. For example, when confronted with the argument that "a single protester would not make any difference," German antinuclear participants responded to the effect that "if everybody thought so, there would be no protest" (1989, p. 236). Conversely, if everybody volunteers, the group can make a difference, and this success, in turn, seems to have convinced participants that their efforts were worthwhile.

The Internalization of Prosocial Norms

While social learning theory can be applied to behavior that is highly pragmatic, one of the attractions of the approach is that it applies equally well to collective actions motivated by normative commitments. Unlike rational choice, learning theory does not impose any assumptions about the capacity for reflective thought. Adaptive responses need not be oriented towards anticipated consequences, but may, instead, be rule-governed and habitual, shaped by unintended outcomes that sanction the legitimacy of the associated rule. Social learning theory thus extends the analytic boundaries to actors who feel their efforts are worthwhile, based on normative rather than instrumental tests.

Normative solidarity differs from instrumental behavior in that the contribution to public goods is "an end in itself, regardless of its status as a means to any other end" (Parsons 1968, p. 75n). For example, obligations based on the rule of fairness depend on the actions of others, not the consequences.

Scott's (1971) theory of moral commitment is perhaps the most fully elaborated application of learning theory to the internalization of norms. The attachment to prosocial norms increases when those who comply are repeatedly rewarded and when those who disregard social obligations and disdain collective welfare are penalized. Conversely, the attachment declines when compliance is penalized and deviance is rewarded. When fully internalized, the learned behavior persists "at a spatial or temporal remove from its sanctions" and may be generalized to other social contexts (1971, pp. 88, 107). Hence, strong political and civil associations can serve as what de Tocqueville called workshops in "the art of association," providing "a thousand continual reminders to every citizen that he lives in society…that it is the duty as well as the interest of men to be useful to their fellows" (de Tocqueville 1969, p. 512).

Local Maxima and Social Traps

Although social learning theory promises to expand the boundary conditions imposed by the rational actor approach, the reduction in cognitive costs does not come without a price. Adaptive actors are more prone to become trapped in local maxima than if they were fully rational. The tendency to repeat rewarded behavior can lead to suboptimal fixation or "satisficing" (Simon 1955). Conversely, the tendency to avoid punished behavior can be "dissatisficing," as actors revisit costly alternatives rather than fixate on the least punitive. Reward-seeking, penalty-aversive behavior is also broadly consistent with Tversky and Kahneman's (1990a) laboratory finding that subjects are risk-averse about gambling to increase rewards but risk-seeking when trying to reduce losses.

In short, "backward-looking" actors can be just as self-interested, just as concerned that their efforts are worthwhile, and just as worried about the future, but they look ahead by holding a mirror to the past. To use a metaphor from cybernetics, they pursue their target relentlessly but cannot know its trajectory and therefore cannot intercept it by plotting a shortcut, a higher-order process that requires the capacity to anticipate future moves by the target. Hence, they do not know when to take one step backwards in order to take two steps forwards, one of Elster's (1979) key criteria for rational choice.

Will the shortsightedness of backward-looking decision making then reduce or exacerbate the tension between collective and individual welfare in social dilemmas? The answer cannot be found in the logic of reward and punishment, since individual members of an interest group are wired together in a system of interactive learning with a logic of its own. In interactive learning, "the behavior of one organism serves as a stimulus for the behavior of another… (which) in turn serves as a stimulus for the first" (Scott 1971, p. 64). Hence, the structural logic need not correspond to the behavioral logic of the component actors. Computer simulations based on a stochastic learning model of social dilemmas show how reward-seeking, penalty-aversive actors lead one another away from mutually beneficial outcomes and into social traps where everyone suffers. The simulations also identify the structural conditions for escape: (1) threshold effects by which random events can precipitate a critical mass of volunteers, and (2) a pattern of low-density social ties that facilitate the spread of the necessary chain reactions.

A STOCHASTIC LEARNING MODEL OF COLLECTIVE ACTION

This study elaborates a stochastic learning model used in my previous work (Macy 1989, 1990, 1991a, 1991b). The model consists of three components— a decision algorithm, a production function by which individual decisions are aggregated into outcomes, and a learning algorithm by which outcomes modify individual decisions.

Decision Algorithm

The simulation model differs from most rational choice formulations in that the decision process is stochastic rather than determined. Formal models developed by Oliver and Marwell (1988), Granovetter (1978), Heckathorn (1989), and others assume that decisions are strictly determined by the calculus of marginal costs and benefits. Participation occurs at "the point where the perceived benefits to an individual of doing the thing in question...exceed the perceived costs" (Granovetter 1978, p. 1422). A stochastic learning model assumes instead that behavior is shaped by its consequences, not determined. The outcomes raise and lower propensities, but choices remain uncertain. In short, anything can happen, but not with equal probability.

The decision algorithm assumes each member j has some propensity to cooperate p_{ij} representing the probability that j chooses to volunteer ($V_{ij} = 1$) at iteration i. Each choice is determined by the magnitude of p_{ij} relative to a random number n_{ij} from a uniform distribution, such that $V_{ij} = 1$ if $p_{ij} \geq n_{ij}$ and $V_{ij} = 0$ if $p_{ij} < n_{ij}$.

Divisible and nondivisible contributions

Most simulation studies of collective action have assumed nondivisible contributions. Members of the group must either participate fully in the collective action or not at all. For example, players in a Prisoner's Dilemma must decide whether to cooperate or defect, or members of a committee must choose whether to raise their hands in response to a call for volunteers.

However, as Elster has observed, "although the assumption of a dichotomous independent variable—the decision to cooperate—is convenient for many purposes, it is often unrealistic. Often, the problem facing the actor is not *whether* to cooperate, but *how much* to contribute" (1989, p. 25).

Divisibility of contributions can be introduced by respecifying p_{ij} not as the propensity to volunteer, but as the proportion of j's relevant resources contributed to the cause. Divisible contributions introduce a problem of interpretation: is the glass half full (and undesired outcomes signal greater caution next time), or half empty (and negative outcomes signal less)? It seems reasonable to assume that as the proportion of individual resources contributed increases, so, too, does the probability that the investment will be regarded by the actor as a genuine contribution and not a token gesture. Thus, the value of V_{ij} is set in the same way as with binary choices, by the magnitude of p_{ij} relative to a random number from a uniform distribution ($0 < n_{ij} < 1$), such that $V_{ij} = 1$ if $p_{ij} \geq n_{ij}$ and $V_{ij} = 0$ if $p_{ij} < n_{ij}$.

The Production Function

The outcome associated with the decision of whether to contribute to the public goods depends on the subjective benefit relative to the cost of contribution. The total cost of producing the public goods depends on the jointness of supply and the size of the group, and the share of this total that falls on each member depends on the distribution of relevant resources within the group and on the individual willingness to volunteer. The benefit depends on the group's contribution level (as given by the benefit function), and the subjective evaluation of the benefit depends on the preferences, perceptions, and expectations of the actor (as given by the utility function).

Jointness of supply

The total cost of providing a public good depends on the size of the group and the jointness (or nonrivalness) of supply. Pure public goods have perfect jointness: if the public goods are available for anyone, then they are also available for everyone else, at no additional cost and with no reduction of anyone's enjoyment. Recall the example of a lighthouse whose cost does not increase, and whose benefit to the local taxpayers does not diminish, no matter how many visitors to their harbor may benefit as well. In contrast, the cost of providing and maintaining an adequate (uncrowded) beach area beside the lighthouse increases with the size of the village.

More formally, we can define the cost of supplying one unit of public goods to every member of the group as N^{1-J}, where N is the size of an interest group and J is the jointness of supply ($0 \leq J \leq 1$). If $J = 1$, the goods have pure jointness of supply; group size then has no effect on unit cost, creating enormous social advantages for large groups. The cost of supplying N members with one unit each will be 1. Consider a group of 100 members, each of whom contributes equally to the production of one unit of public goods. With pure jointness, the cost to each member (.01 units) is trivial compared to the benefit that each enjoys in return, one unit, creating an extremely mild social dilemma. Volunteers have almost nothing to lose. This remains true even in the extreme case, where one stalwart must shoulder the entire cost alone.

At the other extreme, if $J = 0$ (zero jointness), the cost of supplying N members with one unit each is N (i.e., the cost increases with the number who benefit). Again, the social dilemma evaporates, but for the opposite reason. Even if everyone contributes equally, the cost to each of supplying everyone with one unit of public goods will be one unit. Hence, cooperation is pointless; everyone might as well go their separate way.

Oliver and Marwell (1988, p. 4) argue that most public goods have at least some rivalry, while zero jointness generally implies excludability and is thus not usually applicable to public goods. This suggests the plausibility of a midrange value. Suppose $J = .5$, that is, the cost of supplying one unit of public goods to every member increases with the square root of N. With $N = 100$, the cost of compliance is .1, an amount that is nontrivial relative to the optimal benefit, yet not so large as to make collective action pointless. In short, partial jointness is not only more empirically plausible than either pure or zero jointness, it is also more interesting theoretically since it creates a more compelling social dilemma.

Given that the total cost of producing the public goods is N^{1-J}, j's share of the total depends on the distribution of relevant resources within the group and on j's willingness to volunteer. More formally,

$$C_{ij} = \frac{R_j V_{ij}}{N^j} , \tag{1}$$

where C_{ij} is j's cost, R_j is j's share of the total resources in a group of N members, and J is the jointness of supply ($0 \leq J \leq 1$).

Resource inequality

Following Marwell and Oliver (1993), I assume that some volunteers have more resources to contribute than others. Relevant resources include time, money, talent, and freedom from competing private demands (families, careers, etc.). Having chosen to volunteer ($V_{ij} = 1$), the magnitude of each contribution is limited by that individual's resource endowment, R_j. For simplicity, we may assume members of the group average one unit of resources. Hence, C_{ij} ranges from 0 to $\sqrt{N}$ (assuming $J = .5$). $C_{ij} = 0$ if j has no resources ($R_j = 0$) or if j is waiting to see what others will do ($V_{ij} = 0$). $C_{ij} = \sqrt{N}$ if all resources are concentrated in a single benefactor who must shoulder the burden alone

($R_j = N$). The simulations assume that resources are normally distributed with a mean of one.

The benefit function

The level of production of public goods is a nonlinear function of the level of contribution. As Elster (1989) observed, "the benefit function can be expected to be S-shaped, with marginal benefits first rising and then decreasing" as contributions become increasingly redundant toward the end (p. 32). Hence, the marginal impact of each additional contribution will vary with the overall level of participation. Oliver and Marwell consider both accelerating and decelerating functions, which pose start-up and follow-up problems respectively. For convenience, an S curve includes both, with relatively flat slopes at both the low end (the start-up problem) and the high end (the follow-up problem). The S-shape was obtained by modeling the level of public goods, L_i, as a cumulative logistic function of P_i, the rate of contribution,[3] such that

$$L_i = \frac{1}{1 + e^{(.5-P_i)10}}. \tag{2}$$

Equation (2) limits the benefit level to $0 < L_i < 1$. L_i approaches 0 when no one volunteers ($P_i = 0$), $L_i = .5$ when $P_i = .5$, and L_i approaches unity when $P_i = 1$.[4]

The utility function

As with rational choice models, it is useful to distinguish between objective outcomes and subjective utility. Learning theory adds an additional distinction between positive and negative utility, depending on the value of L_i relative to some reference point (or expectation level). As Tversky and Kahneman note, "our perceptual apparatus is attuned to the evaluation of changes or differences rather than to the evaluation of absolute magnitudes. When we respond to attributes such as brightness, loudness, or temperature, the past and present context of experience defines an adaptation level, or reference point, and stimuli are perceived in relation to this reference point (Helson 1964). Thus, an object at a given temperature may be experienced as hot or cold to the touch depending on the temperature to which one has adapted" (1990b, p. 155).

To complicate the problem further, "there are situations in which gains and losses are coded relative to an expectation or aspiration level that differs from the status quo" (1990b, p. 164). For example, at room temperature, a glass of milk seems warm, while a mug of cocoa, at the same temperature, seems cold.

These two reference points apply to satisfaction with the level of participation in collective action. Positive and negative utility is modeled as a function of the level of production of the public goods (L_i), evaluated relative to the status quo (a reference point that is updated through time) and to the expectations of the actor (a reference point that is fixed by prior experience, aspirations, and preferences). These functions are specified by equations (3), (4a), and (4b). Equation (3) models the status quo as an unweighted moving average of the level of production of public goods in the three most recent periods, and then measures current outcomes against this reference point, with

$$\Delta L_i = L_i - \frac{\sum\limits_{k=1}^{3} L_{i-k}}{3} .$$

(3)

Equations (4a) and (4b) then give the utility (U_{ij}) of a given level of public goods (L_i) as a function of perceptions (ΔL_i), expectations (X_j) and interests (I_j). If $\Delta L_i \geq 0$, then

$$U_{ij} = \left[L_i + \Delta L_i (1 - L_i) - \frac{1 - X_j}{2} \right] I_j ,$$

(4a)

and if $\Delta L < 0$, then

$$U_{ij} = \left[L_i + \Delta L_i (L_i) - \frac{1 - X_j}{2} \right] I_j .$$

(4b)

ΔL_i is the difference between the current level of public goods, L_i, and a moving average of the three most recent values, as given by equation (3). Equation (4a) means that the subjective utility of a given benefit level may be substantially enhanced if it represents a noticeable improvement ($\Delta L_i >$ 0). However, the enhancement quickly dissipates if benefits stabilize ($\Delta L_i =$ 0). Note also that as L_i increases (and 1-L_i approaches 0), further increases have less effect on utility. (In other words, L_i approaches unity when $P_i = 1$ and can not be enhanced beyond that upper limit by its magnitude relative to prior levels.)

Conversely, equation (4b) means that an otherwise satisfactory benefit level might signal the need to alter course if it represents a sufficiently sharp reduction ($\Delta L_i < 0$). However, as before, the discounting dissipates once the reduction levels off, and as L_i approaches 0, further deterioration also has less effect.

I_j is j's interest in the public goods, relative to the resources required to produce them, with I_j normally distributed and $\overline{I_j} = 1.$[5] Although the value of the public goods to j is assumed to equal the value of the total bundle of private resources invested in their production (hence $\overline{I_j} = 1$), not every member of the group necessarily agrees. Some may value the public goods more highly

($I_j > 1$), while others place greater weight on the resources that must be contributed ($I_j < 1$).

X_j is from the distribution of actors' expectations for L_i, where $-1 < X_j$ < 1. Assuming (for illustration) that $I_j = \bar{I}_j = 1$ and $\Delta L_i = 0$, as X_j increases, the range of U_{ij} shifts from $-1 < U_{ij} < 0$ to $0 < U_{ij} < 1$. $X_j \approx -1$ implies an extreme "malcontent" whose high expectations make j difficult to satisfy, even with very high levels of production. Restricting the range of U_{ij} to negative values means j responds to all outcomes aversively, some more strongly than others. No matter how well the group is doing, it seems never to be enough.

Conversely, "accommodators" ($X_j \approx 1$) have very low aspirations and are easily resigned to the social costs of rampant egoism. While malcontents see the glass as half empty, accommodators see it as half full. Hence, accommodators tend to respond to the outcomes positively, that is, within a range of 0 to 1. "Realists'" ($X_j = 0$) expectations fall somewhere toward the middle, giving a range of $-.5 < U_{ij} < .5$.

The simulations generally assume that X_j approximates a normal distribution, with $-1 < X_j < 1$, $X_j \approx 0$, and $s_x \approx .3$. In other words, a few are malcontents, a few are accommodators, and most are dissatisfied when collective interests go entirely unmet and are pleased when production of the public good approaches an optimal level. Two other conditions were also tested: positively skewed (most are disgruntled) and negatively skewed (most are easily accommodated).

The Learning Algorithm

The third component of the model is the learning algorithm by which propensities to volunteer and free-ride are modified by the outcomes associated with those behaviors. The learning stimulus is a trigonometric function of the costs and benefits associated with the decision whether to contribute to the public goods, adjusted by a learning-rate parameter, and constrained to an absolute value less than unity, such that

$$S_{ij} = \sin(\arctan\left[(U_{ij} - C_{ij})\, E_j'\right]) \tag{5}$$

S_{ij} is the stimulus and E_j is a learning rate parameter, with $0 < E_j < 1$. $E_j \approx$ 0 means that the reinforcements have no effect on j's behavior, while $E_j \approx 1$ means that j is highly unstable, capable of veering from extreme boldness to extreme caution following a highly aversive outcome. Rapid learning suggests pragmatic error correction, while slow learning may indicate habitual or norm-guided behavior based on rules or routines that take somewhat longer to change. Slow learning might also reflect a lack of interest in the outcomes, a noisy environment, or time lags that weaken the association between outcome and behavior. The simulations assume that E_j approximates a normal distribution, with a few maximally sensitive actors, a few who adhere to their routines regardless of the outcome, and most distributed between these extremes.

Since costs and benefits can exceed unity in absolute value when interests and resources are heterogeneous, a trigonometric transformation was used to limit S_{ij} to the required interval (reinforcements cannot exceed ± 1). Note that the transformation is trivial for $|(U_{ij} - C_{ij})E_j| < .5$, and it constrains all absolute values above unity to the interval $\sqrt{5} \leq S_{ij} \leq 1$.

The Bush-Mosteller model

The learning algorithm is adapted from a conventional Bush-Mosteller stochastic learning model (Bush and Mosteller 1955). The algorithm gives the increase over time in the propensity to volunteer, p_{ij}, or

$$p_{i+1,j} = p_{ij} + [S_{ij}(1 - p_{ij}^{(1/|s_{ij}|)}) V_{ij}] - [S_{ij}(1 - p_{ij}^{(1/|s_{ij}|)})(1 - V_{ij})] \quad (6)$$

The propensity is reinforced when prosocial behavior seems to pay—the reward to volunteers is added to the propensity when $V_{ij} = 1$ and $S_{ij} > 0$. The propensity is also reinforced when noncooperation is costly—the penalty for shirkers is subtracted when $V_{ij} = 0$ and $S_{ij} < 0$, again causing the propensity to increase. Conversely, if feckless behavior pays off or solidarity is suckered, then the propensity to free-ride $(1-p)$ is reinforced, i.e., $1-p$ is substituted for p on both sides of equation (6) and $1-V$ is substituted for V.

Finally, the algorithm assumes that the reinforcement decays as propensities approach their natural limits and that the larger the stimulus, the more rapid this decay, as modeled by the exponential expression $1/|S_{ij}|$. The larger the reward, the faster the decay as the propensity increases. Hence, large rewards change behavior more than small ones, but ten applications of a small reward have greater cumulative efficiency than five applications of a reward of twofold magnitude.

SIMULATION RESULTS

Computer simulations show that these assumptions generate a bistable system with cooperative and noncooperative equilibria (Macy 1990). In the cooperative equilibrium (or "lock-in"), contributions are sufficiently widespread to make participation seem worthwhile, regardless of whether one's contribution is actually necessary at the margin. In the noncooperative equilibrium, prosocial behavior is swamped by rampant individualism. Free-riders may learn that competition is self-destructive, but volunteers also learn that their efforts are wasted and they "burn out" as fast as they are replaced.

Between the equilibria lies a critical point (the "critical mass"), below which the system slides back into noncooperation, and above which contributions become self-reinforcing, leading to lock-in.

Figure 1 reveals that the social trap is a stable and suboptimal equilibrium into which actors tend to gravitate in response to the social cues generated by their interaction. The simulation assumes an interest group of 100 members ($N > 100$ provides little improvement in reliability at a substantial computational cost), moderate jointness ($J = .5$), binary choice (either cooperate or defect), and normally distributed (and uncorrelated) interest ($\bar{I}_j = 1$), resources ($\bar{R}_j = 1$), expectations ($\bar{X}_j = 0$), and learning rates ($\bar{E}_j = .5$). Three trials are depicted, with start values chosen to illustrate the equilibria. In the upper series, the starting propensity is $p_{1j} = .7$. This is the critical mass at which volunteers are sufficiently numerous that their positive reinforcement just offsets the larger benefit for shirkers. Lock-in is then obtained with the level of contribution at $P_e = .7$ (or 70 percent of possible resources committed to the public goods). This is close to optimal, given the flatness of the benefit function at high levels.

In the middle series (indicated by the shaded line), the start value is again .7, but this time the random walk takes the group below the critical mass and they are unable to climb back. Although the actors seek pleasure and avoid pain, their choices lead them to gravitate away from mutually beneficial outcomes toward a punitive equilibrium at $P_e = .38$. At this level of participation, the social costs of widespread deviance are almost as large as the benefits enjoyed by the group (in the upper series) that attains critical mass.

In the lower series, no one is willing to contribute initially ($p_{1j} = 0$), imposing severe social costs that induce numerous members to alter course. Nevertheless, these early volunteers are unable to attain the level of contribution needed to sustain their effort. Instead, they learn that their sacrifice has been for naught and they quickly lose heart. A stable equilibrium is eventually obtained at about .38, with contributions rotated among the group.

The noncooperative equilibrium is like a hole with a slippery and progressively steeper wall that traps any group that wanders in too far. Escape is difficult but not impossible. From the bottom of the hole, the higher the group is able to climb, the easier it becomes to climb higher still. The trick is getting enough people to scramble at the same time to prevent those remaining behind from pulling everyone else back down.

More formally, the noncooperative equilibrium is stable but is not an absorbing state. Learning continues, with each player's propensity fluctuating around the balancing point. If the fluctuations are sufficiently large, the group will sooner or later climb far enough out of the hole to escape its "pull." It is all a matter of timing. Collective action thus benefits from unexpected crises and other nodal points on the social landscape that tend to galvanize a coordinated response.

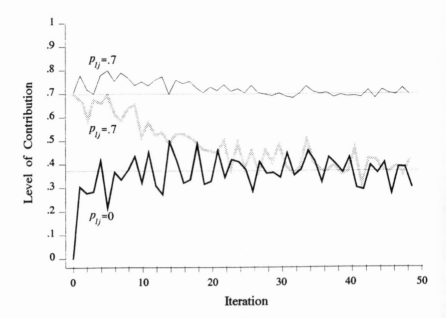

Figure 1. Cooperative and Noncooperative Equilibria in Collective Action. ($N = 100$, $J = .5$, $C_{ij} = [0,1,]$ $\bar{E}_j = .05$, $\bar{X}_j = 0$, normally distributed interest, resources, expectations, and learning rates)

The Coordination Problem

The problem in Figure 1 is that a minimum of 70 members must volunteer to create a critical mass. With equilibrium at .38, it will be quite a long wait before random fluctuations in the overall contribution level pull in a critical mass of volunteers.

The wait may be reduced by increasing the size of the fluctuations or reducing the gap between equilibrium and critical mass. The size of the fluctuations increases with the smallness of the group, but so, too, does the critical region that must be crossed. This is the dilemma of group size. A small group is easier to coordinate (i.e., the fluctuations around the equilibrium tend to be larger) but a broader base gives volunteers a larger payoff, reducing the interval that must be crossed. The tradeoff will thus depend on the jointness of supply. The more that the cost of the public goods increases with the number who benefit, the greater the advantage to small groups.

Holding N constant, the gap between equilibrium and critical mass declines as J increases. The odds of finding a critical mass can be further improved by bringing malcontents and accommodators into the fold. Malcontents fail to consolidate a commitment, while accommodators are content with even badly suboptimal outcomes. The closer to zero the mean and variance in the distribution of expectations, the easier it becomes to cross the gap.

Figure 2 shows how critical mass can be attained in a group identical to that in Figure 1 except that the public goods now have pure jointness of supply ($J = 1$) and everyone in the group is a "realist" ($X_j = 0$ for all j). In order to chart the divergence between volunteers and defectors, Figure 2 displays mean cooperative propensities ($\bar{p}_{ij}$), rather than the level of contribution (P_i).

With $J = 1$, equilibrium rises to $P_e = .47$, with 52 volunteers needed to make cooperation self-reinforcing, a gap of only 5 members, substantially less than the 32 additional volunteers needed in Figure 1. Thus, it does not take long for random fluctuations to wander into critical mass (at $i = 22$).

Once this occurs, the structure of the group changes dramatically. The group splits into permanent classes of producers and exploiters as equilibrium gives way to an absorbing state (or lock-in) in which learning terminates. Exploitation does not occur at noncooperative equilibrium, since the burden of supplying public goods is rotated among the membership. At lock-in, however, the classless interaction is replaced by a permanent system of exploitation of the many (61 percent) by the few (39 percent), but with a marked improvement in the net benefits from collective action (given the steepness of the benefit function in the region between the equilibria). While equilibrium obtains at a level well below the optimal outcome, and while the payoff to exploiters is about four percent larger than that for producers, the improvement may be sufficient for what Przeworski (1985) calls "the material basis of consent." Better to be comfortably exploited than to return to a state of anomie where everyone loses equally.

This solution to the coordination problem depends on the assumptions $J = 1$, $X_j = 0$, $\bar{E}_j = .5$, and $C_{ij} = [0,1]$. Low jointness increases the gap between equilibrium and critical mass, while divisible contributions and low learning rates reduce the magnitude of the fluctuations at equilibrium, lengthening the odds that the group will wander out. Finally, low contribution levels become self-reinforcing for accommodators (the start-up problem), while high contribution levels fail to become self-reinforcing for malcontents (the follow-up problem). Yet, few interest groups are without their share of malcontents, accommodators, and slow learners. Moreover, few public goods have pure jointness of supply and nondivisible contributions (either cooperate or defect).

Clearly, the test of the model is not its ability to explain collective action under all circumstances. On the contrary, the point is to disclose the conditions necessary for achieving critical mass. Nevertheless, high jointness, binary contributions, and uniformly realistic expectations seem overly constraining, given what we know about the prevalence of collective action. In sum, the simulations results reported in Figures 1 and 2 appear to be more useful in diagnosing the social trap than in identifying the way out.

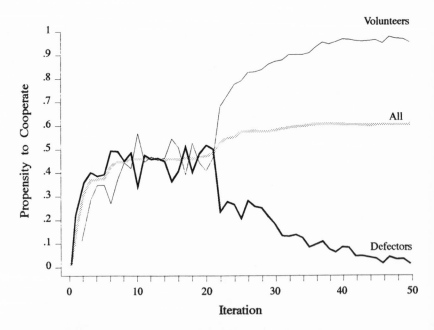

Figure 2. Critical Mass by Random Event. ($N = 100, J = 1$,
$C_{ij} = [0,1,] \bar{E}_j = .5, X_j = 0$, normally distributed interest,
resources, and learning rates, homogeneity of expectations)

Threshold Effects

The coordination problem in collective action calls attention to Granovetter's (1978) threshold model as a more promising solution. His rational choice formulation assumes that the marginal return on investments in public goods increases with the overall level of contribution. Members with stronger preferences for the public goods can be expected to invest earlier than others. Their efforts, in turn, raise the MPCR to the higher level required by their less-interested counterparts, producing a chain reaction in which each new contribution triggers others. These chain reactions are a plausible mechanism by which contributions might be coordinated sufficiently to escape a noncooperative equilibrium.

Granovetter's model seems intuitively much more plausible than game-theoretic models in which the contestants must choose "in parallel," without knowledge of others' strategies. The threshold model assumes *serial* rather than parallel choices: Each actor looks around to see how many others are participating before deciding whether to join in. Granovetter cites several empirical illustrations of this pattern of serial interaction, for example, women in Korean villages who appear to be "wary of adopting birth control devices and wait to do so until some proportion of their fellow villagers do" (1978, p. 1423). Similarly, "workers deciding whether to strike will attend carefully to how many others have already committed themselves." Granovetter's model is also applicable to the spread of rumors, migration, or the decision to go to college or to walk out of a boring lecture (1978, pp. 1423-1424).

Although Granovetter's model assumes rational choice, thresholds need not be determined by calculations of changes in the MPCR across contribution levels. In his study of participation in the civil rights movement, Chong (1991) found that as the proportion of volunteers increases, so too does the moral and social pressure on the remaining members of the group, the obligation to do one's fair share, and the feeling of strength (and safety) in numbers. These pressures vary across members of the interest group. Those with high thresholds may be thick-skinned and relatively immune to social pressure, while those with low thresholds may succumb more easily to feelings of guilt or social obligation when they see even a few other members trying to shoulder the load by themselves.

The incorporation of thresholds into a stochastic learning model is then quite straightforward—we need only replace the propensity to volunteer with the threshold rate of contribution. In other words, the tendency to free-ride is manifested as a high threshold for participation instead of a low propensity to volunteer. The thresholds are then adjusted in the same manner as were the propensities in the previous formulation—each actor's threshold moves up or down in response to the success or failure of collective action and the costs of participation. For example, those with high thresholds are moved to help out only after most members have already done so. Their reluctance may be reinforced if they learn they can count on habitual volunteers to step forward quickly. However, if everyone thinks that way, free-riders may come to believe "if I don't do it nobody else will," causing their thresholds to drop. Unsuccessful instigators, left holding the bag, become less adventurous (their low thresholds rise), while those whose efforts trigger sufficient imitation may become more confident (their thresholds drop further).

Thresholds alter the stochastic learning model in a subtle but highly consequential way. Even in the simulations in Figure 1, based on parallel interaction, the actors learn how many contributed and with what outcome, and this information ultimately shapes the probability that each will volunteer. However, this happens only after they have already committed themselves; hence, the outcomes can only alter the willingness of each actor to volunteer next time. Serial interaction introduces a short circuit, allowing the actions of others to influence *current* decisions. The hypothesis is that this short circuit will greatly facilitate the coordination of responses, making it much more likely that the process will generate a critical mass.

More formally, the decision to participate in collective action is a stochastic function of the distance between the actor's threshold and the current rate of participation by the rest of the group. Let T_{ij} represent j's threshold and P_i the participation rate by $N - 1$, each with a range of 0 (no one contributes) to 1 (full participation). The propensity to volunteer, p_{ij}, can then be modeled as a cumulative logistic (S-shaped) function of the gap between T_{ij} and P_i, such that

$$p_{ij} = \frac{1}{1 + e^{(T_{ij} - P_i)5}} \cdot \tag{7}$$

The sigmoid function means that the probability of participation accelerates as the contribution rate approaches the threshold, reaching $p_{ij} = .5$ when $T_{ij} = P_i$. As the contribution rate rises above the threshold, the probability quickly approaches unity and then decelerates as the gap widens further, with participation all but guaranteed when $T_{ij} = 0$ and $P_i = 1$. As the contribution rate drops below j's threshold, j rapidly becomes unlikely to volunteer, with participation effectively precluded when $T_{ij} = 1$ and $P_i = 0$.

The relevant population for determining thresholds need not be all $N - 1$ members of the group. Four conditions were tested, in order of the extensiveness of social influence: (1) *parallel decision making,* in which actors respond to the costs and benefits of participation and free-riding without regard to what others are doing (the assumption in Figures 1 and 2); (2) *high-density networks,* in which members concern themselves only with those in social proximity, with little overlap among reference groups (the condition sometimes assumed in studies of crowd behavior; see McPhail 1991); (3) *low-density networks* that link socially distant actors (the condition typified by groups with weak ties); and (4) *groupwise serial interactions,* in which members take their cues from the contribution level of the group as a whole.

Thresholds are then modified in the same way that propensities were modified in equation (6), except that thresholds move in the opposite direction, that is, the threshold increases when prosocial behavior is punished ($V_{ij} = 1$ and $S_{ij} < 0$) or noncooperation is rewarded ($V_{ij} = 0$ and $S_{ij} > 0$). This gives

$$T_{i+1,j} = T_{ij} - [S_{ij}(1 - T_{ij})^{(1/|S_{ij}|)}) \, V_{ij}] + [S_{ij}(1 - T_{ij})^{(1/|S_{ij}|)}) \, (1 - V_{ij})] \, . \tag{8}$$

Extensive simulations reveal a stable pattern of results under widely varied assumptions about jointness, learning rates, divisibility of contributions, expectations, and group heterogeneity (see Macy 1991b). To preview the principal finding, when members of a group take their cues from the actions of others, they can easily trap themselves in a vicious circle in which no additional members will contribute until someone else does, as Granovetter has emphasized. What Granovetter's static model fails to reveal, however, is that there is virtue in the vicious circle. The impasse causes the distribution of thresholds to evolve towards a critical state, precipitating chain reactions of various magnitudes. Eventually, one of these chain reactions may be long enough to shift the system into a new stable state with a self-reinforcing level of contribution. Moreover, this does not require pure jointness of supply, binary contributions, or positive reinforcement for volunteers. The bandwagon effect is so strong and so persistent, under a wide range of model parameters, that it suggests a promising solution to the timing problem in collective action.

Figure 3 illustrates threshold effects under conditions that would preclude critical mass if choices were made without regard to what others were doing. Recall how two of the three simulations in Figure 1 converged towards a punitive equilibrium. The conditions here are even more inauspicious: divisible contributions, moderate jointness of supply ($J = .5$), heterogeneity of expectations, and an average learning rate one-fourth as high ($\bar{E_j} = .125$). The lower learning rate corresponds to the tendency for actors to follow norms or heuristics that are modified over time, in contrast to the pragmatic error correction allowed in Figure 1. (The lower rate also slowed the process down sufficiently to better illustrate the logic.)

As indicated in the upper series (with normally distributed expectations), when members of an interest group take their cues directly from the actions of others, the group has little trouble locking in very high levels of contribution. The contrast with the lower series of Figure 1 is striking. In the latter, the level of participation initially rises very quickly but then levels off. In the upper series of Figure 3, on the other hand, members are initially much more reluctant to respond, with virtually no participation during the first 10 iterations. This affirms Granovetter's finding that the distribution of thresholds can block contributions even by those with a strong interest in the public goods. However, the standoff is part of a dynamic that eventually propels the group into lock-in. The prudent tendency to wait for others allows the distribution of thresholds to evolve toward a critical state in which a minor and random event triggers an avalanche of cooperation. It is this avalanche that jolts the system out of its noncooperative state and into lock-in.

The process illustrates the principle of "self-organized criticality" that has proved somewhat successful in predicting earthquakes, and has been applied as well to market collapses, biological evolution, and turbulence in fluids (Bak and Chen 1991). Complex interactive systems do not tend towards permanent equilibria that can only be disturbed by exogenous shocks. Rather, the normal workings of these systems cause them to "evolve toward a critical state in which a minor event can lead to a catastrophe" (1991, p. 46). In the case of collective action, the catastrophe may be benevolent.

The negatively skewed expectations in the middle series mean that most members of the group are malcontents who get little positive reinforcement even at high contribution levels ($\bar{X_j} = -.86$). The results show that positive reinforcement is not necessary for a successful chain reaction, but the system

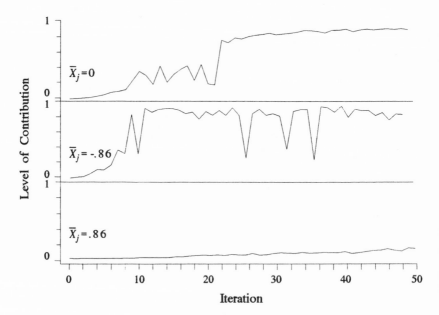

Figure 3. Threshold Effects in Collective Action. ($N = 100$, $J = .5$, $C_{ij} = [0...1]$, $\bar{E}_j = .13$, normally distributed interest, resources, and learning rates)

will fail to differentiate permanent classes of producers and exploiters and equilibrium will be highly unstable.

Conversely, with expectations skewed toward accommodation, threshold effects remain unable to induce critical mass, as indicated in the lower series. The profusion of accommodators dampens aversive responses to the social costs of collective action failures ($\bar{X}_j = .86$). Start-up problems thus appear to be caused not by the low MPCR to early investors, as rational choice theorists have argued, but by the tendency for actors to become discouraged by failure and to accommodate the social costs of rampant anomie.

If organizers can raise members' expectations, the flatness of the benefit function does not cause a start-up problem, despite the low MPCR to early investors. The flatness simply means that thresholds continue to fall despite the rash of frustrated but futile outbursts. Eventually, the distribution of thresholds will fan out into a pattern that produces the necessary chain reaction.

Once this occurs, an organizer's emphasis should be redirected to the benefits achieved through collaborative effort. Otherwise the group may have the follow-up problem evident in the middle series of Figure 3. As Alinsky counsels, once cooperation gains a foothold, "the organizer knows that his biggest job is to give the people the feeling they can do something" (1971, p. 113), or as

Fireman and Gamson put it, "try to keep tangible, though perhaps small, victories coming..." (1979, p. 30).

The Social Structure of Collective Action

Granovetter's is one of the few rational choice models that goes beyond the microfoundations of collective action to consider the effects of social structure, specifically, the social ties through which members find out what others are doing. A structural analysis suggests that collective action may depend not only on the strength of collective interests, but also on the network of social ties that channel the necessary chain reactions. Hence, threshold models should be useful for understanding how participation in collective action might follow the contours of social structure, for example, how standing ovations spread throughout an audience or how crowd reactions may depend on the location of instigators and the density of social ties.

However, structural analysis is badly handicapped by Granovetter's rational choice assumption that thresholds derive from the tendency of the marginal return to vary with the overall level of investment. This means that individual participation is triggered by the contribution rate of the group as a whole. The choices of those in one's immediate circle have no special relevance to the current rate of return and, hence, should not be expected to be of much interest to rational actors.

The structure of social ties is much more relevant if the actions of others shape individual behavior directly rather than as a byproduct of an underlying interest in the MPCR. This influence may take the form of social pressure to do one's fair share or the desire simply to imitate one's close associates. The overall contribution level then loses any special significance; involvement will be triggered by the participation of just those members the actor is willing or able to attend to, be it a small circle of friends or the group at large. It follows that the spread of chain reactions through social space will depend on the network of social ties linking the members of the collectivity.

The higher the network density, the fewer the mediations required for information to pass between any two randomly selected nodes. At maximum density, every node is tied to all others, permitting cues to travel from any point to any other point without mediation. This condition enables each actor to choose, based on knowledge of what everyone else is doing. At the other extreme, all ties are disconnected and the network disappears. This is the condition assumed with parallel decision-making in which each actor chooses without knowledge of or concern for the choices of others.

Between these extremes, each member is linked to a subset of significant others. For example, the actors may have greater awareness of and concern for the actions of familiar or nearby persons. There is now very strong empirical evidence that social contacts are important conduits for recruitment into organizations, interest groups, crowds, and social movements (Snow, Zurcher, and Eckland-Olson 1980; McPherson and Smith-Lovin 1987; McPhail 1991; and McAdam 1989). Suppose members of a collectivity do not choose in isolation from one another, nor are they aware of what everyone else in the group is doing. Rather, they take their cues from those in social or spatial proximity—friends, family members, neighbors, or colleagues. Although he did not address the issue in his analysis of threshold effects, Granovetter's argument for the "strength of weak ties" seems relevant here. The facility with which chain reactions spread through the network will depend on social ties that bridge the boundaries of local clusters. "The stronger the tie between A and B, the larger the proportion of individuals in S to whom they will both be tied....This overlap in their friendship circles is predicted to be least when their tie is absent, most when it is strong, and intermediate when it is weak" (1973, p. 1362).

The important point is not the strength of the ties, but their density. Strong ties tend to be associated with high-density interactions, such as where all members of A's reference group are tied to each other—B and C influence A, and B influences C. Hence, information passes easily within each circle, but poorly between circles. Weak ties, on the other hand, have the "strength" of their typically low density: few members of A's reference group refer to one another, choosing instead those outside A's circle—B and C influence A but D does not, while D, and not B, influences C. Granovetter's work suggests that chain reactions will spread much more easily when density is low. If so, then weak ties, by bridging greater social distances, may facilitate the coordination of contributions needed for a critical mass.

Both high- and low-density configurations were simulated. The high-density configuration was created by randomly sorting the membership into 18 circles. Each member interacts with all other members of the circle. The size of the circles varies to give each member between three and six interactants, with a mean of 4.5. The size was limited in order to accentuate the contrast with groupwise interactions in which each member responds to the choices of all other members of the group.

Low-density ties were created by having each actor randomly select three interactants from the full population. By random chance, in a group of 100, two percent of the members can be expected to choose the same interactant twice, which doubles this person's influence, and three percent choose to interact with themselves. Note that everyone chooses, but not everyone is chosen. About 37 percent of the actors are overlooked each time everyone networks. Anyone who is chosen must then add to her reference group any persons who picked her as one of their interactants. For example, someone picked by three people will have six interactants, the three people she picked, plus the three who picked her. Someone overlooked entirely will have only the three interactants he selected, and his ties will be asymmetric (A influences

B but B does not influence A). Hence, the reference groups for each actor range in size from three to six, the same as in the high-density configuration. The critical difference is that an actor is not required to pick those whom his other interactants are paired with, nor do actors necessarily attend to those who attend to them. The absence of cliques means that information can diffuse much more easily across social space.

The two series charted in Figure 4 confirm that low-density interactions are more conducive to collective action. The simulations are identical to the upper series of Figure 3 except that each member of the group responds only to the choices of a subset of interactants and ignores everyone else. In the low-density network, cooperative equilibrium is obtained with a .68 contribution level, about optimal. In contrast, the high-density group (indicated by the shaded line) performs only slightly better than those in which the actions of others are irrelevant (e.g., the lower series of Figure 1). The group remains trapped in a noncooperative equilibrium with roughly a .43 contribution level. "Cliquishness" thus appears to have much the same effect as social isolation, dampening the chain reactions needed to reach critical mass.

Given the wide disparity in low-density configurations, the simulations results showed much greater variability than in groupwise simulations, necessitating statistical tests of the difference in performance between the two structures. Based on 60 trials (with small within-group variations in the model but no between-group variations), the differences between groups were highly significant. Lock-in was attained in 17 of 30 trials with low density, and only 4 of 30 trials with high density ($p < .001$). In addition, five trials with low density led to partial lock-in, with slow decay. The low-density group failed to escape equilibrium in eight of the 30 trials. The high-density group attained partial lock-in six times, and failed in 20 of the 30 trials.

The level of contribution was also much higher at lock-in for the groups with low density. The average contribution level was .72, with a range of .52 to .87. For the groups with high density, the average contribution level at lock-in was only .54, with a range of .49 to .59.

Finally, lock-in took twice as long in the groups with high density. With low density, the median time to lock-in was 20 iterations, with a mean of 26 ($s_i = 16$), and a range of 12 to 75. With high density, lock-in required a median of 47 iterations, a mean of 50 ($s_i = 12.2$), with a range of 40 to 65. These differences were statistically significant at $p < .001$.

Although the difference between the two configurations is striking, an important caveat is in order. The simulations assumed that social influence does not vary with network density. Hence, it is unclear whether the structural advantage is sufficient to compensate for the characteristic weakness of socially distant ties. The threshold effects that are critical for collective action depend decisively on the willingness of actors to attend to the choices of others. If actors in low-density networks are less sensitive to social pressure, the "strength"

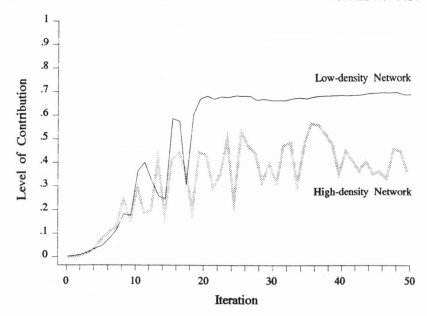

Figure 4. The Strength of Weak Ties in Collective Action.
$(N = 100, J = .5, C_{ij} = [0...1], \bar{E}_j = .13, X_j = 0$, normally distributed interest, resources, and learning rates)

(or structural advantage) of weak ties may be offset by the dampening of the threshold effect.

SUMMARY AND CONCLUSION

Human history has no shortage of tragedies, and many of these take the form of social dilemmas. How can people with mutual interests learn to avoid self-defeating competition, despite the risk that socially responsible behavior may be exploited by others? The social leverage generated by combination does not necessarily provide an explanation for collective action, for two reasons. First, an even greater advantage is enjoyed by free-riders, and second, the marginal impact of individual contributions is usually hard to appreciate.

Functionalist explanations center on the internalization of norms needed for social order and cohesion. The selective incentives argument is closely parallel, except that compliance is not internalized, but must be secured by an instrumental interest in the associated sanctions. Both arguments encounter a similar difficulty: *the capacity of a group to monitor, sanction, or socialize compliance with cooperative norms does not explain collective action, it presupposes it.*

Recent contributions have aimed to restore the centrality of collective interests as more than an unintended byproduct. The Oliver-Marwell studies show how a core group may be willing to shoulder the burden themselves, no matter how many others might enjoy the benefit for free. However, the solution is circumscribed by the assumption that symbiotic behavior issues from an instrumental assessment of marginal returns. The problem is that MPCR cannot be directly observed but must be calculated, and this is no easy task. Many collective action problems arise among actors who lack the necessary temperament or information.

Learning theory allows us to relax the cognitive and information demands of the rational actor model and suggests an alternative interpretation of the efficacy problem in collective action. The key assumption is that actors judge whether their efforts are worthwhile not by their marginal impact, but by what the group is able to accomplish. Simply put, their enthusiasm for participation reflects the experience of collective rather than personal efficacy.

This hypothesis can be formally modeled as a stochastic learning model of contributions to public goods. The model assumes that symbiotic behavior emerges in response to the signals generated through social interaction, without the capacity or disposition for forward-looking calculation. The study begins with the game-theoretic convention that decisions to cooperate are made in parallel, independently of others' moves. Extensive simulations reveal a bistable system with both noncooperative and cooperative equilibria, and show how stable cooperation can evolve out of anomic Hobbesian conflict, and how it can degenerate as well. While the actors gravitate toward behavior that is more rewarding and less punitive, the simulations show that the search process can lead away from rewarding outcomes and into a punitive equilibrium or social trap.

The probability of escape depends on the number of moves that must be stochastically synchronized to attain critical mass. This is a function of the size of the group (small is better), the distance that must be traversed between equilibrium propensity and the threshold for escape, and the distance covered by each move. The distance to be traversed is a function of the jointness of supply and group size (big is better). The opposing effects of group size pose a dilemma: small groups tend to be easier to coordinate, while large groups can provide much greater social leverage on individual efforts.

This coordination problem identifies the decisive importance of threshold effects. Collective action is much more likely to flourish when each member of the group takes into account what others are doing before deciding whether to join in. Hence, a strong interest in the public good is not enough; even highly interested actors may pitch in only if a sufficient number of others participate as well. However, this constraint on individual participation has the opposite effect on the group. The interdependence facilitates the coordination of responses needed to escape a noncooperative equilibrium, even with only modest jointness of supply, slow learning, divisible contributions, and outcomes that provide little positive reinforcement for volunteers.

This threshold effect depends not only on the susceptibility of the actors to social influence, but also on the structure of social ties through which chain reactions can spread throughout the group. Collective action may fail due to cliquishness or provincialism that isolates local clusters of interactants. Conversely, ties that link socially distant actors can promote collective action, as long as the distance does not dissipate the social pressure on which threshold effects depend.

To conclude, the behavioral assumption of "backward-looking" or adaptive behavior resonates with the bedrock sociological principle that the whole is more than the sum of its parts. The rationality of the group may not depend on the capacity of individual members to grasp the systemic logic of their interaction; rather, this intelligence may be hard-wired in the structural arrangements that constrain individual choices. Yet, these structures are themselves created by the choices of individuals. Hence, computer simulations of adaptive processes seem especially inviting as a tool to explore the emergence of social structure out of the choices it constrains.

ACKNOWLEDGMENT

This study elaborates a stochastic learning model used in my previous work, integrates material presented piecemeal in separate articles, and provides a more developed theoretical rationale for this approach. The model has grown in complexity and refinement over several years, and this paper reaffirms earlier simulation results using a more recent formulation. The author wishes to thank those whose patient and careful criticisms of my work have prompted many of these revisions, especially Barry Markovsky, Michael Hechter, Pamela Oliver, Gerald Marwell, Douglas Heckathorn, and John Orbell.

NOTES

1.   Heckathorn (1989) proposes a provocative solution to the second-order free-rider problem based on "hypocrisy," the willingness of first-order free-riders to support sanctions against people like themselves. Heckathorn shows that under some circumstances it can be rational to contribute to social controls, even though it may not be rational to abide by them. He offers the example of corrupt sheriffs in the Old West who nevertheless "increase the level of order" (1989, p. 97). Computer simulations of a learning-theoretic reformulation show that Heckathorn's solution obtains even with greatly relaxed behavioral assumptions (Macy 1993).

2.   For Bandura (1971), "social learning" refers to vicarious reinforcement, or modeling behavior observed to be rewarding. I use the term more broadly to include other forms of social influence. The most important is interactive learning, in which reinforcements are generated by choices that are, in turn, influenced by the rewards and penalties imposed by the choices of others. I also model threshold effects in which actors wait to see what others will do before choosing to follow.

3. Assuming resource homogeneity, P_i equals the proportion of the group who volunteer. With resource inequality, P_i equals the proportion who volunteer weighted by their relative resource endowments.

4. The multiplier 10 in the exponent of equation (2) alters the shape of the curve so as to approximate a cumulative normal function. As the multiplier is reduced, the function approaches linearity.

5. The simulations also assume that the distribution of interest in the public goods is orthogonal to the distribution of resources. Elsewhere, I vary the distributions and correlation of interest and resources; see Macy 1991b.

REFERENCES

Alinsky, Saul. 1971. *Rules for Radicals*. New York: Random House.Arrow, Kenneth J. 1986. "Rationality of Self and Others in an Economic System." Pp. 201-216 in *Rational Choice: The Contrast Between Economics and Psychology*, edited by R. Hogarth and M. Reder. Chicago: University of Chicago Press.

Axelrod, Robert. 1984. *The Evolution of Cooperation*. New York: Basic.

Bak, Per and Kan Chen. 1991. "Self-Organized Criticality." *Scientific American* 264(1):46-53.

Bandura, Albert. 1971. *Social Learning Theory*. Morristown, NJ: General Learning Press.

Baum, William M. 1979. "Matching, Undermatching, and Overmatching in Studies of Choice. *Journal of Experimental Analysis of Behavior* 32:269-281.

Bush, Robert and Frederick Mosteller. 1955. *Stochastic Models for Learning*. New York: Wiley.

Chong, Dennis. 1991. *Collective Action and the Civil Rights Movement*. Chicago: University of Chicago Press.

Coleman, James S. 1986. "Social Structure and the Emergence of Norms Among Rational Actors." Pp. 55-84 in *Paradoxical Effects of Social Behavior*, edited by A. Diekmann and P. Mitter. Heidelberg: Physica-Verlag.

Cook, Karen and Margaret Levi. 1991. *The Limits of Rationality*. Chicago: University of Chicago Press.

Cross, John. 1983. *A Theory of Adaptive Economic Behavior*. Cambridge: Cambridge University Press.

de Tocqueville, Alexis. 1969. *Democracy in America*. New York: Doubleday.

Elster, Jon. 1979. *Ulysses and the Sirens: Studies in Rationality and Irrationality*. Cambridge: Cambridge University Press.

———. 1989. *The Cement of Society*. Cambridge: Cambridge University Press.

Fireman, Bruce and William A. Gamson. 1979. "Utilitarian Logic in the Resource Mobilization Perspective." Pp. 8-44 in *The Dynamics of Social Movements*, edited by M. Zald and J. McCarthy. Cambridge, MA: Winthrop Press.

Granovetter, Mark. 1973. "The Strength of Weak Ties." *American Journal of Sociology* 78(6):1360-1380.

———. 1978. "Threshold Models of Collective Behavior." *American Journal of Sociology* 83(6):1420-1443.

Hardin, Russell. 1982. *Collective Action*. Baltimore: Johns Hopkins University Press.

Hechter, Michael. 1987. *Principles of Group Solidarity*. Berkeley: University of California Press.

Heckathorn, Douglas D. 1989. "Collective Action and the Second-Order Free-Rider Problem." *Rationality and Society* 1:78-100.

Helson, H. 1964. *Adaptation-Level Theory*. New York: Harper.

Herrnstein, Richard. 1981. "A First Law of Behavioral Analysis." *The Behavioral and Brain Sciences* 4:392-395.

————. 1988. "A Behavioral Alternative to Utility Maximization." Pp. 3-60 in *Applied Behavioral Economics*, edited by S. Maital. London: Wheatsheaf.

Homans, George C. 1974 [1961]. *Social Behavior: Its Elementary Form*, 2nd ed. New York: Harcourt Brace Jovanovich.

Macy, Michael W. 1989. "Walking Out of Social Traps: A Stochastic Learning Model for Prisoner's Dilemma." *Rationality and Society* 1:197-219.

————. 1990. "Learning Theory and the Logic of Critical Mass." *American Sociological Review* 55:809-826.

————. 1991a. "Learning to Cooperate: Stochastic and Tacit Collusion in Social Exchange." *American Journal of Sociology* 97:808-843.

————. 1991b. "Chains of Cooperation: Threshold Effects in Collective Action." *American Sociological Review* 56:730-747.

————. 1993. "Backward-Looking Social Control." *American Sociological Review* 58:819-836.

Marwell, Gerald and Ruth Ames. 1979. "Experiments on the Provision of Public Goods I: Resources, Interest, Group Size, and the Free Rider Problem." *American Journal of Sociology* 84:1335-1360.

————. 1980. "Experiments on the Provision of Public Goods II: Provision Points, Stakes, Experience, and the Free Rider Problem." *American Journal of Sociology* 85:926-937.

Marwell, Gerald and Pamela Oliver. 1993. *The Critical Mass in Collective Action*. Cambridge, England: Cambridge University Press.

McAdam, Douglas. 1989. "The Biographical Consequences of Activism." *American Sociological Review* 54:744-760.

McPhail, Clark. 1991. *The Myth of the Maddening Crowd*. New York: Aldine De Gruyter.

McPherson, J. Miller and Lynn Smith-Lovin. 1987. "Homophily in Voluntary Organizations: Status Distance and the Composition of Face to Face Groups." *American Sociological Review* 52:370-379.

Mueller, Dennis. 1986. "Rational Egoism Versus Adaptive Egoism as Fundamental Postulate for a Descriptive Theory of Human Behavior." *Public Choice* 51:3-23.

Oliver, Pamela. 1980. "Rewards and Punishments as Selective Incentives for Collective Action: Theoretical Investigations." *American Journal of Sociology* 86:1356-75.

————. 1984. " 'If You Don't Do It, Nobody Else Will': Active and Token Contributors to Local Collective Action." *American Sociological Review* 49:601-610.

Oliver, Pamela., Gerald Marwell, and Ruy Teixeira. 1985. "A Theory of Critical Mass. I. Interdependence, Group Heterogeneity, and the Production of Collective Action." *American Journal of Sociology* 91(3):522-556.

Oliver, Pamela and Gerald Marwell. 1988. "The Paradox of Group Size in Collective Action: A Theory of Critical Mass. II." *American Sociological Review* 53:1-8.

Olson, Mancur. 1965. *The Logic of Collective Action*. Cambridge, MA: Harvard University Press.

————. 1990. "Mancur Olson," Pp. 167-185, in *Economics and Sociology*, edited by R. Swedberg. Princeton: Princeton University Press.

Opp, Karl-Deiter. 1989. *The Rationality of Political Protest. A Comparative Analysis of Rational Choice Theory*. Boulder: Westview Press.

Parsons, Talcott. 1968. *The Structure of Social Action*. New York: The Free Press.

Przeworski, Adam. 1985. *Capitalism and Social Democracy*. Cambridge: Cambridge University Press.

Rachlin, Howard, Ray Battalio, John Kagel, and Leonard Green. 1981. "Maximization Theory in Behavioral Psychology." *The Behavioral and Brain Sciences*. 4:371-417.

Rapoport, Anatol and Albert Chammah. 1965. *Prisoner's Dilemma: A Study in Conflict and Cooperation*. Ann Arbor: University of Michigan Press.

Scott, John Finley. 1971. *Internalization of Norms: A Sociological Theory of Moral Commitment*. Englewood Cliffs, NJ: Prentice-Hall.

Simon, Herbert. 1955. "A Behavioral Model of Rational Choice." *Quarterly Journal of Economics.* 63:129-138.

Snow, David, Louis Zurcher, and Sheldon Eckland-Olson. 1980. "Social Networks and Social Movements: A Microstructural Approach to Differential Recruitment." *American Sociological Review* 45:787-801.

Stinchcombe, Arthur. 1990. *Information and Organizations.* Berkeley: University of California Press.

Tversky, Amos and Daniel Kahneman. 1990a. "Prospect Theory: An Analysis of Decision Under Risk." Pp. 140-170 in *Rationality in Action*, edited by P. Moser. Cambridge, England: Cambridge University Press.

————. 1990b. "Rational Choice and the Framing of Decisions." Pp. 60-89 in *The Limits of Rationality,* edited by K. Cook and M. Levi. Chicago: The University of Chicago Press.

Winter, Sidney. 1986. "Comments on Arrow and on Lucas." Pp. 243-250 in *Rational Choice: The Contrast Between Economics and Psychology,* edited by R. Hogarth and M. Reder. Chicago: University of Chicago Press.

van de Kragt, Alphons, John Orbell, and Robyn Dawes. 1983. "The Minimal Contributing Set as a Solution to Public Goods Problems." *The American Political Science Review* 77:112-122.

————. 1988. "Are People Who Cooperate 'Rational Altruists'?" *Public Choice* 56:233-247.

CROSS-NATIONAL EXPERIMENTAL INVESTIGATIONS OF ELEMENTARY THEORY: IMPLICATIONS FOR THE GENERALITY OF THE THEORY AND THE AUTONOMY OF SOCIAL STRUCTURE

David Willer and Jacek Szmatka

ABSTRACT

This paper examines the methodology of cross-cultural experiments and presents extensive experimental results on variety of small social structures which were studied in the U.S. and Poland. Results for exchange, coercive and profit point structures in both settings are reported. These were cross-national replications and, as explained in the methodological discussion, since the results from the two settings are similar, the generality of the Elementary Theory is supported. Beyond supporting the theory, these results suggest that social structures investigated were autonomous from particular variations of time and place. Further implications of this autonomy are considered.

Advances in Group Processes, Volume 10, pages 37-81.
Copyright © 1993 by JAI Press Inc.
ISBN: 1-55938-280-5

INTRODUCTION

This paper investigates models drawn from Elementary Theory (Willer and Anderson 1981, Willer 1987) and reports experimental results from Polish and American settings. Though the two settings differ on a number of important political, economic, linguistic, and cultural conditions, our research does not treat any of these differing conditions as independent variables. Instead, our interest focuses on the similarity of experimental observations between the two settings. Far from wanting to find that Poles and Americans differ, we prefer to find that they act the same for, as we will show, similarity supports the generality of Elementary Theory (ET).

Social theory must make the apparently preposterous claim that its scope of application is not limited by the specifics of time and place. Both theory and the scope statements which govern its application must be "comprised of constructs that do not refer to particulars of time and place" (Walker and Cohen 1985). Call this the principle of universality which is, of course, common to all sciences.

Theories conform to the principle of universality for practical reasons. Theories are formulated and tested so that they can explain and predict. That is to say, theories must be formed so that they can be applied today, in the future and in the past. For example, toward the end of the fourth century A.D. the Roman Empire, being divided into largely autonomous eastern and western parts, was invaded by barbarian tribes. Among these tribes were the Visigoths, who settled for some years in the province of Illyricum, which was located between East and West. Unlike most Germanic tribes, the Visigoths maintained their autonomy from both Romes for some time. By 397 the Visigoths invaded the Eastern Empire, gaining plunder and forcing a settlement which granted the Visigoth king the title of "Master of Soldiers of Illyricum." Over the next ten years Visigoth attacks concentrated on the West, extracting tribute and ransom. By 412 they were induced to Spain, then to Gaul, and, by 418, were settled in Aquitania as Roman allies with dependent status (Bury [1904] 1967 and Lind 1987). Why was Visigothic autonomy, so long successfully defended, lost after their move to the western provinces? We will offer an explanation based on our experimentally tested theory.

Though formal theory and experimental research are frequently paired, the goals of both must transcend the limits of the laboratory. Readers of this Annual need not be reminded that experiments are the best tests for theory and that theory is the best design for experiments. But no theory is developed merely to be tested in a laboratory. We hope that the cross-national experiments which we will report here are significant in themselves. To us, however, their significance most importantly lies in the paths they open toward applications outside the laboratory, including application to cases like the Visigothic one to which we later return.

One of the tasks of this paper is to show how our cross-national experimental tests might justify wide ranging applications of ET for explanation and prediction. The task has two parts. First we must demonstrate what all social theorists claim and few, if any, have substantiated—that predictions drawn from ET hold independent from particulars of time and place. We begin by reviewing the methodology of cross-cultural replications departing from Foschi's superb discussion (1980) and then introduce the basic concepts and procedures of the theory. The core of the paper consists of an array of contrasting models and experiments, with results from Poland and the United States extending the confirmational scope of the theory.

Second, we consider how experimental results bear on the generality of the phenomena of the theory (Faucheux 1976). We hope to show that, beyond supporting generality, our experimental results also imply that social structures are autonomous from the particulars of different cultures to an important degree. By *social structure* we mean a set of interrelated positions with valued resources which are occupied by actors whose interests, values and beliefs spring from the relations, positions and resources. By *culture* we mean an array of norms, values and beliefs associated with a group of actors. By *the particulars of culture* we mean any specific cultural qualities that do not spring from a social structure. For social structures to have the claimed autonomy means that processes and outcomes are affected by the general conditions of structures and actors, and not by cultural particulars. If autonomy can be substantiated, wide-ranging theory applications will be radically simplified. The paper concludes by reviewing the theory's confirmational status.

THE METHODOLOGY OF
CROSS-CULTURAL EXPERIMENTATION

Whether knowledge is—or could be—universal in social psychology has long been a contended issue. Gergen (1973) asserted that the theories and findings of social psychology are time and place bound, but Schlenker (1974) responded that only empirical generalizations are so limited while universal propositions are, in principle, "transhistorical" (p. 3). Schlenker was unable, however, to cite even one example of a universal proposition which had been successfully tested cross-culturally. Though the number of cross-cultural studies is large (Brislin 1983), the number of cross-cultural experiments is still small (Foschi and Hales 1979), while the number based on formal theory is very small. Thus, the generality of knowledge once conjectural remains so today, and Sell and Martin's observation is still true: "we do not argue that most social psychology is transhistorical and cumulative, but that it could be." (1983).

These limits are not particular to social psychology; the proven generality of knowledge in all social sciences is similarly limited. For example, economic historians question whether neoclassical economic theory can be applied beyond the confines of modern capitalism (Finley 1974). While the comparative research of classical sociological theorists such as Weber ([1918] 1968) ranged across continents and centuries, his theory, like other theories of the era, was quite informal and its applications, though wide ranging, could hardly qualify as tests of universality. In light of these limitations, the disjunction between formal theory and cross-cultural experimentation is particularly unfortunate. When cross-cultural research seeks universality, to cover variations found, even to accurately express them, formal theory is indispensable. For, as Foschi and Hales explain, the results of cross-cultural research "become meaningful only after they have been incorporated into a theory" (1979, p. 246).

Methodological discussions of the relation between theory and cross-cultural experimental research remain largely hypothetical. For example, Foschi could illustrate her points only with (1) cross-cultural exploratory experiments which used no formally developed theory, and (2) theoretically driven experimental designs which could, at some future time, be extended cross-culturally (1980). Nevertheless, Foschi (and Foschi and Hales) have distinguished two crucially different types of cross-cultural experiments, which they differentiate by the way that experimental limitations are addressed.

Experimental limitations are "the particular values assigned by the experimenter to variables which are not part of the theory but which may have an effect on the dependent variable" (Foschi 1980, p. 94). Cross-cultural experimental research can address experimental limitations in two contrasting ways. First, cultural differences can be used to operationalize an independent variable. We will call these *cultural difference* experiments. Second, cultural variations can be used to test for the generality and robustness of a theory. We will call these *cross-national replication* experiments. We discuss the nature and purposes of each type in the following two subsections. In both cases, our discussion begins with Foschi's (and Foschi and Hales') points, to which we add thoughts of our own.

Cultural Difference Experiments

In these experiments, cultural contrasts are used to set values for an independent variable (Foschi and Hales 1979), while the dependent variable is measured just as it would be in any single-culture experiment. Examples of this type include: Berry (1967), Tedeschi, Smith, Gahagan and Elinoff (1972), Tallman and Ihinger-Tallman (1979) and Schwinger (1980).

For example, Tallman and Ihinger-Tallman speculate that people "living in social structures characterized by a scarcity of material goods will place a greater value on material well being" than people of a wealthier society (1979, p. 222). The subjects were from Minnesota (U.S.) and Mexico; the experiment consisted of a game in which they could chose lifestyles emphasizing social relations or material goods. The hypothesis that Mexicans are more materialistic was supported by the results.

The Tallman and Ihinger-Tallman study exemplifies the design of cultural difference experiments in that it begins, not with formal theory, but with an orienting perspective. Two mental states associated with two cultures form the independent variable, while measures associated with the game provide values for the dependent variable. In this design, assignment of subjects to the two experimental groups (analogous to "experimental" and "control") is determined by place of residence and is not random; that is, one experimental group was Mexican and the other was Minnesotan. The Tedeschi et al. design is logically similar.

Tedeschi et al. hypothesize that individuals in more economically "advanced" societies are less cooperative than individuals in less "advanced" societies. Subjects from the U.S. and Ghana played the Prisoner's Dilemma (P/D) game under classic conditions of noncooperation; subjects' choices to cooperate or defect were effectively simultaneous and binding contracts were disallowed. Contrary to their hypothesis they found that Americans were significantly more likely to cooperate than Ghanians (1972). From Tedeschi et al.'s perspective, because Ghanians act more cooperatively in everyday life, a cooperative mental state is formed which generally predisposes behavior. Thus, Ghanians should be more cooperative in the P/D game than are people from the U.S. However, they are not.

Many cultural difference experiments, like the two above and others reviewed by Foschi, rest on assumptions to the effect that (1) there are easily identifiable cultural characteristics that (2) are routinely internalized as mental states generally predisposing behavior, and (3) these states can be measured by an experimental task like the P/D game. While all three assumptions are open to question, the second raises quite general issues for theory. Though the structural perspective of this paper does not deny cultural differences, it does not accept that behavioral differences are necessarily produced by differing mental states. For example, if it is true that Ghanians cooperate more than Americans, we are inclined first to look to social structural conditions. As Axelrod (1984) explains, while defection is the dominant strategy, when playing the P/D game a large number of times with a specific other, Tit for Tat, a reciprocal and cooperative strategy, avoids Pareto suboptimality. Assume many social relations in Ghana and the U.S. are like the P/D game. Then, if Ghanian people encounter specific others more frequently, they should behave more cooperatively than people in the U.S. who encounter specific others less frequently. Because greater cooperation is not due to contrasting mental states, however, playing an experimental P/D game will not find Ghanians more cooperative. That is to say, if the behavioral differences are due to structural differences alone, there will be no contrasting mental states to measure cross-culturally.

While issues of theory and inference are important, the most serious problems associated with cultural difference experiments are methodological. As Foschi and Hales (1979) point out, the results of cultural difference experiments always pose problems of interpretation and always face threats to internal validity, because subjects cannot be randomly assigned to different treatments. As they explain, however, this problem also occurs when parameters such as gender and ethnicity are used as independent variables in experiments that do not cross cultures.

The problems of interpretation and threats to validity are due to *correlated biases*. "Correlated biases refer to those additional characteristics of the people in the two groups which are also related to the...(independent)...variable" (Willer and Willer 1973). As Selvin (1957) pointed out, categories of survey research like rural and urban are not logically like the categories experimental and control. People who are rural are not different from people who are urban *only* by their place of residence. To the contrary, rural-urban differences can be correlated with as many other differences as time and money allow us to find. Therefore, if differences in behavior are found between the two, it cannot be known that these differences are due only to place of residence.

While only random assignment allows inferences connecting independent and dependent variables, the design of cultural difference experiments obviates random assignment. Thus, had Tedeschi et al. found their expected differences, it would be impossible to know that they were due only to differing levels of "development" between U.S. and Ghana and not to one or more of the thousands of other differences between the two settings, differences which are correlated biases and quite impossible to untangle. Similarly, though their hypothesis was supported, it cannot be known that Tallman and Ihinger-Tallman's results were due only to differences in wealth as they suggested. To the contrary, only through random assignment can systematic differences be expected to cancel out, tests of significance be justified and inferences connect independent to dependent variable (Selvin 1957). It is precisely because random assignment justifies the test allowing the inference that experimental research is normally preferred to survey research.

Foschi contends that the problems of cultural difference experiments can be alleviated through the use of theory. Her idea is to substitute formal theory for orienting perspectives and to use "cultural differences to operationalize theoretical variables" (1980, pp. 98-99). Certainly experiments of this kind need much more than a "let's see if there are any cultural differences in *blank*" approach (Sechrest 1977, p. 116). While allowing a more sophisticated statement of findings, formal theory also helps by replacing empiricist experimental design with theory-driven experimental design (Willer and Willer 1973, Willer 1987). Though the use of theory will not eliminate correlated biases in findings, it can eliminate correlated biases at the level of the theory. That is, if a formal theory is subject to intensive development—of which the cultural difference experiment is only a part—threats to validity *of the theory* will decline as the series of tests accumulate and will wash away if not all tests are of the cultural difference type.

Cross-National Replication Experiments and Contextual Contrasts

Cross-national experimentation is intended to vary conditions which are not part of theory and which, according to Foschi, "can be identified by default only" (1980, p. 93). Experiments of this type are intended to discover whether a theory can be applied independent of time and place (Rohner 1977), use a test-retest reliability design (Finifter 1977), and test for universality through contrasting settings.

The experiments presented in this paper are cross-national replications. The same series of experiments was run in both the United States and Poland. Our hope was that Polish subjects would behave the same as U.S. subjects, such that exactly the same contrasts across laboratory social structures would be found. As a test-retest reliability design, one aim was to draw experimental subjects from a population distinct from undergraduates at American Universities. While it is true, as Foschi notes, that conditions like "U.S. undergraduate student" and "Polish graduate student" are not part of the theory, there are structural conditions of both societies comprehended by ET which could affect experimental outcomes. Those conditions will be discussed shortly.

Unlike the cultural difference design, which is bedeviled by correlate biases, cross-national replication prospers from uncontrolled variations. The uncounted and uncountable variety of cultural characteristics, which pose problems of internal validity for cultural difference experiments, contribute to the power of cross-national replication experiments. The reason for this good fortune lies in the following. For cross-national replications, the hope is that experimental results will be *unaffected* by cultural variations; success is measured by the similarity of findings across settings. Therefore, when a cross-national replication succeeds, it does so in spite of the large number of the cultural differences which could be designated. When a replication has succeeded, however, there is no need to designate all of the differences. For many purposes it is enough to know that there are multiple sources of variation that could conceivably threaten the generality of the theory, but have not.

For the experiments to follow, since no attempt is made to measure differences due to culture, these replications do not claim that any specific difference was assuredly present. We do not know whether the two settings of our research, the United States and Poland, are in distinct cultures which produce distinct mental states in our subjects. All that is asserted is that the experiments took place in different settings. Thus we use the term cross-*national* in preference to cross-*cultural* to designate this type.

Applying ET, however, we find that there are important social structural differences between the two settings, which bear upon the confirmational generality of our theory. Because ET is a theory of social structure which is not limited to the laboratory, it points to an array of contrasting conditions in the larger society which could confound results. While not formally measured, the most important of these conditions are identified and discussed below.

The social structural contrasts between the United States and Poland could hardly have been greater. The U.S. has, since its inception, been a capitalist society, and arguably the purest case thereof due to the absence of (1) traditional limits on the workings of private property, and (2) traditional statuses complicating its stratification structure. At the time of the research, Poland was a Soviet-type society with state-owned means of production and distribution. Poland was subject to the inefficiencies of a bureaucratic centralized economy and to a political system which routinely violated human rights, both of which had been imposed on it by the Soviet Union at the end of World War Two. While also imported from outside, capitalism in the U.S. was long enough established to be considered indigenous and produced a significantly higher standard of living than that found in Poland. Human rights were incorporated, at least formally, into the U.S. legal system and were more routinely honored.

During the research period in Poland there were important changes which eventually led to a democratic political structure and, at the time of this writing, toward disestablishing the centralized economy. While these changes worked to bring the two structures closer together, when the research was ongoing there was considerable uncertainty about the direction toward which change would lead. It was not known whether the political-economic revolution—for that is what it was—would succeed or be suppressed internally, by the Communist party and Political Police, or externally, by intervention of the Soviet Union. During the research period in the U.S., American society was politically and economically stable and, far from being in revolution, was in a period of reaction following the more revolutionary times of the 1960s and early 1970s. Since most of the U.S. research was completed in the middle 1970s and all the Polish research in the late 1980s most was separated by as much as fifteen years. The one exception was replication of Markovsky, Willer, and Patton (1988), which occurred three years later, in 1990.

More descriptively, the settings and subjects of the research were also quite distinct. The U.S. research was completed at the University of Kansas, a typical midwestern U.S. university established in the middle of the 19th Century and located in Lawrence, Kansas, a town less than 150 years old at the time of the research. The Polish research was completed at Uniwersytet Jagiellonski, which traces its history to medieval times and is located in Krakow, a commercial center of great antiquity. Polish experimental subjects were all

Polish citizens and all were graduate students. Though most U.S. subjects were U.S. citizens, some few were international students—though none of the internationals were from Soviet-type societies—and all were undergraduates. In both settings, however, no attempt was made to control for gender.

ET, like other theories of structural social psychology, uses very simple actors and more complex formulations for social structure. Since only one simple actor model is used, variations in actors' behavior between structures is due to variations of structural conditions. As a result, two threats to internal validity are particularly pernicious. The first is the possibility that behavior observed in experiments emanates from psychological traits brought to the experiment by the subjects and is completely independent from the social structural conditions modeled. It should be possible to eliminate this threat without cross-national study. Because behavior is expected to vary by structure, the emanation hypothesis is falsified when experiments with contrasting structures draw forth contrasting behavior. Nevertheless, cross-national replications can offer a further test. Assume that behavior in experiments is similar in two settings which are highly contrasting societies. Then it is highly unlikely, perhaps impossible, for that behavior to have emanated from exactly the same complex of psychological traits.

The second threat to internal validity and to generality as well is the possibility that observed behavior is due to an *interaction* between experimental conditions and a complex of psychological traits not modeled by the theory. Unlike the first, this threat does not suggest that behavior is independent of social structural conditions modeled in the experiment. Instead, it maintains that behavior is the joint result of a narrow population of subject strategies and the general social conditions of the experiment. If true, experimental results violate the principle of universality. Cross-national tests appear to offer the best way, perhaps the only way, to test for this threat. The threat fails if, over a set of distinct structural models and experiments and in spite of contrasting social conditions outside the laboratory, the same social conditions inside the laboratory draw forth the same behavior and, irrespective of setting, there is a one-to-one correspondence between each model and its experiment.

Confidence that the threat fails should be greatest when contrasting social conditions outside the laboratory are directly relevant to conditions being investigated. If there is a narrow population of subject strategies not comprehended by the theory which interacts with experimental conditions to produce results, and if that narrowness can violate the principle of universality, then subject strategies cannot be common to all people at all times. Instead, the strategies must be particular to some special social conditions and presumably are produced by them. In fact, contrasting social conditions between the United States and Poland are directly relevant to the social structures we investigated.

Weber defined the state as that organization which attempts to monopolize the means of violence in a given geographical area ([1918] 1968). The state is a centralized coercive structure and, in that regard, is structurally similar to centralized coercive structures which were experimentally investigated. Since both of our subject populations lived in modern states, both had experience dealing with coercion. Coercive relationships in the two settings were, however, quite different, being quite routine and unobtrusive in the U.S. and capricious and obvious in Poland. Thus, coercion was not unknown to both subject pools, but its style and substance was quite different. Part of the research on coercive structures investigated the effects of collective action. While people in the U.S. are relatively politically inert, Polish subjects were experiencing the period of *Solidarnosc* (the Solidarity uprising). We do not claim, however, that the two subject populations had internalized different inclinations to cooperate. We only note that levels of cooperation and collective action then directed toward the state in the two societies differed.

Also investigated experimentally were three contrasting exchange structures. While it should be emphasized that private personal property was common to both societies, the rights of private property which are basic to exchange were far more universally and firmly established in the U.S. Furthermore, the styles of interaction when exchanging in the two settings were not at all alike. Since private property rights were initial conditions of exchange experiments in both settings and since subject negotiations were face-to-face, there was ample opportunity for strategies formed outside the laboratory to interact differentially with experimental conditions.

Finally, one set of experiments investigated networks in which exchange was simulated by pools of profit points. Profit point relations are found in Poland or the U.S. only when they are purposefully produced in sociology laboratories. Because profit points are not found outside the laboratory, experiments utilizing that type of relationship might be immune to interactions with strategies formed outside. If so, results of profit point experiments will be more similar cross-nationally than results of exchange and coercive experiments. Alternatively, because profit point relations and exchange relations have similarly structured payoff matrices, it is conceivable that a complex of strategies produced outside the laboratory could still differentially interact across the two settings. In that case, profit point experiments will also differ across the two settings. Both possibilities will be considered.

The Structure of the Theory and Cross-National Replications

As indicated by its name, the structure of Elementary Theory is strongly influenced by the structure of theory in exact sciences. As explained by Hawking:

A theory is a good theory if it satisfies two requirements: it must accurately describe a large class of observations on the basis of a model that contains only a few arbitrary *elements,* and it must make definite predictions about the results of future observations. (1988, p. 9 [italics added].)

Given in Figure 1 is a schematic for the ET as thus far investigated in the U.S. Under "Exchange Structures" and "Coercive Structures," each of the twenty-two boxes represents an experimental study, five of which are currently in process. ("In process" means that models have been constructed, predictions generated, experimental research completed, and manuscripts are being prepared.) Each study investigates a number of contrasting models, at times as few as three (Markovsky, Willer, and Patton 1988) and sometimes as many as nine (Willer 1987, Chap. 5). The first priority of the program diagramed in Figure 1 is to discover the scope of ET (Toulmin, 1953, p. 112). Thus only a few replications, normally three to six, are run for each model.

The studies replicated in Poland are represented by four boxes of the diagram. The first replication discussed is "centralized exchange networks," found under "exchange structures" and "exclusion," traced to the far left. The second replications are found in two boxes, "centralized coercive networks" and "collective action networks," which are under and to the left of "coercive structures." The third replication is "1-exchange structure [GPI₁]," found under "exchange structures" and "profit point networks," traced to the bottom left.

Altogether, 13 different models were investigated in both settings and, because a few more replications were completed in the U.S., there were a total of 48 Polish and 60 U.S. experiments. While this is a substantial number of experiments, it represents only about 18% of the theoretic models for structures which have been investigated. The significance of cross-national replication, however, is not merely a function of percent of scope investigated. Instead, it is the diversity of scope which is decisive. As indicated by placement in the network schematic, the three represent importantly different parts of the theory's scope. Each investigates a structure composed of a different kind of relation—exchange, coercion, and profit point—which, together, represent all of the types of relation investigated by the theory. While all three studies traced the development of power, one study also investigated how power is countervailed. Both simple and complex network structures are included.

ET has three components: (1) a modeling procedure used to draw diagrams, (2) principles, and (3) laws applied to the diagrams for explanation and prediction. The idea of a theory with three components is copied from physical theories (see Galilei [1665] 1954, Newton [1729] 1966, and Toulmin 1953). Looking ahead to the figures given in the next section, note that the diagrams consist of a few simple elements, arcs with one or two signs which connect nodes into network shapes. Signed arcs stand for acts and are called *sanctions* while the nodes are actors. The signs represent the direction of actors'

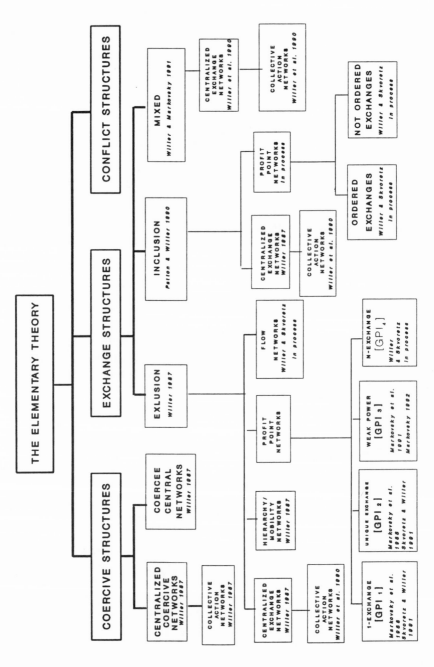

Figure 1. A Schematic For The Elementary Theoretic Research Program

preference change consequent on sanction transmissions and receptions. When an arc has a single sign, it will be adjacent to the receiving actor and indicates direction of preference change on that actor consequent on reception. When the arc has two signs, the sign adjacent to the transmitting actor indicates direction of preference change consequent on transmission. In Figure 3 all sanctions are positive in their effect on receiving actors and the relations are exchanges. In Figure 4 "negative sanctions" trace from the center to peripherals, while "positive sanctions" trace from peripherals to the center, indicating a structure of coercive relations.

Sanctions are flows which can vary quantitatively and the objective of ET is to predict those flows, predictions which can be expressed variously as rates of exchange or coercion for the differing relations. The actors that transmit and receive these sanctions are endowed with preferences and beliefs and are the locus of decision making. Borrowing the term "reflection" from Marx ([1867] 1967), for the simplest models, actors' values and beliefs "reflect" the valued resources which can flow as sanctions in the network. As shown below, applications of two principles and one law govern decisions of the actors and thus predict the rates of flow.

Theory is tested across an array of models by establishing a correspondence between each model and the experiment drawn from it. The correspondence is demonstrated when (1) each model and its experiment are similar and (2) contrasts between experiments correspond to contrasts between the models for them. For example, in discussions below we will compare the array of mean rates of coercion between our two settings to the array of mean rates within each setting. Because all Polish means are within the range of U.S. means, we conclude that results are similar between the two settings.

In the experiments in Poland and the United States, subjects were seated in chairs arranged in a semicircle and resources for negotiation were placed before each. Related subjects were in eye contact, while subjects not directly related were divided by partitions. That is, relations among subjects were laid out like the shapes of the figures, while values of resources which subjects transmitted and received corresponded to the signed flows. Consistent with the models, information was open so all subjects could hear the negotiations of others. The most important experimental conditions were given ostensively by the physical setup of the room and the number and kind of resources. Setting the remaining initial conditions was linguistically simple. Experimental instructions note the time limit imposed on negotiation and the value of each resource in the local currency. To set subjects' preferences they were asked to earn as much as possible and were paid by points.

We have found that paying subjects by points is the only reliable way to produce the desired value system (Willer 1987, Chap. 7). The greatest difficulty in standardizing conditions did not lie in translating instructions, but in standardizing subject payments between U.S. dollars and Polish zloty.

Standardization posed a problem because the official dollar/zloty exchange rate was completely unrealistic and because of significant inflation in Poland. We finally settled on a commodity equivalent, paying subjects a sum which would purchase a nice (but not elegant) lunch or four beers at a local bar.

Demand characteristics are an important threat to experimental validity when experimental conditions give subjects the information needed to infer experimental outcomes desired by the experimenter (Orne 1962, 1969). The danger lies in the possibility that favorable data might be produced by subjects playing the role of the "good subject" (Orne 1962, p. 778) or the "faithful subject" (Weber and Cook 1972, p. 275). Subjects' knowledge of network shape seems an obvious cue. Therefore, it is important to note that, contrary to early Power Dependence Theory (Emerson 1972a, 1972b), shape alone does not determine the resource flows in any of the structures investigated.

Furthermore, rumor from experiments already completed, a second source of demand characteristics (Orne 1962, 1969), provided a poor basis for subjects to infer results. Runs involved a number of models with similar but not quite identical initial conditions and very different predicted rates. Since subtle changes in initial conditions of the experiment strongly affected rates, it would have been particularly difficult for subjects to infer results; and, if desired results cannot be inferred, it is impossible for subjects to produce results simply by playing compliant roles. These and other issues of experimental control are discussed more extensively elsewhere (Willer 1987).

THEORY, MODELS AND EXPERIMENTAL RESULTS

Elementary theoretic actors have preferences and beliefs and are the locus of decision making. In the simplest models (which are the ones most interesting to experimentalists) all actors' values and beliefs are simply *reflections* of the structure containing the actors. As noted above, we adapt the idea of reflective actors from Marx ([1867] 1967) but also from Weber ([1918] 1968) and from widespread practice of historians (Bloch [1942] 1953, Braudel [1949] 1976, Martin 1977).

Actors' decisions follow from two principles and a law. The first principle of the theory holds that

P_1 ALL SOCIAL ACTORS ACT TO MAXIMIZE THEIR EXPECTED PREFERENCE STATE ALTERATION (Willer 1981a, p.28).

The principle is sufficient to govern a variety of decisions, such as actors' choices among networks and between positions in a network, acceptance or rejection of a price, and so forth. The conditions of these decisions are parametric, that is, the payoffs of the alternatives are fixed (Elster 1986, Willer 1992a). The limits of a decision rule like P_1 in exchange have been known since Edgeworth (1881), who explained that, for the dyad, the rate of exchange is indeterminate. The rate is indeterminate because actors with opposed interests cannot jointly maximize. In fact, this indeterminacy is more general than exchange and occurs in any relationship in which actors' interests are opposed but complimentary, including coercion. "Resistance" was developed to predict for relations which are beyond the determinate predictive scope of the rationality principle (Willer 1981b, 1984, Heckathorn 1983).

Resistance in Three Social Relationships

Given in Figure 2 are three relationships with payoffs displayed. Figure 2a is an exchange relation in which A has unit resources, each of which costs A one to transmit and benefits B one when received. B has a single unit resource, costing nothing to transmit and benefiting A ten when received. Because of the differential valuation of B's sanction there will be a gain of ten to be distributed in the relationship; when P_i is the payoff gained by i, $P_A + P_B = 10$. Since no actor will exchange for less than a minimal payoff, which is stipulated here to be one, the best result for which either A or B can hope, $P_{A\,max}$ or $P_{B\,max}$, equals nine, which, principle one asserts, corresponds to A and B's initial offers, respectively. If no further offers are forthcoming, however, the two will not agree and the relation will be in confrontation at no exchange and $P_{A\,con} = P_{B\,con} = 0$, because no sanctions flow.

The concept of resistance asserts that actors are motivated by two interests: (1) $P_{A\,max} - P_A$, the interest in gaining the best payoff and (2) $P_A - P_{A\,con}$, the interest in avoiding confrontation. Resistance of A is defined as the quotient of the two interests, as follows:

$$R_a = \frac{P_{A\,max} - P_A}{P_A - P_{A\,con}} \ .$$

And, according to the second principle:

COMPROMISE OCCURS AT EQUIRESISTANCE FOR UNDIFFERENTIATED ACTORS IN A FULL INFORMATION SYSTEM (Willer 1981b).

Thus, for the exchange relationship in Figure 2a,

$$R_A \ = \ \frac{9 - P_A}{P_A - 0} \ = \ \frac{9 - P_A}{P_B - 0} \ = \ R_B \ .$$

a. Exchange

b. Coercion

c. Profit Point

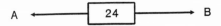

Figure 2.   Three Types of Relations

Since, as noted above $P_A + P_B = 10$, we solve for equiresistance: $P_A = 5$ and $P_B = 5$. In terms of sanction flows, resistance predicts that B transmits its one resource, worth ten, to A in exchange for A transmitting five resources, each at a loss of one to A and a gain of five for B. Resistance explains this prediction by the balance of the two interests between the two actors.

The relation in Figure 2b is coercive and, applying principle one, C (the coercer) has an interest in extracting valued resources from D (the coercee) by threatening to transmit the negative. If the threat succeeds, D will transmit some positives to C; and C, avoiding the cost of transmission, will not send the negative. Since C transmits the negative only if the threat fails, in coercion and unlike exchange, only one and not both sanctions flow. Since we assume reflective beliefs, we assume that D's beliefs fit the relation as modeled. That is, the model contains C's negative, so the assumption of reflection implies that C's threat is believed by D. (Alternatively, had D not believed the threat, D's believed structure when drawn would contain only D's and not C's sanctions— and for that belief D has no interest in transmitting any sanctions to C.) Since the threat is believed, D will transmit some number of resources between zero and nine to C. Thus $P_{C_{max}} = 9$, $P_{D_{max}} = 0$, and $P_C = - P_D$.

As in exchange, principle one alone does not solve the relation and principle two and resistance equations are used to find the predicted compromise rate. When agreement does not occur the negative is transmitted such that $P_{C_{con}} = -1$ and $P_{D_{con}} = -10$. Thus

$$R_C = \frac{9 - P_C}{P_C + 1} = \frac{0 - P_D}{P_D + 10} = R_D$$

with $P_C = 4.5$, and $P_D = -4.5$. Thus, resistance predicts that D will send (on the average) 4.5 sanctions to C in order to avoid C's negative. Resistance explains this prediction by the balance of interests in gaining the best payoff and avoiding confrontation—a balance occurring in *both* actors.

In the profit point relations of 2c, actors negotiate over division of a pool which, in this case, contains 24 points. Thus $P_A + P_B = 24$ and, applying principle one, $P_{A_{max}} = 23$ and $P_{B_{max}} = 23$. Since no profit points are gained in the absence of agreement, $P_{A_{con}} = 0$ and $P_{B_{con}} = 0$. Applying resistance,

$$R_A = \frac{23 - P_A}{P_A} = \frac{23 - P_B}{P_B} = R_D$$

and $P_A = 12$, $P_B = 12$ and a $12/12$ division is predicted. It is important to note, however, that the prediction is $12/12$, not because $12 = 12$, but because $12/12$ is the point of equiresistance.

Ordinary language definitions of power usually refer to a condition in which one actor is advantaged at the expense of another (Willer 1992b, Willer et al. 1989, p.314ff). When interests are opposed, as in the relations just discussed, however, arguably all rates advantage one actor at the expense of the other and all settlements are power events. Thus, the problem before us is to differentiate power events in which actor *i* is exercising more and actor *j* less power from the reverse. To do so we will stipulate an equipower baseline rate; then the actor receiving better than the baseline rate is exercising (more) power. The stipulation is straightforward in exchange and profit point relations. For both, equiresistance represents a balance of interests and that rate is the stipulated equipower baseline. For example, in the Figure 1a exchange relation, if A pays less than 5 for B's sanction, A benefits more, B benefits less, and A and is exercising power, whereas if A pays more than 5, B benefits more, A benefits less, and B is exercising power—and similarly for deviations from the $12/12$ baseline of profit point relations.

For coercion, however, the equiresistance rate cannot be the stipulated equipower rate, because it is inconsistent to treat equiresistance in both exchange and coercion as equipower. This can be seen by adding a negative sanction to an exchange relation. The addition produces a mixed coercive/exchange relation in which the equiresistance point has been shifted favor to the coercer. For example, allowing A of the exchange relation in Figure 2a the same negative as C in the coercive relation of Figure 2b shifts the equiresistance rate from $P_A = 5.00$ to $P_A = 5.77$, which advantages A and disadvantages B. That is, the shift is a power exercise in the meaning adopted above. Thus we adopt the following nominal definition:

EQUAL POWER occurs at the equiresistance rate when $P_{A_{con}} = P_{B_{con}}$.

One result of this stipulation is, consistent with ordinary language discussions, all coercive relations are differential power relations (cf. Bierstedt 1950, p.162ff). That is, for Figure 2b, since $P_{A_{con}} > P_{B_{con}}$, A is exercising power over B, and similarly for all other coercive relations.

Stated now in terms of power, for the three dyadic relations resistance predicts equipower settlements in exchange and profit point relations and predicts that A will exercise power over B in the coercive relation. The subsections which follow trace the power conditions of structures composed of each of these relations and report cross-national experimental results.

Centralized Exchange Structures: Models, Predictions and Cross-National Results

Exclusion, not shape, determines power in exchange networks. That is, all three exchange networks of Figure 3 are centralized at A, but only in Figures 3b and 3c, where some B's are excluded, and not in Figure 3a, where no B is excluded, is power predicted to be centralized at A. Let N_i be the number of relations in which i can exchange and M_i be the maximum number in which i can exchange under given conditions. Then if $N_i = M_i$, i is null connected, and if $N_i > M_i$, i is exclusively connected. In the Figure 3 networks, $N_A = 3$ in Figure 3a, $N_A = 4$ in Figure 3b, and $N_A = 5$ in Figure 3c; but for all three we have set $M_A = 3$. Using the N_i/M_i convention describing branch networks, in Figure 3a the 3/3 network $N_A = M_A$ and A is null connected; but Figures 3b and 3c are 4/3 and 5/3 branches and, because $N_A > M_A$, A is exclusively connected in both. (See Willer 1992b.)

The 3/3 network is constructed by connecting three of the 1a exchange relations at A and multiplying A's resources by three. For the network as a whole,

$$R_A (3/3) = \frac{3(9 - P_A)}{3P_A}$$

a. A 3/3 null connected structure

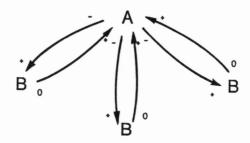

b. A 4/3 exclusionary structure

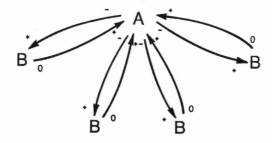

c. A 5/3 exclusionary structure

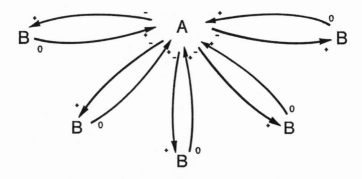

Figure 3. Three Exchange Structures

and, since $R_A(3/3) = R_A$, null connection has no effect on rates of exchange. Thus, resistance predictions for the 3/3 network of Figure 3a are exactly the same as for the dyad; in each relationship of Figure 3a, A will transmit five sanctions in exchange for B's sanction.

In exclusionary networks like 4/3 and 5/3, terms of the resistance equation which had been constants are altered by the structure in the following way. Assume for 3b that A and three of the B's have negotiated the equiresistance rate and, before finalizing the three agreements, A turns to the last B. Since three B's have already offered to exchange their resources for five from A, the last B cannot hope to receive nine. That is $P_{B_{max}}$ is no longer 9, but has declined to 4, one less than the offers of the other B's.

Furthermore, A has already negotiated three exchanges and, since $M_A = 3$, can only exchange with three. Failing to exchange with the fourth B, A will receive 5 from another B and $P_{A_{max}}$ is no longer zero, but has increased to five, such that

$$R_A = \frac{9 - P_A}{P_A - 5} = \frac{4 - P_B}{P_B - 0} = R_B$$

with $P_A = 7.5$ and $P_B = 2.5$. Exclusion having produced A's power exercise over the fourth B will subsequently produce a similar power exercise over two more B's. But now with three offers of $P_A = 7.5$, $P_B = 2.5$, a new resistance analysis shows that the remaining B must adjust its offer further to $P_A = 8.4$ and $P_B = 1.6$—and similarly for further adjustments by the other B's. The end point of the power process—for that is what exclusion produces—is offers for which $P_A = 9$ and $P_B = 1$, the maximum power exercise. The development of power in 1c is faster than 1b because the 5/3 structure has a higher proportion of exclusions than does the 4/3 structure.

Experiments were constructed using the three models of Figure 3. Following from the exchange relation, subjects in A were allocated ten resources worth one to A or B and B was allocated one resource worth ten only to A. These resources took the form of poker chips with dollar or zloty value. Because the 3/3 structure connects three exchange relations and A can exchange in all, A's resources were multiplied by three such that A subjects initially held thirty resources and each B held one. Though the 4/3 structure connects four relations at A, A was limited to exchanging only in three so A was again allocated thirty resources and similarly for the 5/3 structure. A and B subjects were seated face-to-face in the experimental room, but the peripheral B's, not being connected by exchange relations, were separated by dividers. Dividers eliminated eye contact but allowed all to hear the offers of each.

Prior to the experiment subjects read instructions outlining the values of resources and rules of negotiation. Experiments were organized into rounds, periods, and runs. For each structure there were three runs at each location, each with a different group of subjects. There were as many periods per run as there were subjects in the structure. Thus the 3/3 structure had four periods; the 4/3, five periods; and the 5/3, six periods. Each period was divided into four rounds; within each round, subjects negotiated and exchanged. Each round ended when A had completed three exchanges or, failing that, when five minutes had elapsed. At the end of each period each subject was moved to a new position in the structure, with the experiment ending when each subject had occupied each position for four rounds. Rotation introduces systematic control for subject differences. If some subjects are more effective negotiators than others and gain better rates, due to rotation through the structure, these differences cancel when means are calculated.

There were a number of important similarities among the three experiments. In all three A held thirty resources, each B held one and A exchanged with as many as three Bs. Predicted experimental outcomes were, however, quite different. In the 3/3 structure, A was predicted to gain fifteen, five in each of three relations and each B to gain five. Both 4/3 and 5/3 were predicted to have power processes beginning at rates like the 3/3 and moving toward the extreme favoring A. At that extreme A was predicted to gain 27 and the three B's not excluded each to gain the minimum of one.

Let us now turn to experimental results. Given in Table 1a are average exchange rates for the Figure 3a structure, from United States and Poland. Exchange rate (Ex) is defined as the number of sanctions transmitted by A divided by the total gain possible for the relation. It was already established that $P_A + P_B = 10$, so the total gain in the relation is ten and, since at equipower A transmits five sanctions, the equipower exchange rate is Ex $= 5/10 = .50$. When Ex $> .5$, the B's are exercising power and when Ex $< .5$ the A is exercising power. As can be seen, the mean, mode, and median values for the Polish data fit the resistance predictions slightly better than do those from the U.S trials. In both cases, the 3a rate of exchange slightly favors the B's over A indicating a small power exercise from periphery to center which is particularly weak in the Polish data.

When the United States rates were first published, it was speculated that the power exercise by the Bs was due to the open face-to-face relations within which the three B's could jointly pressure the A to gain better rates. This speculation was subsequently supported. Later studies of 3/3 branches found little or no power exercise when subjects' communications were limited (Skvoretz and Willer 1991). Thus, current evidence favors the inference that the observed Ex $> .50$ power exercise was due to the B's jointly pressuring the A. But why United States subjects applied more pressure when acting as B's, or were more sensitive to the pressures when acting as A's we cannot say.

Turning now to the structures of Figures 3b and 3c, average rates given in Tables 1b and 1c indicate substantial power exercise in both Poland and the

Table 1a.   Centralized Exchange Structure with No Exclusion
GPI = 1
(Three Experimental Groups)

RATE OF EXCHANGE	U.S. DATA	POLISH DATA
MEAN	0.69	0.55
MEDIAN	0.70	0.60
MODAL	0.70	0.50

Table 1b.   Centralized Exchange Structure with One Exclusion
GPI = 4/3
(Three Experimental Groups)

RATE OF EXCHANGE	U.S. DATA	POLISH DATA
MEAN	0.33	0.26
MEDIAN	0.30	0.30
MODAL	0.20	0.30

Table 1c.   Centralized Exchange Structure with Two Exclusions
GPI = 5/3
(Three Experimental Groups)

RATE OF EXCHANGE	U.S. DATA	POLISH DATA
MEAN	0.22	0.24
MEDIAN	0.20	0.20
MODAL	0.10	0.20

Source: of U.S. Data: Willer 1987, p. 73

United States for both structures. These mean, mode, and median values are consistent with the predicted power process. As predicted by the resistance analysis, more power was exercised in the 5/3 than in the 4/3 structures for both U.S. and Polish cases. Examination of Polish and United States data indicate power processes in *all* experimental runs with rates moving to the extreme of Ex = .1.

More generally, the averages given in Table 1 indicate a similarity within structural type regardless of location. That is, only in the Figure 3a structure were B's able to exercise any power over A, and they did so, in both Poland and the United States. Furthermore, in both settings the mean rates for the 4/3 and 5/3 exclusionary structures are quite distinct from the mean for the 3/3 null structure. Comparing between power structures, the mean for the Polish Figure 3b structures is closer to the parallel United States structures than it is to the Polish Figure 3c structures, and the mean for the Figure 3b structure in the United States is closer to the Polish Figure 3b structure than it is to the United States Figure 3c structure—and similarly for comparisons across settings and structures for the data in Table 1c. Interestingly, though the theory was first developed and applied in the United States, the poorest fit between theory and data is the U.S. 3/3 structures.

These results show what we hope to find in the rest of the experiments. They show that, regardless of setting, theoretically similar experimental structures have empirically similar rates. Predictions are supported by the results in both settings If and only if results continue this pattern across a variety of further models and experiments should it be concluded that the theory is supported across settings.

Centralized Coercive Structures: Models, Predictions and Cross-National Results

Power exercise can be traced either to (1) conditions of relationships or (2) conditions of structures. In the exchange structures studied above, because the relationships did not produce power, all power exercise was due to structure. By contrast coercive relations are power relations. Thus, a *structure* with coercive power can be constructed by simply connecting coercive relations, as in Figure 4. If so, there is power *in* the structure which is *not due to* the structure, but *due only to each of the relations*.

In fact, one of the three coercive structures investigated had relational power and no structural power. Call it the A structure, which was built by connecting five coercive relations like the one discussed above. As displayed in Figure 4, five relations are connected at the coercer. As in the dyad, each D initially holds ten resources and each resource is worth one to C or D. C holds negatives which cost C one to transmit and cost D ten when received. As in the coercive dyad, each D is predicted to transmit 4.5 sanctions to the C. Let the rate of coercive exploitation, C_x, be defined as the ratio of resources transmitted to resources initially held by the coercee. Then, for the dyad and for each relation of the A structure, $C_x = .45$ is predicted at equiresistance. $C_x = .45$ is due to relational power alone. To produce $C_x > .45$, structural power analogous to the exclusionary exchange branches will be required.

Because both sanctions do not flow in coercive relations, it is not necessary to arm C in the A structure with five sanctions. As can be seen in the following

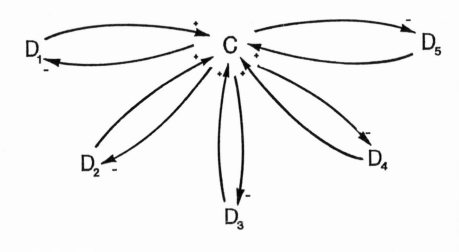

Figure 4.   A Coercive Structure

example, if C's threats are believed, only one negative sanction will serve. Assume that C has only one negative sanction and let C negotiate first with D_1, concluding negotiations by receiving 4.5 resources. Now A still holds its one negative sanction and turns to D_2, again concluding negotiations by receiving 4.5 sanctions—and similarly for the remaining D's. That is, for the predicted outcomes, the transmission of 4.5 sanctions by each of the five D's, C transmits *no negatives whatever*. Since exactly the same outcome is predicted for the A structure regardless of the number of negatives initially held by C, in the experiments C will be allocated two. The reason for two and not some other number will become evident as we turn to the B and C structures.

The coercive A structure is analogous to the null connected exchange branch because structure has no effect on rates. We now construct coercive structures which are analogous to the Figure 3b and 3c structures—that is, coercive structures which, like exclusionary exchange structures, have structural power. In fact, they will have both relational power *and* structural power. Structural power occurs because of the structural allocation of costs of confrontation. Looking back to the 4/3 exchange branch, A was powerful because one of its four relations had to end in confrontation, a confrontation which had no cost to A but which deprived one of the Bs of any gain from exchange. In the 5/3 branch A was powerful because two of its five relations had to end in confrontation at no cost to A but depriving two of the Bs of any gain from exchange. More generally, all exclusionary exchange branches are power structures because (1) exclusion is confrontation in exchange and (2) the costs of that confrontation are borne not by the central actor, but only by one or more peripherals.

In coercion, transmitting the negative is confrontation and structural power will be produced by negative sanctions which the coercer has an interest in transmitting. One such sanction is here called a *con sanction*, for its effect is to confiscate the resources of any D receiving it. Each con sanction is worth one to C and, when transmitted, moves all D's resources to C. When received by D, $P_D = -10$, which is exactly the same effect on B as the negative sanction of the A structure. Its net payoff to C, however is $P_C = -1 + 10 = 9$, not $P_C = -1$, as in the negative sanction of the A structure. Thus, C has an interest in transmitting the con sanction unless at least nine of D's sanctions are received. Since D prefers to transmit nine sanctions, for which $P_D = -9$, than to receive the con sanction, for which $P_D = -10$, we predict Ex = .9 and no con sanction transmissions when information is perfect and complete. Conversely, when the con sanction is transmitted, D's sanctions are confiscated and are *not* extracted by force threat. Thus, when calculating rates of coercive exploitation, no resource movement from D to C produced by confiscation is counted. At issue is the effect of the con sanction on the relations in which it is not used.

A coercive structure analogous to the 5/3 exchange structure is built by allocating two con sanctions to a C connected to five D's as in Figure 3. Call this the C structure. As in exchange, two confrontations can occur which are costless to C, but costly to two D's. Also like the 5/3 exchange structure, the negotiations in each branch are not independent. On the contrary, each D knows that the C has an interest in transmitting the two con sanctions to the two D's making the worst offers to C, just as the B's in the exchange structure know that the two making A the worst offers will be excluded. Thus, rates of coercion should increase over time to approach $C_x = .9$ in all of C's relations. At that point, at least three D's will be transmitting nine sanctions to C. For the remaining two relations, C will be indifferent between transmitting the con sanction and receiving nine sanctions from those D's. Also investigated is a B structure in which C is allocated one of each type of negative sanction. The B structure is analogous to a 5/4 exclusionary exchange structure and, like the C structure, is predicted to move to the extreme $C_x = .9$ rate though more slowly.

There are a number of important similarities among the three coercive structures. In all, there are five D's with each initially allocated ten resources, while the C is allocated two negatives, each of which costs a D ten points when received. Differences among the structures are due to the presence or absence of structural power. Structural power is absent in the A structure when C's sanctions are like those of the relationship modeled in Section 3.1 and $C_x = .45$

is predicted. Structural power is present in the B and C structures due to the presence of con sanction(s) and C_x is predicted to approach the $C_x = .9$ extreme. The coercive experimental design was similar to the design of the exchange experiments in many regards. Subjects read instructions prior to the experiment which gave the values of the resources as traced above and resources were represented again by poker chips, in this case, red for C's negative and white for D's positives. Like exchange experiments, subjects were paid by points earned. Again experiments were organized into rounds, periods and runs. As displayed in Tables 2a and 2b, there were eight runs for each structure in the United States and six in Poland. Since each experiment contained six subjects, each had six periods. Like the exchange experiments, between periods subjects moved, such that each occupied each position for one period. Again each period was divided into four rounds of negotiation and sanction transmission—each maximally five minutes. Also like the exchange experiments, peripheral subjects not connected by relations—in this case the D's—were separated by dividers, but C and the D's were face to face.

Consider, first, average rates of coercive exploitation by experiment and by type. For all A structures in Poland the mean $C_x = .538$, and for the U.S., the mean $C_x = .536$. While the two were gratifyingly similar, both indicate a greater coercive power exercise by the C than the predicted $C_x = .45$ equiresistance rate. Interestingly, there is a greater range of rates between experiments in the United States, ranging from .476 to .601 while the Polish data cluster more closely, ranging only from .519 to .561. The overlap of means is complete, in that all Polish means fall inside the range of the U.S. means. While this emphasizes the similarity of results between settings, we offer no speculations concerning why the U.S. runs were more variable.

Moving to data for the C structure, means for both settings closely approach the extreme rates of $C_x = .90$, but Polish means were slightly higher at .877 than those for the United States trials, at .841. Looking at averages by experiment, there was less overlap between settings than for the A structure. In the United States the highest $C_x = .878$; two of the six Polish experiments were lower than that rate. By contrast, comparing across the B structures we find that, like the A structures, the overlap of means is complete; all Polish means lie between the U.S. extremes of .749 and .883.

The number of negatives transmitted in each experiment was less regular. For the Polish A structure in one experiment, only three negatives were transmitted and at most nine were transmitted, while, for the A structures in the United States the range was from three to sixteen. Ranges are comparably large in B and C runs in both settings. In both settings, there appears to be no relation between transmitting negatives and receiving positives. More specifically, transmitting more negative sanctions did not increase the average rate of coercive exploitation. This result would be surprising from an operant

Table 2a. Data for Part One of the Experiments on Weak and Strong Centralized Coercive Structures: U.S. Data

Type of Structure	A								B								C							
Group Number	2	3	7	6	4	5	1	8	5	2	3	4	8	6	1	7	7	2	5	8	6	4	3	1
Average Rate of Coercive Exploitation	.476	.477	.494	.529	.547	.577	.599	.601	.749	.761	.785	.789	.813	.835	.878	.883	.734	.827	.831	.850	.865	.870	.876	.878
Number of Negative Sanctions Transmitted	9	6	16	5	3	2	6	5	3	4	25	12	11	0	23	26	22	45	34	26	10	28	9	12
Average Rate of Coercive Exploitation by Type	.538								.812								.841							
Average Number of Negative Sanctions Transmitted by Type	6.5								13.0								23.3							

Source: Willer 1987, p. 132

63

Table 2b. Data for Part One of the Experiments on Weak and Strong Centralized Coercive Structures: Polish Data

Type of Structure	A						B						C					
Group Number	3	5	6	1	2	4	1	3	6	4	2	5	4	1	2	5	3	6
Average Rate of Coercive Exploitation	.519	.525	.529	.530	.552	.561	.778	.782	.796	.834	.842	.882	.860	.864	.880	.881	.886	.888
Number of Negative Sanctions Transmitted	9	8	9	6	7	3	2	24	17	19	0	11	15	13	22	22	9	23
Average Rate of Coercive Exploitation by Type	.536						.819						.877					
Average Number of Negative Sanctions Transmitted by Type	7.0						12.2						17.3					

point of view—that is, from a point of view that saw positives as being extracted by the application of negatives.

Our models, however, see positives as extracted, not by negatives, but by threat of negatives (see above). Furthermore, positives extracted by confiscation are not part of the calculated rates. Thus, D's who believe C's threat will transmit positives and will not await the arrival of negatives to do so; and only if the threat is not believed will the negative be transmitted. Since subjects' propensity to believe threats is not controlled experimentally, the numbers of negative sanction transmissions can vary widely. Nevertheless, there is a rough overall relation between cost of transmission of the negative and numbers transmitted for both settings. That is, transmission of any negative in the A structure always cost C one while transmission in the C structure always gained C nine, with either being a possibility in the B structure, depending upon the type transmitted. Thus, it is hardly surprising that the largest numbers of negatives were usually transmitted in the C, the smallest usually in the A, and the B structure was typically between.

The results given in Table 2 are for a total of 42 experimental runs, more than twice the number reported for the exchange experiments. For both settings, B and C means were near the predicted extreme of $C_x = .90$. Less gratifying is the fit between the A structure rates and the predicted $C_x = .45$ equiresistance rate. Very satisfying, however, are the similarities between settings for each structural type. Clearly the United States and Polish A structures are very similar to each other and much more similar than they are to the B and C structures within either setting. The B structures are also very similar between settings for, like the A structures, all Polish means were within the range of the United States means. Only the C structure rates range apart between the two settings, but again there are greater similarities by type than differences by setting. That is, the Polish C structures are more like the U.S. C structures than the Polish B structures and similarly for the United States.

Collective Action and Countervailing Structural Power

The coercive experiments were extended to a second part. In the second part, the D's actions, which had previously been independent, are allowed to become collective. Whereas collective action is known to countervail power (Olson 1965), we will predict that only structural power and not relational power is countervailed. Though finer distinctions will be made below, the general effect of collective action on the A structure is nil. That is, since the A structure has only relational power, the D's can do no better acting collectively than they did individually. By contrast, the general effect of collective action on the B and C structures is to eliminate structural power, rolling coercive exploitation back to lower rates.

Prior to the beginning of part one (as reported in the previous section), subjects were told that the study had two parts. However, the nature of the second part was not revealed at that time. At the conclusion of part one, a half sheet of new instructions was handed out setting the conditions for either collective group or normative group formation among the B's. Either formation allowed B's to act collectively, but normative group structure could become more cohesive (see Willer 1987, p.144ff). It must be emphasized, however, that the instructions only modified experimental conditions such that group formation was possible. No mention was made of forming a group and no suggestions put forward regarding the purposes joint action could serve. Nevertheless, in all but one case, subjects formed groups and attempted collective action.

There are three types of coercive structure, and two types of collective action for each, giving a total of six experimental types. Each of the six subsections to follow briefly discusses one of the types. Resistance is applied to predict the rate of coercive exploitation and the prediction is compared to data from the two settings. (For a more extensive discussion of each type, see Willer 1987.) For part two, the following experimental conditions were common. Subjects were no longer rotated and the subject scoring the most points in part one was made C for the rest of the run. Selection of the top scoring subject was preferred to random selection for it assured at least a minimal level of competence at the C position. Ten more rounds of negotiation and sanction transmission followed then the experiment concluded and subjects were paid for points earned in both parts.

The A Structure with Collective Group

When D's act individually, as in the first part of the experiment the C sequentially extracted value by threat. As a result, C's two negative sanctions had the same effect as five, one for each relation. Since a collectivity cannot share the costs of confrontation, C's two sanctions applied in sequence still do the work of five. Aggregating resources and payoffs across the collectivity, $P_{C_{max}} = 45$, $P_{C_{con}} = -2$, $P_{D_{max}} = 0$, and $P_{D_{con}} = -50$. Thus,

$$R_C = \frac{45 - P_C}{P_C + 2} = \frac{0 - P_D}{P_D + 50} = R_D$$

and $P_C = 23$, which means that each D transmits 4.6 sanctions for a $C_x = .46$ rate in each relationship. This is effectively the same as the rate for the A structure with D's acting independently.

As given in Tables 3a and 3b, the grand means that for the two settings bracket the predicted rate: the Polish rate, $C_x = .481$, is higher and the U.S.

rate, $C_x = .393$, is lower. Contrary to our prediction that collective action would not affect rates, however, both are substantially lower than rates from the first part when D's acted independently. We offer no explanation for this anomaly: either something more is needed in theory or our experiments did not exactly fit the models, or both. We note, however, that the result is very robust: the mean rates for all Polish and all United States groups were more favorable to the D's after collective group formation than before.

The B Structure with Collective Group

Collective group formation stops structural power such that only relational power remains. Relational power for the B structure is calculated exactly as it was for the A, but now C holds one sanction of each type and $P_{C_{con}} = 10 - 1 - 1 = 8$. That is, C's "cost" of confrontation is now a benefit and

$$ R_C = \frac{45 - P_C}{P_C - 8} = \frac{0 - P_D}{P_D + 50} = R_D $$

with $P_C = 26$ and $P_D = -26$ meaning that each D transmits, on the average, 5.2 sanctions and $C_x = .52$ is predicted. This rate is substantially lower than the $C_x = .90$ extreme predicted when the D's act individually.

Again the grand means for the two settings bracket the prediction; as given in 4a and 4b, $C_x = .562$ for the Polish data and $C_x = .468$ for the United States data. Note that there is one U.S. run in which the rate is very low, at $C_x = .115$. Ignoring that case, the U.S. grand mean is .583, very close to the Polish values. (We will discuss cases of very low C_x rates like the $C_x = .115$, below.) In all Polish cases, collective action was effective at lowering the mean rate of coercive exploitation; however, collective action was successful in only three of the four U.S. cases. Of the 28 cases in which collective action was predicted to countervail power, only in B2 was the rate of coercive exploitation higher in part two than part one. The failure to countervail power is clear evidence that collective action failed in that case.

The C Structure with Collective Group

Given that the collective group eliminates structural power in the C structure, the equiresistance rate can be calculated as it was for the B structure. In fact, all values are the same but for $P_{C_{con}} = 18$, because C holds two con sanctions. Plugging the values into the resistance equation we find that the predicted rate is $C_x = .58$, slightly higher than the B structure because of C's increased benefit from confrontation. The grand mean for the U.S. data is $C_x = .588$, very close to the predicted rate and lower than the Polish value of $C_x = .724$. The near

Table 3a. Data for Part Two of the Experiments on Weak and Strong Centralized Coercive Structures: U.S. Data

Type of Structure	A								B								C							
Type of Resistance Group	Collective				Normative				Collective				Normative				Collective				Normative			
Group Number	3	1	2	4	6	5	7	8	1	3	4	2	6	8	5	7	4	2	1	3	8	5	6	7
Average Rate of Coercive Exploitation	.282	.391	.434	.464	.135	.181	.184	.261	.115	.400	.567	.789	.151	.250	.313	.327	.313	.581	.680	.776	.000	.044	.200	.265
Number of Negative Sanctions Transmitted	5	3	3	5	7	3	2	6	12	10	2	0	19	8	12	15	19	19	15	5	20	18	18	10
Average Rate of Coercive Exploitation by Type	.393				.190				.468				.260				.588				.127			
Average Number of Negative Sanctions Transmitted by Type	4				4.5				6				13.2				14.2				16.2			

Source: Willer 1987, p. 147

68

Table 3b. Data for Part Two of the Experiments on Weak and Strong Centralized Coercive Structures: Polish Data

Type of Structure	A		B		C	
Type of Resistance Group	Collective	Normative	Collective	Normative	Collective	Normative
Group Number	5 4 6	1 3 2	5 6 4	3 2 1	1 3 2	6 5 4
Average Rate of Coercive Exploitation	.464 .478 .502	.171 .191 .268	.424 .514 .748	.149 .253 .314	.603 .768 .800	.000 .141 .436
Number of Negative Sanctions Transmitted	5 4 1	1 3 0	8 2 0	3 7 8	20 6 15	20 16 8
Average Rate of Coercive Exploitation by Type	.481	.210	.562	.239	.724	.192
Average Number of Negative Sanctions Transmitted by Type	3.3	1.3	3.3	6	13.6	14.6

69

identity of United States and predicted rates is coincidental, for the U.S. data show one outlying case in which $C_x = .313$. The remaining U.S. cases have rates very similar to Polish cases. Though in most cases observed mean rates are higher than predicted, in *all* C structures of both settings collective action did succeed in lowering the rate of coercive exploitation below that observed when D's acted independently.

The A structure with Normative Group

The D's in the normative group can share resources after the C has transmitted negatives. Therefore, C's two sanctions can no longer do the work of five. In fact, they can only do the work of two. Thus now $P_{D_{con}} = -20$. Aggregating across the D group,

$$R_C = \frac{45 - P_C}{P_C + 2} = \frac{0 - P_D}{P_D + 20} = R_D$$

and $P_C = 13.5$, $P_D = -13.5$. Thus, each D transmits 2.7 sanctions and $C_x = .27$ at equiresistance.

That C_x value will not, however, be treated as the predicted mean. Instead, we treat $C_x = .27$ as the upper bound. $C_x \leq .27$ because the D's can jointly adopt the strategy of offering C nothing. Offering no sanctions to C is a reasonable strategy, because then the D's can share the costs of confrontation. If they do—even if C always transmits both negatives—each D will only lose 4 resources. Thus our prediction is $C_x \leq 2.7$. In fact, one case from each setting is close to the predicted maximum while the remaining cases are substantially lower. As predicted, all normative groups in both settings were more successful than any collective group from either setting, with all mean rates indicating that power was substantially countervailed.

The B structure with Normative Group

As in the normative group above, the D's can share the costs of confrontation and C's two sanctions do only the work of two. Since one of C's sanctions is a con sanction $P_{C_{con}} = 8$, and other values are as above. Thus, equiresistance occurs when $C_x = .316$. Again, offering no sanctions to C is an alternative strategy and our prediction is $C_x \leq .316$. Both United States and Polish results show a similar range, from slightly above the predicted range to a low of .151 and .149, respectively.

The C structure with Normative Group

Again, the D's can share the costs of confrontation but now the C has two con sanctions and $P_{C_{con}} = 20$. Thus, $C_x = .38$ at equiresistance. But if the D's offer no resources whatsoever and C transmits both sanctions, the D's can share the thirty resources remaining. In that case each will earn six resources per round, which is only slightly less than $10 - 3.8 = 6.2$ earned at the equiresistance rate. Thus, a fully revolutionary strategy by the B's is more reasonable here than in any other scenario.

The zero rates indicate that the revolutionary strategy of offering C nothing was adopted from the beginning, while rates like .044, .141, and so forth indicate that the D's adopted that strategy after a few sanctions were sent. With only one exception, in both settings all had very low mean rates of coercive exploitation. The exception was a Polish case where $C_x = .436$, which is higher than the predicted rate but lower than $C_x = .860$ for that group prior to coalition formation.

Summarizing the results of part two of the experiment, in both settings collective action countervailed power to rates like those predicted by resistance. For the forty-two experiments only in one case did subjects fail to successfully organize. Both settings evidenced similar contrasts among types, contrasts which corresponded well to predictions. Rates within experimental type were quite similar between settings. These results strongly support the cross-national validity of ET.

A Complex Profit Point Network

Prior to its replication in Poland, the "T" network shown in Figure 5 provided what may be the first decisive critical test between any pair of social structural theories. The theories are the vulnerability procedure of Power-Dependence Theory and Elementary Theory's GPI procedure (Cook et al. 1983, Willer 1986, Cook, Gillmore, and Yamagishi 1986, Markovsky, Willer, and Patton 1988, Willer, Markovsky, and Patton 1989, Markovsky, Willer, and Patton 1990, and Yamagishi and Cook 1990). Contrary to Lakatos (1970), critical tests can provide a decisive basis for choice between theories. The basis for choice is decisive when it can be shown that one theory but not the other predicts incorrectly due to contradictions. This method of falsification was first used by Galilei ([1665] 1954). Cross-national replication allows us to check further the contrasting predictions of the two theories.

Vulnerability defines *network flow* as the number of profit points which are divided, and RMF as the *reduction from maximum flow* resulting from the removal of a relation. When RMF = RMF$_{max}$, the position is high in power. In the T network of Figure 5, all positions are limited to a single exchange and the maximum flow for the network occurs when B divides with either A

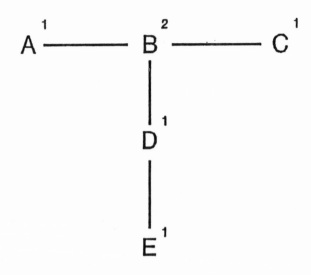

Figure 5. A 1-Exchange Network with Initial GPI Values

or C and D divides with E, such that $24 + 24 = 48$. Removing either A or C (but not both), RMF = 0 because two divisions are still possible. That RMF = 0 indicates that A and B are low in power. Removing B, only D and E can divide and RMF = 24, indicating that B is high in power. Removing D, only B can divide profit points (with either A or C, but not both), and RMF = 24, indicating that D is high in power. Removing E, only B can divide profit points (with only one of A, C, or D), RMF = 24, and E is high in power. Since high power positions gain more than half of their divisions, vulnerability predicts that B, D, and E all gain more than one half of the profit points between them, a mathematical impossibility. Nevertheless, this prediction will be checked against data from both settings.

GPI assigns a numerical index to each position by counting nonintersecting paths from a position. Odd-numbered paths which are advantageous are added while disadvantageous even-numbered paths are subtracted. Assigning a GPI index to each position allows the order of profit point divisions among positions to be inferred. A second step is needed to determine which positions will exercise power over which others. It is an axiom of the GPI procedure that all actors seek their best exchanges. Like all other positions in "T," D is limited to a single exchange. But D has two potential partners, each at a different power level. Thus, the theorist using GPI can infer that D will prefer to exchange with the equal-power E than to exchange with the higher-power B. Space will not allow us to further discuss and apply the GPI procedure, its equations, associated axioms and method of application, but all are available elsewhere (Markovsky, Willer, and Patton 1988, Willer, Markovsky, and Patton 1989, and Willer 1992b).

For the T network, GPI predicts that B is high in power and will exercise power over only A and C, gaining substantially better that 12/12 divisions. There will be no B-D exchanges. D and E will divide equally at 12/12. In both settings the experiment was set up much like the others discussed above. Again, only subjects connected by relations were in eye contact, while all others were separated by dividers. Since these were profit point relations, a pool of 24 counters was placed between each pair of positions connected in the T and instructions explained that all were limited to a single division per round. Four rounds made up a period and five periods made up a run. In both settings, subjects were rotated between periods so that each occupied each position during one period. In the U.S. there were five runs, and in Poland there were three.

All GPI predictions are supported in both settings. The results for each setting, displayed in Table 4, indicate that B exercised power over A and C. B did not exercise power over D in Poland because, as predicted, there were no B-D exchanges. If B exercised power over D in the U.S. it was trivial because there were only three B-D exchanges out of 100 opportunities, significantly fewer than could be attributed to chance. Furthermore, D-E divisions were very close to 12/12 in both settings. (We note that E did slightly better than D in both settings, but attempt no explanation.) These results also support vulnerability for profit points gained by A, B, and C. Mean profit points gained by D and E in both settings indicate that neither is a high-power position, however. GPI is clearly favored and vulnerability is falsified.

Were experimental results for this structure more similar cross-nationally than others? In Section 2.2 it was explained that, because profit point relations do not occur outside of laboratories, they may be less affected by cross-national variations than exchange or coercive relations. That is to say, if the particulars of culture affect actions in structures, the effect should be least for structures with profit point relations. In fact, the results for all experiments were quite similar, independent from the type of relation. Since experimental results do not support the view that the particulars of culture strongly interact, they do not support the view that cultural differences interact with different degrees of effect.

Table 4. One Exchange Networks

POSITION	PROFIT POINTS	
	U.S. DATA	POLISH DATA
A	4.81	4.64
B	19.21	18.95
C	4.91	4.36
D	11.83	11.67
E	12.12	12.36

Source: of U.S. Data: Markovsky, Willer, and Patton 1988, p. 227

THE AUTONOMY OF SOCIAL STRUCTURE

In sociology and other social sciences, points of view have been influential even when not supported by the slightest shred of evidence. For example, Homans summarized his reductionist view in the following way: "the institutions, organizations and societies that sociologists study can be analyzed, without residue, into the actions of men." (1971, p.378). Since no term of the referenced sentence is precisely defined by Homans, the statement may be unassailable due to lack of determinant meaning. Alternatively, it may have some unassailable meaning; it may mean, for example, that sociologists only study organizations peopled by humans and do not study organizations peopled by Martians. As received at the time, however, Homans' reductionism was not thought to be meaningless or trivial (Szmatka 1989a, 1989b, and 1989c).

Instead it was thought Homans was asserting that all of the conditions which determine behavior are inside individuals—a meaning contradicted in a number of ways by this research. For example, the conditions of structural power are not in any actor but in the connection of relationships. When a structure is analyzed into its component parts, relationships are disconnected. Since conditions of connection occur only in the structure and not in its parts, they must count as residue and Homans was wrong. The proof that he was wrong lies in findings such as those reported above. We doubt, however, that these proofs will deter future reductionists from equally groundless claims.

In this section we briefly consider the bearing of our cross-national studies on another commonly held view which, like reductionism, has importantly inhibited the growth of general knowledge in the social sciences. The view holds that particular cultural traits are internalized and influence (or produce) behavior independent from the structures in which it occurs. Harris (1968) calls this view *historical particularism* and traces its development through Dilthey, Rickert, and Windelband to Boas and U.S. anthropology. Historical particularism appears to be a very commonly held position and is found, for example, in the belief that Japanese organizations produce more effectively than U.S. (or Polish) organizations because Japanese culture strongly emphasizes cooperation (or obedience, or any of a number of similar traits). Historical particularism is a descriptively rich form of reductionism and, like Homans' more abstruse form, seems free from serious evidential support.

There is considerable evidence from field studies that does not support historical particularism, however. For example, two outstanding studies have offered explanations of Japanese productivity and both specifically deny historically particular appeals to Japanese culture. Womack, Jones, and Roos (1990) show that much of the advantage enjoyed by Japanese factories lies in a system of lean production, the effectiveness of which is importantly due to the greater responsibility given workers. Dore (1973) relates workers' willingness to take responsibility in the Japanese factory to its system of recruitment, retention, and promotion and to its much lower levels of stratification and profit. That is, Japanese workers holding lifetime jobs and being less exploited are consequently more responsible.

The experiments also do not support historical particularism. We discussed two possible threats to internal validity of experiments: (1) that there is a complex of psychological traits which alone produce experimental behavior, and (2) that there is a complex of psychological traits which interact with structural conditions to produce observed behavior. Either could be a form of cultural particularism. As already discussed, prior to the cross-national studies we were satisfied that the first was wrong because variations in structures between experiments produced variations in behavior. Prior to the studies, however, we could not address the second claim: that historically particular cultural/psychological traits *interact with* structural conditions to produce behavior. In fact, our cross-national studies offer no support whatsoever to the second claim. The striking quality of the array of findings reported above is not differences between settings and similarities between types of structures as historical particularism would expect. To the contrary, the striking quality is the very great similarity between settings and the practical identity of contrasts between structural types. The evidence is consistent enough that, in our view, cross-national experiments falsify the claims of historical particularism.

Cross-national replications support a position very different from historical particularism, which we term *structural autonomy*. Stated as a thesis, structural autonomy asserts that when actors' beliefs accurately reflect the conditions of the social structures in which they act, their acts are not affected by the particulars of culture. (For the issue of "reflection" see discussions above.) The structural autonomy thesis amounts to the denial that any culture independently produces a unique population of strategies which condition behavior in structures. At least one anthropologist, Malinowski (1959), specifically denies the existence of special strategies varying by culture.

Experiments are intended to produce reflective social structures by determining subjects' values through experimental payments and subjects' beliefs through information about the structure and its parameters. But the experiments do not control or determine the strategies used by subjects. Thus cross-national replications test for structural autonomy. Our contention is that, if there were such strategy differences, they would have been produced by U.S.-Poland differences and would have systematically pulled results of the two settings apart. But such was not the case. If the structural autonomy thesis is right and strategy does not differ by culture, future studies replicating the structures investigated here in new settings will find results like those we found in the United States and Poland.

The cross-national studies raise the possibility that application of ET can be very parsimonious; however, that parsimony rests to an important degree upon the extent to which structures determine the values and beliefs of people acting in them—that is, the extent to which structures are reflective. While we believe this determination is routinely realized in the laboratory, at issue is determination outside. Arguing for a simple reflective view is the far greater salience of sanctions outside the lab. Since reflective structures are easily produced in the laboratory where payoffs are almost trivial, reflection should also be expected in the field where the preference effect of sanctions is so much greater. It may be that structures in the larger society are reflective and yet they overlap such that values produced by one structure carry over to affect others. (cf. Willer 1992a.) If so, ET applications will be more complex. Determining whether applications outside the laboratory will be simple or complex, however, is clearly beyond the limits of this paper.

Nevertheless, we can answer the question asked in the introduction: "Why was Visigoth autonomy so long successfully defended?" ET's explanation uses an exclusionary coercive structure. As long as the Visigoths were physically between the two Empires they gained the advantages of structural power in that they could strike East or West, but not both—and benefited from either. By threat they could and did extract concessions from both Romes, one result of which was ongoing autonomy. Only when the Visigoths left that structurally favored position and moved far to the west did they form a conflict dyad with western Rome, the result of which was their allied and dependent status. The case suggests that the Visigoths, while able to employ a structural power advantage, were not aware of the conditions necessary to reproduce it. That is to say, the Visigoths were canny and opportunistic, as were the Romans. There is no reason to suppose, however, that Visigoths (or Romans) understood the general structural conditions under which they acted.

CROSS-NATIONAL EXPERIMENTS
AND CONFIRMATIONAL STATUS

In the studies reported above, the design of each experiment and the contrasts between experiments flowed from models drawn in relation to the theory. These were not empiricist experiments of the familiar experimental-control-group design as invented by Fisher (1935, 1956), but scientific experiments the design of which was developed in its modern form by Galilei ([1665] 1954). (Also see Willer 1987.) The theory was tested, not by finding differences between experimental and control groups, but through a series of predictions across its scope of application (Toulmin 1953, Lakatos 1970).

The ET investigations diagramed in Figure 1 are inquiries into a series of related problems. In answering earlier problems new problems come to be recognized. For example, the study of structural power in exchange structures led to the discovery of structural power in coercive structures. The study of exclusionary and null connection lead to the discovery of inclusionary connection and associated mixed forms. Investigation of profit point relations under the 1-exchange condition led to extensions to n-exchange conditions, while the limitations of power exercise in complex profit point networks led to the investigation of flow networks. In this way, the scope of the theory was systematically extended.

The Polish experiments were replications, being within an already established scope of the theory and using the face-to-face design. As the Polish studies were in process, a second experimental design was being developed in the U.S., called ExNet. ExNet places subjects in separate rooms, allowing them to negotiate through connected PCs. Skvoretz and Willer (1991) compared results from a number of structures between face-to-face and ExNet settings. Though minor differences were observed, as when power developed more slowly in ExNet than in the face-to-face setting, it was found that the theory predicted well in both designs. That results were similar even though the two experimental designs were not at all alike adds support to the Polish replications by demonstrating that the predictive power of the theory is not limited to a single experimental design. At issue now is what the cross-national replications add to the theory's confirmational status.

As noted in the introduction, theory is useful in prediction and explanation only if its application is not limited by specifics of time and place. That is to say, theory must claim universality. While any well-stated theory is universal in form, there should also be evidence that it is universal in application. The issue of universality is quite different from the issue of generality of scope and the two can vary quite independently. A theory which has been applied over a wide range of theoretic variation, but only to a limited population in one society, has broad scope but lacks demonstrated universality. In contrast, a theory that successfully spans an array of settings has universality, even if all applications were restricted to one narrowly limited theoretic scope.

Unlike the investigation of scope, which, as we have seen, proceeds systematically and directly, guided by theory, at times theory offers little guidance for the investigation of universality. In many cases, as Foschi pointed out, the conditions being contrasted, such as U.S. undergraduate and Polish graduate, are not part of the theory, yet might represent important limitations of prior work. Since there are an unlimited number of implicit conditions to be varied, does universality demand a large number and range of studies? Because each cross-national replication can vary a large number of implicit conditions, however, universality does not necessarily demand a large number and range of replications. For these replications, as we have shown, ET gave an important degree of direction to the investigation of universality. ET is a theory of social structures and, outside the laboratory, its applications include large power structures. Since many contrasts between prerevolutionary Poland and the United States are relevant to the experimental structures, if special conditions of time and place can confound results, those specific conditions should have confounded them. But they did not.

Therefore, we conclude that universality was supported while noting that the studies represented an almost unique opportunity for finding that support. We fervently hope that recent changes in Poland and elsewhere will decisively improve the quality of life. Whether they do or not, because those changes move in a capitalist direction, they are certainly eliminating many of the theoretically relevant contrasts which held through our study. Within this narrow window of opportunity, ET survived cross-national replications which the theory suggests might well have confounded it. It produced its phenomena quite similarly in spite of important differences between the two settings, thereby providing evidential basis for its universality.

ACKNOWLEDGMENTS

The first author acknowledges the National Science Foundation for support under Grant # SES9010888 and both authors acknowledge the National Science Foundation for support under the NSF/ASA program. The second author acknowledges the Jagiellonian University for support under Grant BW/II/IS/1/91. We would like also to express our thanks to Izabella Uhl, Jagiellonian University for organizing and running part of the Polish experimental work.

REFERENCES

Axelrod, Robert. 1984. *The Evolution of Cooperation.* New York: Basic.
Berry, J. W. 1967. "Independence and Conformity in Subsistence-Level Societies." *Journal of Personality and Social Psychology* 7:415-418.
Bierstedt, Robert. 1950. "An Analysis of Social Power." *American Sociological Review* 15:161-84.
Bloch, Marc. [1942] 1953. *The Historian's Craft.* Translated by P. Putnam. New York: Vintage.

Braudel, Fernand. [1949] 1976. *The Mediterranean and the Mediterranean World in the Age of Philip II.* Translated by S. Reynolds. New York: Harper.

Brislin, Richard W. 1983. "Cross-Cultural Research in Psychology." *Annual Review of Psychology* 34:363-400.

Bury, J. B. [1904] 1967. *The Invasion of Europe by the Barbarians.* New York: Norton.

Cook, Karen S., Mary Gillmore, and Toshio Yamagishi. 1986. "Power and Line Vulnerability as a Basis for Predicting the Distribution of Power in Exchange Networks." *American Journal of Sociology* 92: 445-448.

Cook, Karen S., Richard M. Emerson, Mary R. Gillmore, and Toshio Yamagishi. 1983. "The Distribution of Power in Exchange Networks: Theory and Experimental Results." *American Journal of Sociology* 89:275-305.

Dore, Ronald. 1973. *British Factory/Japanese Factory.* Berkeley: University of California Press.

Edgeworth, Frederik Y. 1881. *Mathematical Psychics.* London: Kegan Paul.

Elster, Jon. (ed.) 1986. *Rational Choice.* New York: New York University Press.

Emerson, Richard M. 1972a. "Exchange Theory, Part 1: a Psychological Basis for Social Exchange." In *Sociological Theories in Progress*, Vol. 2, edited by Joseph Berger, Morris Zelditch, Jr., and Bo Anderson. Boston: Houghton Mifflin.

―――. 1972b. "Exchange Theory, Part 2: Exchange Relations and Network Structures." In *Sociological Theories in Progress*, Vol. 2, edited by Joseph Berger, Morris Zelditch, Jr., and Bo Anderson. Boston: Houghton Mifflin.

Faucheux, Claude. 1976. "Cross-Cultural Research in Social Psychology." *European Journal of Social Psychology* 6:269-322.

Finifter, Bernard M. 1977. "The Robustness of Cross-Cultural Findings." Pp. 151-184 in *Issues in Cross-Cultural Research*, edited by Leonore Loeb Adler. New York: New York Academy of Sciences. 151-84.

Finley, M.I. 1974. *The Ancient Economy.* Berkeley: University of California Press.

Fisher, Ronald A. 1935. *The Design of Experiments.* Edinburgh: Oliver and Boyd.

―――. 1956. *Statistical Methods and Scientific Inference.* Edinburgh: Oliver and Boyd.

Foschi, Martha. 1980. "Theory, Experimentation and Cross-Cultural Comparisons in Social Psychology." *Canadian Journal of Sociology* 5:91-102.

Foschi, Martha and William M. Hales. 1979. "The Theoretical Role of Cross-Cultural Comparisons in Experimental Social Psychology." In *Cross-Cultural Contributions to Psychology*, edited by L. Eckensberger, W. Lonner, and Y.H. Poortinga. Lisse: Swets and Zeitlinger.

Galilei, Galileo. [1665] 1954. *Dialogues Concerning Two New Sciences.* Translated by Henry Crew and Alfonso deSalvio. New York: Dover.

Gergen, K. J. 1973. "Social Psychology as History." *Journal of Personality and Social Psychology* 26:309-320.

Harris, Marvin. 1968. *The Rise of Anthropological Theory.* New York: Thomas Y. Crowell.

Hawking, Stephen W. 1988. *A Brief History of Time.* New York: Bantam.

Heckathorn, Douglas. 1983. "Extensions of Power-Dependence Theory: The Concept of Resistance." *Social Forces* 61:1248-1259.

Homans, George C. 1971. "Commentary." In *Institutions and Social Exchange*, edited by H. Turk and R. Simpson. Indianapolis: Bobbs-Merrill.

Lakatos, Imre. 1970. "Falsification and the Methodology of Scientific Research Programs." In *Criticism and the Growth of Knowledge*, edited by Imre Lakatos and Alan Musgrave. Cambridge: Cambridge University Press.

Lind, Joan. 1987. "Exchange Processes in History." *Sociological Quarterly.* 28:223-246.

Malinowski, Bronislaw. 1959. *Crime and Custom in Savage Society.* Paterson, NJ: Littlefield, Adams.

Markovsky, Barry, David Willer and Travis Patton. 1988. "Power Relations in Exchange Networks." *American Sociological Review* 53:220-236.

_____. 1990. "Theory, Evidence and Intuition." *American Sociological Review.* 55:300-305.

Martin, Rex. 1977. *Historical Explanation.* Ithaca, NY: Cornell University Press.

Marx, Karl. [1867] 1967. *Capital.* New York: International Publishers.

Newton, Isaac [1729] 1966. *Principia.* Translated by A. Motte. Berkeley: University of California Press.

Olson, Mancur. 1965. *The Logic of Collective Action.* Cambridge: Harvard University Press.

Orne, M. T. 1962. "On the Social Psychology of the Psychological Experiment: With Particular Reference to Demand Characteristics and their Implication." *American Psychologist* 17:776-783.

_____. 1969. "Demand Characteristics and the Concept of Quasi-controls." In *Artifact in Behavioral Research,* edited by R. Rosenthal and R. L. Rosnow. New York: Academic.

Rohner, Ronald P. 1977. "Why Cross-Cultural Research?" In *Issues in Cross-Cultural Research,* edited by Leonore Loeb Adler. New York: New York Academy of Sciences.

Schlenker, Barry R. 1974. "Social Psychology as Science." *Journal of Personality and Social Psychology* 29:1-15.

Schwinger, T. 1980. "Just Allocations of Goods: Decisions Among Three Principles." In *Justice and Social Interaction,* edited by G. Mikula. New York: Springer-Verlag.

Sechrest, L. 1977. "On the Dearth of Theory in Cross-Cultural Psychology: Is there Madness in Our Method?" In *Basic Problems in Cross-Cultural Psychology,* edited by Y.H. Poortinga. Amsterdam: Swets and Zeitlinger.

Sell, Jane and Michael W. Martin. 1983. "An Acultural Perspective on Experimental Social Psychology." *Personality and Social Psychology Bulletin* 9:345-350.

Selvin, Hanan C. 1957. "A Critique of Tests of Significance in Survey Research." *American Sociological Review.* 22:519-527.

Skvoretz, John and David Willer. 1991. "Power in Exchange Networks: Setting and Structure Variations." *Social Psychology Quarterly.* 54:224-238.

Szmatka, Jacek. 1989a. "Individualism, Holism, Reductionism." *International Sociology.* 4:169-186.

_____. 1989b. *Male Struktury Spoleczne. Wstep Do Mikrosocjologii Strukturalnej.* [Small Social Structures. Introduction to Structural Microsociology.] Warszawa: PWN.

_____. 1989c. "Reduction in the Social Sciences: The Future or Utopia?" *Philosophy of the Social Sciences* 19:425-444.

Tallman, Irving and Marilyn Ihinger-Tallman. 1979. "Values, Distributive Justice and Social Change." *American Sociological Review* 44:216-235.

Tedeschi, J.T., R.B. Smith, III, J.P. Gahagan, and J. Elinoff. 1972. "Economic Development and Social Conflict: A Cross-Cultural Study of Americans and Ghanaians." *Human Relations* 25:65-76.

Toulmin, Stephen. 1953. *The Philosophy of Science.* New York: Harper and Row.

Walker, Henry A. and Bernard P. Cohen. 1985. "Scope Statements: Imperatives for Evaluating Theory." *American Sociological Review* 50:288-301.

Weber, Max. [1918] 1968. *Economy and Society.* Berkeley: University of California Press.

Weber, Stephen J. and Thomas D. Cook. 1972. "Subject Effects in Laboratory Research: An Examination of Subject Roles, Demand Characteristics and Valid Inference." *Psychological Bulletin* 77:273-295.

Willer, David. 1981a. "The Basic Concepts of Elementary Theory." *Networks, Exchange and Coercion,* edited by D. Willer and B. Anderson. New York: Elsevier/Greenwood.

_____. 1981b. "Quantity and Network Structure." In *Networks, Exchange and Coercion,* edited by D. Willer and B. Anderson. New York: Elsevier/Greenwood.

————. 1984. "Analysis and Composition as Theoretic Procedures." *The Journal of Mathematical Sociology* 10:241-270.

————. 1986. "Vulnerability and the Location of Power Positions." *American Journal of Sociology* 92:441-444.

————. 1987. *Theory and the Experimental Investigation of Social Structures.* New York: Gordon and Breach.

————. 1992a "The Principle of Rational Choice and the Problem of a satisfactory Theory." In *Rational Choice Theory: Advocacy and Critique,* edited by J.S. Coleman and T.J. Fararo. Newbury Park, CA: Sage.

————. 1992b. "Predicting Power in Exchange Networks: A Brief History and Introduction to the Issues." *Social Networks* 14:187-211.

Willer, David and Bo Anderson (eds.). 1981. *Networks, Exchange and Coercion.* New York: Elsevier/Greenwood.

Willer, David, Barry Markovsky, and Travis Patton. 1989. "Power Structures: Derivations and Applications of Elementary Theory." In *Sociological Theories in Progress,* Vol. 3, edited by Joseph Berger, Morris Zelditch, and Bo Anderson. Newbury Park: Sage.

Willer, David and Judith Willer. 1973. *Systematic Empiricism: Critique of a Pseudoscience.* Englewood Cliffs, NJ: Prentice-Hall.

Womack, James P., Daniel T. Jones, and Daniel Roos. 1990. *The Machine That Changed the World.* New York: Harper.

Yamagishi, Toshio and Karen Cook. 1990. "Power Relations in Exchange Networks: A Comment on "Network Exchange Theory." *American Sociological Review* 55:297-300.

POLITICS AND CONTROL IN ORGANIZATIONS

Gerald R. Ferris, Julianne F. Brand, Stephen Brand,
Kendrith M. Rowland, David C. Gilmore, Thomas R. King,
K. Michele Kacmar, and Carol A. Burton

ABSTRACT

A theoretical model of perceptions of organizational politics and control in organizations is proposed and tested in two separate studies. The model suggests that organizational politics perceptions are associated with a number of work outcomes, including stress, satisfaction, and withdrawal. However, these relationships are believed to be moderated by perceptions of control over politics; perceived control, it is suggested, affects the interpretation of politics by translating it as a threat or an opportunity. Two empirical studies, involving five organizations, are reported and provide support for the model. Implications for theory and research are discussed.

Advances in Group Processes, Volume 10, pages 83-111.
Copyright © 1993 by JAI Press Inc.
All rights of reproduction in any form reserved.
ISBN: 1-55938-280-5

INTRODUCTION

Since the early works of Weber (1947) and Marx (1963), the organizational pathologies that occur due to political behavior have been discussed extensively by organizational researchers. Indeed, whereas politics often reflects simply the way things get done in organizations, the common-sense notion of political behavior typically is a negative one. Yet, while the belief that organizations are inherently political is rarely questioned, empirical research that explores how their members perceive and react to the existence of organizational politics is limited. Thus, systematic efforts to better understand the nature of political behavior in organizations are needed. A conceptualization is proposed in this paper of the consequences of organizational politics perceptions. Part of that conceptual model characterizes perceptions of organizational politics being associated with job anxiety, job satisfaction, and organizational withdrawal, but not in all cases. Rather, this linkage is believed to be moderated by perceived control over the nature of politics in the work environment, and perceived control is proposed to affect the interpretation of politics as a threat or an opportunity. Two studies were conducted to test this conceptual model.

Nature of Political Behavior in Organizations

Organizational scientists have developed different notions of what constitutes political behavior. Some have defined organizational politics in terms of the behavior of interest groups to use power to influence decision making (e.g., Pettigrew, 1973; and Tushman, 1977), or through coalition building and bargaining (Bacharach and Lawler, 1980). Others have focused on the self-serving and organizationally nonsanctioned nature of individual behavior in organizations (e.g., Burns, 1961, Porter, 1976; Farrell and Petersen, 1982; Mayes and Allen, 1977; Schein, 1977; Gandz and Murray, 1980). Still, others have characterized organizational politics as a social influence process with potentially functional or dysfunctional organizational consequences (Allen, Madison, Porter, Renwick, and Mayes, 1979; Ferris, Russ, and Fandt, 1989; Porter, Allen, and Angle, 1981), or simply the management of influence (Madison, Allen, Porter, Renwick, and Mayes, 1980). While subscribing to aspects of several of these definitions, Pfeffer (1981) more directly established the linkage between politics and power, and conceived of organizational politics as "the study of power in action" (p. 7). Finally, Mintzberg (1983) referred to politics as "individual or group behavior that is informal, ostensibly parochial, typically divisive, and above all, in the technical sense, illegitimate—sanctioned neither by formal authority, accepted ideology, nor certified expertise (though it may exploit any one of these)" (p. 172). While the foregoing do not exhaust all possible definitions of organizational politics, they provide a representative sample.

Indeed, the foregoing definitions appear to reflect considerable diversity, which has led at least one source to suggest that the construct of organizational politics is largely unknown (Vrendenburgh and Maurer, 1984). We would not be quite as pessimistic but, instead, look for common themes that emerge from the existing definitions and research evidence in an effort to make sense out of this phenomenon.

Several recent conceptualizations of organizational politics have been proposed, one investigating a single set of political behaviors (i.e., ingratiation) (Liden and Mitchell, 1988), and others examining a number of different tactics and strategies (Ferris and Judge, 1991; Ferris et al., 1989; Gardner and Martinko, 1988; Porter et al., 1981; and Tedeschi and Melburg, 1984). Additionally, empirical research has investigated a variety of different types of political behaviors (e.g., ingratiation, opinion conformity, other enhancement, entitlements, excuses, apologies, justifications, etc.) on outcomes such as employment interview decisions (Baron, 1986; and Gilmore and Ferris, 1989), performance evaluations and salary increases (Dreher, Dougherty, and Whitely, 1988; Ferris, Judge, Rowland, and Fitzgibbons, in press; Kipnis and Schmidt, 1988; Kipnis, Schmidt, and Wilkinson, 1980; and Wayne and Ferris, 1990), career progress (Gould and Penley, 1984), and managerial action recommendations (Wood and Mitchell, 1981). Furthermore, political behaviors have been studied as reactions to stress (Mayes and Ganster, 1988), and job dissatisfaction (Farrell, 1983).

Perceptions of Organizational Politics

In reviewing the research examining organizational politics in addition to the definitions cited, several issues seem to emerge and can be collectively organized in an effort to develop a more informed understanding of organizational politics perceptions. First, there appears to be an assumption of intentionality in the demonstration of political behavior. While this may be a less critical issue for some, we believe it is important for how perceivers cognitively evaluate the observed event (i.e., the political behavior exhibited).

A second issue emerging from the existing evidence pertains to level of analysis. It appears that political behavior is demonstrated by individuals, groups, and organizations. Furthermore, Ferris et al. (1989) argued that because similar types of political behaviors are likely used by individuals, groups, and organizations, they saw no particular theoretical advantage at present in distinguishing among the various levels in their definition of organizational politics.

A third notion that can be extracted is that the very nature of organizational politics reflects a social influence process, both in terms of the conditions under which political behaviors are likely to occur (Fandt and Ferris, 1990; Ferris and Mitchell, 1987; Ferris and Porac, 1984; and Pfeffer, Salancik, and

Leblebici, 1976), and the actual demonstration and consequences of such behavior (Allen et al., 1979; Ferris et al., 1989; and Porter et al., 1981).

Finally, a fourth issue concerns the extent to which political behavior is necessarily good or bad. In general, it is fair to say that individuals tend to interpret the term "organizational politics" negatively, perhaps due to the imputation of malignant intent to the individual engaging in such behaviors. However, the existing theory and research would imply that politics can be either good or bad, in perception and outcome. Of course, the more precise issue is good or bad for whom; do the most benefits accrue to the individual, the group, or the organization? Ferris et al. (1989) have proposed at least three possible reactions of individuals to perceptions of organizational politics, some of which would be interpreted as negative and some positive (this is discussed in more detail later in this paper).

Thus, a number of relevant issues seem to emerge from theory and research in organizational politics that permit some convergence on just what is going on out there in the work environment that people would perceive as political behavior. In suggesting directions for research on politics in organizations, Ferris et al. (1989) proposed that more work needed to be done on the conditions under which political behavior occurs, as well as the types of political behaviors that are demonstrated and their consequences. A third area of research also was proposed which, to date, has been virtually ignored; that is, the determination of antecedents and consequences of individuals perceiving a work environment as political. This area of research is a bit different than the other two, because it suggests that organizational politics is a subjective perception, not necessarily an objective reality. Whereas one might assume that there is a strong correspondence between actual political behavior (i.e., to the extent that an indication of so called objective political behavior could be obtained) and behavior that is perceived as political, it must be acknowledged that this assumption may be unfounded, that perceptual differences occur, and that it is important to try to better understand how and why this happens.

Organizational scientists have long argued the distinction between objective and perceived work environments (e.g., James and Jones, 1974; Naylor, Pritchard, and Ilgen, 1980; and Schneider, 1975). More recently, arguments have been made that work environments are molded by the types of people attracted and granted entry (Schneider, 1987), that both selection and socialization processes contribute to political environments (Ferris et al., 1989; and Ferris, Fedor, Chachere, and Pondy, 1989), and that some organizational environments are more political than others (Riley, 1983).

However, in this area of research, we agree with Gandz and Murray (1980), who suggested that rather than exclusively an objective state, it is appropriate to conceive of organizational politics as a subjective evaluation, and, thus, as a state of mind. Many years ago, Lewin (1936) suggested the very important notion that people respond on the basis of their perceptions of reality, not reality per se, and, later on, Porter (1976) argued that perceptions are important to study and to understand, even if they are misperceptions of actual events, with particular reference to organizational politics. Furthermore, researchers interested in other aspects of work environments (e.g., organizational climate), in discussing true versus perceived attributes, have argued for a definition of work environments based on perceived attributes (James and James, 1989; Naylor et al., 1980; and Schneider, 1975).

In summary, then, the interest we have in organizational politics for purposes of the present paper is to conceptualize and investigate the nature of *perceptions* of organizational politics. Themes and issues emerge from existing theory and research concerning just what constitutes political behavior in organizations. We are interested in the cognitive evaluation and subjective experience of those behaviors and events occurring in the work environment that seem to constitute political behavior.

Furthermore, the empirical research on perceptions of organizational politics has been quite limited (Ferris and Kacmar, 1992; Gandz and Murray, 1980; and Madison et al., 1980), perhaps due in part to a lack of theory to guide research. Empirical results have tended to demonstrate a negative relationship between perceptions of organizational politics and outcomes, but job satisfaction is the predominant outcome investigated. Furthermore, without theory to guide research in this area, meaningful interpretation of results is quite difficult. What is needed, then, is a systematic conceptualization that articulates the consequences of organizational politics perceptions, and the conditions under which they are likely to be observed.

POLITICS PERCEPTIONS AND WORK OUTCOMES

Employees may respond several ways when they perceive the organization to be political in nature. Affected will be their levels of organizational withdrawal, job involvement, job anxiety, and job satisfaction. Ferris et al. (1989) suggested that at least three potential responses to politics perceptions would be to withdraw from the organization, remain a member of the organization but do not become involved in the politics, and to remain a member of the organization and become involved in the politics. These responses appear similar in nature to Hirschman's (1970) exit, loyalty, and voice, respectively.

One potential response of an employee who views the work environment as political is to withdraw from the organization. Frost (1987) also has made this suggestion, arguing that employees may leave to avoid engaging in an organization's political games. Withdrawal can take one of two forms, absenteeism or turnover. This type of behavior is expected when perceivers do not wish to become involved in the political games of the organization. While some employees will have the luxury of external mobility, others will not be able to leave due to constraints, or because of other organizational features they find appealing. Of those who elect to stay, there may be an increase in their absenteeism rate. With the decision to stay comes several alternative outcomes.

First, if organizational politics is perceived negatively, employees might immerse themselves in their work in an effort to ignore the surrounding political behavior. With this in mind, the positive relationship between job involvement and organizational politics perceptions becomes obvious. A second alternative is to become involved in the political process. If this option is selected, both job anxiety and job satisfaction can be influenced. An inverse relationship between perceptions of politics and job satisfaction has been established by prior research. Further, when employees engage in political activities, the perceptions of others concerning the political nature of the organization may increase. This, in turn, may increase job anxiety due to a more uncertain or ambiguous environment fueled by the increased political behavior. It might be useful to take a more careful look at some of these work outcomes.

While it seems logical that employee perceptions that a work environment is political in nature might have negative or stress-inducing consequences, and that perceptions of politics might in fact represent a work-related stressor, this notion has been formally stated only quite recently. Matteson and Ivancevich (1987) identified a number of potential stressors associated with the work environment, including politics. However, several definitions and conceptualizations of stress in organizations have provided opportunities to view politics as relevant and inclusive subject matter. Three key features serving to integrate politics and stress in these conceptualizations are their perceptual nature, uncertainty or ambiguity regarding processes or outcomes, and the threat or opportunity status of politics and stress.

Like the discussions of politics perceptions presented by others (Ferris et al. 1989; and Gandz and Murray, 1980), a number of stress researchers have tended to view stress as an individually experienced phenomenon rather than a characteristic of the environment (McGrath, 1976; Schuler, 1980; and Schuler and Jackson, 1987). As noted above, people respond on the basis of their perceptions of reality, not reality per se (Lewin, 1936), so the key to both stress and politics, as discussed here, are individual perceptions.

A second feature that serves to integrate stress and politics is the central role of uncertainty. Uncertainty or ambiguity represents one of the principal conditions under which political behavior occurs, as well as a major influence on perceptions of politics (Fandt and Ferris, 1990; and Ferris et al., 1989). McGrath (1976) emphasized perceived uncertainty as a primary determinant of individual stress reactions, which led others to also highlight the role of uncertainty of processes and/or outcomes in their definitions of stress (e.g., Beehr and Bhagat, 1985; Beehr and Schuler, 1982; and Schuler, 1980). In fact, Schuler and Jackson (1987) have argued that stress is uncertainty that occurs in the work environment, and that a more informed understanding of stress can be gained by more careful examination of the construct of uncertainty.

The third feature that might serve to integrate politics and stress is the identification of each as threat (constraint) or opportunity. Nearly two decades ago, Caplan, Cobb, French, Harrison, and Pinneau (1975) defined stress in terms of environmental characteristics that pose threats to individuals. Shortly after, McGrath (1976) characterized stress in terms of both constraints and opportunities. More recently, Schuler (1980) conceived of stress as dynamic conditions in which opportunities, constraints, or demands are placed on people. Politics in organizations seem to provide similar options and, thus, can be construed in a comparable manner. It seems quite reasonable to conceive of politics as providing situations of potential gain (i.e., opportunity) as well as situations of potential loss (i.e., threats). Schuler (1980) suggested that opportunity stress should be positively related to affective psychological outcomes, including reduced anxiety and increased job involvement, and one could see opportunity politics as reflecting similar relationships. In addition, it would be reasonable to expect threat-oriented stress and politics to be negatively related to such psychological outcomes. What seems less clear, however, are the processes by which perceptions of organizational politics are translated into opportunities or threats.

Perceptions of Control

One could construe the perceived control over one's work environment as a mechanism which translates stressors (e.g., politics) into opportunities or threats, and thus moderates relationships between stressors and outcomes. Sutton and Kahn (1986) proposed such a conceptualization relative to stress research and the stress-strain relationship, and Tetrick and LaRocco (1987) provided some supporting evidence. Figure 1 presents a similar conceptualization of how control moderates the perceptions of organizational politics-work outcomes (e.g., job anxiety, job satisfaction, or organizational withdrawal) relationships.

Control represents the extent to which people can exercise influence over their environment and is a concept that, either implicitly or explicitly, underlies most theories of individual behavior in organizations (Ganster and Fusilier, 1989). In a seminal review of the control literature, Averill (1973) distinguished among behavioral, decisional, and cognitive types of control. Whereas behavioral and decisional control referred to conditions in which individuals had direct influence over their environment, cognitive control involved one's interpretations or perceptions of control, which may or may not have an objective basis. Although many studies fail to make the distinction between objective and perceived control, most studies investigating this construct use individuals' beliefs concerning the amount of control they exercise (Thompson,

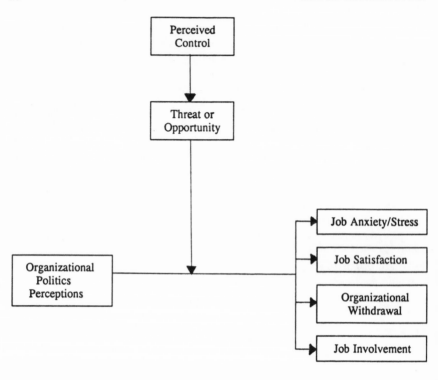

Figure 1. Model of Perceived Control Influences on the
Organizational Politics Perceptions-Work Outcomes Relationship

1981). Indeed, when examining stress and other mental and physical health outcomes, perceived control has been found to be more of a direct causal factor than the objective forms (Ganster, 1989).

Perceived control has been extensively investigated (Thompson, 1981) and operationalized in a number of ways including autonomy, supervisory status, and decision latitude (Adelmann, 1987). Karasek (1979) found that more decision latitude (i.e., many opportunities to make decisions concerning the job) reduced the negative impact of job demands relative to mental strain. Other research has suggested that being or feeling in control is something people actively seek (Greenberger & Strasser, 1991), and increases the probability that environmental uncertainty will be interpreted as challenging opportunities, rather than as threats from which protection should be sought (Lerner, 1980; Schuler, 1980). Furthermore, perceived control has been found to be most effective when it leads to uncertainty reduction (Averill, 1973). Information control (i.e., the feelings of control that are derived from obtaining information about an event or situation) not only helps reduce ambiguity, it enables individuals to more effectively employ their own coping strategies in obtaining desired outcomes (Fiske & Taylor, 1984).

Therefore, in light of the foregoing discussion and the conceptualization presented in Figure 1, it is suggested that the relationships between organizational politics perceptions and job anxiety, job satisfaction, and organizational withdrawal are moderated by perceived control. More specifically, if people perceive that politics go on in the work environment, and if they perceive they have little control over the process, politics can be interpreted as a threat and would be expected to lead to more negative outcomes. However, if employees believe that they have a high degree of control over the political process and outcomes, less negative and even more favorable outcomes should result. In this sense, organizational politics takes on the role of a potential environmental stressor which has more or less negative consequences to the extent individuals perceive they can control the source of stress.

These notions of threats and opportunities are derived from recent theoretical and empirical work by Dutton and Jackson (Dutton & Jackson, 1987; Jackson & Dutton, 1988). They argued that how issues or phenomena are interpreted is a function of their characteristics, and particular characteristics influence whether issues are interpreted as opportunities or threats. When situations are interpreted as threats, individuals tend to cope by relying on faith, resigning their futures to fate, or engaging in wishful thinking (McCrae, 1984). We would argue that these reactions all relate to a perceived lack of control. In fact, Jackson and Dutton found that issues classified as opportunities tended to be identified as positive with high feelings of control. Conversely, issues classified as threats were identified as negative and involved a lower perceived degree of control.

Present Research

In the present research, two studies were conducted to systematically investigate perceived control as a moderator of the organizational politics perceptions-work outcomes relationships. In Study 1, the focus is on the moderating influence of perceived control on the politics perceptions-job anxiety relationship. In this study, supervisory status was used as a proxy for control in order to circumvent potential problems of method covariation producing spurious results. Mixing objective and subjective measures can help to reduce this problem (Schmitt, Colligan, and Fitzgerald, 1980); however, there are tradeoffs concerning the legitimacy of the proxy employed. In this study, it was believed that supervisory status was a legitimate proxy for the control construct for several reasons. Prior research has used supervisory status (i.e., whether one supervises others) as an operationalization of the control construct (Adelman, 1987). Furthermore, the use of supervisory status as a proxy for perceived control has received some empirical support recently. Tetrick (Personal Communication, April 11, 1989) reported that one's position in the organizational hierarchy was significantly and positively correlated with control (i.e., control increased as one went up the hierarchy).

Study 2 extends Study 1 by examining how perceived control moderates the relationship of organizational politics perceptions with other outcome measures specified in Figure 1. Perceived control was measured more directly, using self-report responses, and job satisfaction and organizational withdrawal were examined as work outcome variables.

In both studies, the hypothesis is that organizational politics perceptions should be associated with more positive outcomes (i.e., less anxiety, lower withdrawal, and greater job satisfaction) when employees perceive a greater degree of control over their work environments and thus the process of politics. However, under conditions of low perceptions of control, politics perceptions should be associated with negative outcomes (i.e., greater anxiety and withdrawal, and lower job satisfaction).

STUDY 1: METHOD

Sample

An organizational survey was administered to a total of 310 employees in four different organizations. The sample consisted of 89 middle- and lower-level line and staff managers from a large heavy-equipment manufacturer; 81 registered nurses from a large (800-bed) hospital; 94 nursing service employees from a skilled nursing care facility operated by a county government; and 46 customer-service representatives from a savings and loan company. Employees present at work on that day completed questionnaires in meeting rooms on the organizations' time and in the presence of a member of the research team. Participants were asked to respond as accurately as possible, and they were assured of complete anonymity and confidentiality of responses. Survey feedback sessions, reporting aggregate data, were conducted at each of the organizations approximately three months after data collection.

Characteristics of the total sample and the four separate organizations are reported in Table 1.

Measures

Organizational Politics Perceptions

Organizational politics was measured by summing five items designed by Ferris and Kacmar (1992) to assess the extent of political behavior respondents

Table 1. Sample Characteristics (Study 1)

Variable	Organization #1 (Manufacturing Managers)	Organization #2 (Hospital Nurses)	Organization #3 (Nursing Home Aides)	Organization #4 Savings and Loan Representatives)	Total
Sample Size (N)	89	81	94	46	310
Age					
Mean	45.55	38.88	33.46	34.04	38.48
Standard deviation	8.11	14.74	13.57	10.49	13.08
Education					
1. Less than high school	0	2	15	1	18
2. High school diploma	11	17	36	1	65
3. Some school after high school (no degree or certificate)	27	30	18	7	82
4. Associate degree	0	2	1	2	5
5. 3 years of college or post-secondary schooling	5	4	4	1	14
6. Bachelor's degree	32	17	9	4	62
7. Master's degree	13	9	2	1	25
8. Doctoral degree	1	0	0	0	1
9. Other	0	0	7	29	36
Sex					
Male	89	45	9	2	145
Female	0	36	85	44	165
Supervisory Status					
Supervise	80	52	30	24	186
Do not supervise	9	29	61	21	120
Organizational Politics					
Mean	11.03	15.26	12.70	13.94	13.07
Standard deviation	2.39	3.63	3.91	4.22	3.85
Job Anxiety					
Mean	65.79	63.31	35.33	61.21	55.25
Standard deviation	7.50	8.26	11.64	9.30	16.19

93

perceived in their work environment. Each of the items was measured on a five-point Likert-type scale (1 = Strongly Disagree; 5 = strongly agree) and they were summed to create an index which ranged in value from 5 to 25, with higher total scores reflecting a greater degree of perceived organizational politics. The five items were:

1. Favoritism rather than merit determines who gets ahead;
2. There is no place for yes-men around here; good ideas are desired even when it means disagreeing with supervisors (reverse scored);
3. You can get along around here by being a good guy, regardless of the quality of your work;
4. Employees are encouraged to speak out frankly even when they are critical of well established ideas (reverse scored); and
5. There are "cliques" or "in-groups" which hinder the departments' effectiveness.

The coefficient alpha internal consistency reliability estimate for the organizational politics perceptions scale was .76. This measure was developed with the objectives of designing a concise, unidimensional, construct-valid measure of overall organizational politics perceptions that possessed acceptable psychometric properties. Ferris and Kacmar (1992) demonstrated, through factor analyses and assessment of relationships with other measures across two samples, that these objectives were met. Furthermore, whereas politics has been defined in different ways by different researchers, as noted earlier in this paper, the present measure tends to focus more on the dark side of politics.

Job Anxiety

Job anxiety was measured with the State-Trait Anxiety Inventory (STAI) developed by Spielberger, Gorsuch, Lushene, Vagg, and Jacobs (1983). The essential qualities evaluated by STAI involve feelings of tension, nervousness, worry, and apprehension. The scale consists of 20 items such as "I feel tense," "I feel nervous," "I feel anxious." Respondents were asked to indicate the extent to which each statement described how they felt in connection with their job or work at the present. Thus, the focus was on state (not trait) anxiety. The responses were measured using a four-point, Likert-type scale (1 = not at all; 2 = somewhat; 3 = moderately; 4 = very much so), so STAI scores potentially could range from 20 to 80. The STAI has known psychometric properties reflecting a median coefficient alpha value of .93 based on data consisting of nearly 5000 cases, and reflects careful development from both theoretical and methodological perspectives (Spielberger et al., 1983). The STAI has been used more extensively than any other anxiety measure in psychological research over the past decade (Buros, 1978).

Perceived Control

Perceived control, the extent to which people believe they can exercise influence over their environment, was operationalized as supervisory status, as suggested by Ferris et al. (1989) and reported by Adelmann (1987). Supervisory status was assessed by asking respondents to indicate as either (1) true or (2) false, whether they presently supervise other regular paid employees.

Background Data

Information was gathered from respondents concerning their age, sex, and education level. For education level, respondents checked one of the following categories:

1. Less than high school;
2. High school diploma;
3. Some school after high school (no degree or certificate);
4. Associate degree;
5. 3 years of college or post-secondary schooling;
6. Bachelor's degree;
7. Master's degree;
8. Doctoral degree;
9. Other.

Data Analysis

In order to test portions of the conceptual model presented in Figure 1, hierarchical moderated regression analysis was conducted. The interest here was in examining variation in job anxiety that can be explained by perceptions of politics and perceived control variation, as well as the interaction. Furthermore, it was important to be able to interpret the results as freely of potential confounds or competing explanations as possible. Therefore, it was decided to control the criterion variance due to potential organization differences. Specifically, the organization to which each employee belonged was controlled by inserting organization, as a dummy coded variable, hierarchically on the first step in the regression analysis. This has the effect of disentangling the criterion variance due to differences in organizational practices, and thus providing the opportunity for a cleaner interpretation of the results.

STUDY 1: RESULTS

Means and standard deviations of all variables under investigation are presented in Table 1, and intercorrelations of the variables are presented in Table 2. Of particular importance to note is that the zero-order correlation

Table 2. Intercorrelations of All Variables (Study 1)

Variable	1	2	3	4
1. Organizational Politics Perceptions	—			
2. Job Anxiety	.01	—		
3. Organization	.33*	.35*	—	
4. Perceived Control (Supervisory Status)	.14*	−.22*	.08	—

* $p < .01$

between perceptions of organizational politics and job anxiety is near zero $(r = .01)$.

Hierarchical moderated regression results demonstrated that after controlling for organizational differences, and removing the criterion variance due to the main effects, the interaction of politics perceptions and perceived control accounted for a significant increment in job anxiety variance explained. The results are presented in Table 3.

Under conditions of low perceived control (i.e., for employees who did not supervise others), the relationship between organizational politics perceptions and job anxiety was significant and positive $(r = .17, p < .05, n = 120)$. However, under conditions of high perceived control (i.e., for employees who did supervise others), the relationship between politics and anxiety was nonsignificant but inverse in direction $(r = -.06, \text{n.s.}, n = 186)$.

The results of Study 1 provided support for the hypothesis that perceived control would moderate the relationship between perceptions of organizational

Table 3. Hierarchical Regression Results Examining the Moderating Effects of Perceived Control on the Organizational Politics Perceptions-Job Anxiety Relationship, Controlling for Organization (Study 1)

Variables	β	R^2	ΔR^2	F (Step)	df
Job Anxiety					
Control Variable Organization[a]		.13	.13	23.07**	3,298
Predictor Variables Organizational Politics					
Perceptions (A)	−.16	.15	.02	7.32*	1,297
Perceived Control (B)	−.22	.20	.05	16.64**	1,296
A × B	1.33	.27	.07	27.62**	1,295

[a] The organization variable was dummy coded and entered as a nominal variable.
* $p < .01$
** $p < .001$

politics and job anxiety. However, a limitation of this study was the operationalization of the perceived control moderator variable. Supervisory status was used as a proxy for perceived control. Although this has been suggested in prior research, the question remains as to whether supervisory status is a legitimate operationalization of perceived control. In Study 2, perceived control is measured more directly.

STUDY 2: METHOD

Sample

An organizational survey was administered to 386 employees of a large health-care organization located in the Midwest. Subjects were selected by stratified random sampling from each of 28 administrative and clinical services. Most of the participants were permanent, full-time employees (92%), and the average employee surveyed had worked at the organization for nine years. A majority of the survey respondents were white (80%) and about two-thirds were women. The modal level of education attainment was "some college," and the ages of the respondents ranged from 20 to 73 years old ($M = 39$).

The survey was administered to employees during paid working time in meeting rooms away from their regular work station. Members of the university research team were responsible for all phases of these data collection meetings. Employees were assured that participation in the survey was voluntary and that individual responses would be kept in the strictest confidence. The results of data analyses involving groups of employees were provided in a feedback report to the health care organization.

Measures

The survey assessed employee's perceptions of the frequency with which organizational politics affected decision making and the degree to which the employee could control various aspects of his or her work situation. Furthermore, employee job satisfaction and organizational withdrawal measures also were assessed.

Organizational Politics Perceptions

To measure perceptions of organizational politics, employees were asked directly to indicate the extent to which politics influenced specific decisions concerning employment and working conditions (e.g., promotions, pay, layoffs, work schedules, and workload). Employees responded to each of 12 items on a five-point Likert-type scale (1 = Never; 5 = Always). To obtain an index of politics perceptions, the average of these items was computed. A principal axis factor analysis with Varimax rotation yielded one factor with an eigenvalue greater than 1.00, on which all items had loadings greater than .30, suggesting that the index taps one dimension underlying ratings of organizational politics. The coefficient alpha internal consistency reliability estimate for the politics perceptions scale was .93. This scale has been used previously in our research, and it has been found to correlate with the politics perceptions scale used in Study 1 to a substantial degree (i.e., $r = .60$).

Perceived Control

To assess the employees' perceptions of control, the survey employed the Influence at Work scale from a battery of occupational stress measures developed by the National Institute of Occupational Safety and Health (Hurrell and McGlaney, 1988). Employees were provided with a definition of influence at work, which included control over what others do at work as well as autonomy and independence. Employees were then instructed to rate the degree to which they believe they could influence each of 16 aspects of their work situation (e.g., availability of supplies and equipment, and decisions as to when things will be done). For each item, subjects indicated the extent of their influence on a five-point Likert-type scale (1 = Very little, 5 = Very much). To obtain an overall index of perceived control, the average of these ratings was computed. This overall index was found to have an internal consistency reliability of .90.

Job Satisfaction

The survey included scales to assess four specific facets of job satisfaction. Scales from the Michigan Organizational Assessment Questionnaire (Seashore, Lawler, Mirvis, and Camman, 1982; and Mirvis and Lawler, 1977) were utilized to measure Satisfaction with Extrinsic Reward (4 items: e.g., "Pay"), Satisfaction with Decision Making (4 items: e.g., "Chances to take part in decisions"), Satisfaction with Performance Evaluations (3 items: e.g., "Feedback you get from the performance evaluation process"), and Satisfaction with Labor-Management Relations (3 items: e.g., "Benefits and rights provided by the union contract"). For each item, employees indicated the level of satisfaction or dissatisfaction on a seven-point Likert-type scale (1 = Very Dissatisfied; 7 = Very Satisfied). To obtain indices of job satisfaction, for each scale, the average of ratings across items was computed. The coefficient alpha internal consistency reliability estimates for the satisfaction scales were: Extrinsic Rewards = .77; Decision Making = .84; Performance Evaluations = .89; Labor-Management Relations = .77. In addition, initial studies have supported the construct validity of the index of satisfaction with Extrinsic

Rewards, demonstrating that lower scores on these scales are associated with lower levels of job involvement and stronger intentions to turn over (Seashore et al., 1982).

Organizational Withdrawal

The survey included measures of three dimensions of organizational withdrawal: negative evaluation of the decision to take the current job, intentions to leave, and absences. To assess negative evaluation of the decision to take the current job, three items were utilized from the employment preferences section of the National Institute of Occupational Safety and Health stress battery (Hurrell & McGlaney, 1988):

1. "Knowing what you know now, if you had to decide all over again whether to take the type of job you have now, what would you decide?";
2. "If you were free right now to go into any type of job you wanted, what would your choice be?"; and
3. "If a friend of yours told you he/she was interested in a job like yours, what would you tell him/her?."

For each item, employees were instructed to select one of three answers ranging from favorable to unfavorable evaluation of their current job. To obtain an index of negative evaluation, the average rating for these three items was computed, and the coefficient alpha reliability estimate was .78.

To assess intentions to leave current job, two items from the Hurrell and McGlaney (1988) battery were utilized:

1. "How likely is it that you will be actively looking for a new job in the next year?"; and
2. "How likely is it that you will be quitting your job at the (name of organization) within the next year?."

For both items, employees responded on a four-point likelihood scale (1 = Not at all likely; 4 = Extremely likely). To obtain an index of intentions to leave, the average of the ratings for these two items was computed. The coefficient alpha reliability estimate for the intentions to leave scale was .73.

Finally, to assess absences from work, employees were instructed to indicate the number of sick days that they had utilized in the past 12 months. Because the number of absences exhibited a strong negative skew, and this variable has an absolute zero point, a logarithmic transformation of this variable was utilized in the following analyses (cf. Cohen and Cohen, 1975). Specifically, absences were computed as equal to [$Ln(1 + absences)$]. This transformation resulted in a closer approximation to the normal distribution, and reduced the influence of extreme observations.

Table 4. Means, Standard Deviations, and Intercorrelations of All Variables (Study 2)

Variables	1	2	3	4	5	6	7	8	9
1. Politics Perceptions									
2. Perceived Control	-27c								
Satisfaction									
3. Extrinsic	-44c	29c							
4. Performance	-51c	27c	43c						
5. Decisions	-46c	41c	50c	63c					
6. Labor-Man.	-35c	24c	47c	46c	50c				
Withdrawal									
7. Evaluation	28c	-27c	-36c	-33c	-45c	-35c			
8. Leaving	20c	-10	-27c	-19c	-27c	-22c	-26c		
9. Absences	23c	-16b	-15b	-21c	-12a	-11a	18c	05	
Mean	2.92	2.69	4.54	4.27	4.59	4.10	1.63	1.43	1.58
S.D.	0.93	0.80	1.33	1.62	1.33	1.31	0.55	0.74	0.89

Tabled values are correlation coefficients (decimals are omitted). Sample sizes for specific correlations ranges from 341 to 367 due to pairwise deletion of missing data.
[a] $p < .05$
[b] $p < .01$
[c] $p < .001$

100

STUDY 2: RESULTS

Means, standard deviations, and intercorrelations of all variables are presented in Table 4.

The correlates of organizational politics perceptions are of particular interest, because it is intuitively thought that low levels of job satisfaction and higher levels of organizational withdrawal are important consequences of organizational politics. The findings of the correlational analyses were largely consistent with these predictions.

Overall, in stage one of the analyses, indices of satisfaction and withdrawal were related in the expected direction to organizational politics perceptions. However, the model also suggests that the strength of the relationship between organizational politics perceptions and these indices should be moderated by perceived control. Thus, the second stage of analyses examined perceived control as a moderator of the relationship between organizational politics perceptions and job satisfaction and withdrawal. These analyses followed hierarchical moderated regression procedures (Cohen and Cohen, 1983; and Pedhazur, 1982).

Job Satisfaction

As can be seen in Table 4, the four measures of satisfaction are highly intercorrelated. Thus, it seemed more appropriate to create a composite job satisfaction measure on which to examine the interaction of politics perceptions and perceived control, than to run four separate regression analyses. Therefore, the four measures of satisfaction were summed to create a composite measure of job satisfaction.

Hierarchical moderated regression analysis demonstrated that the interaction term accounted for a significant increment in job satisfaction variance explained over the main effects model. These results are presented in Table 5.

Under low levels of perceived control, organizational politics perceptions exhibited a strong negative correlation ($r = -.63$, $p < .01$) with job satisfaction as predicted. In contrast, under conditions of high perceived control, markedly weaker, but still statistically significant and negative correlations were found between organizational politics perceptions and satisfaction ($r = -.39, p < .01$). Overall, these findings are consistent with the hypothesis that organizational politics perceptions are of greater importance for job satisfaction when levels of control are perceived to be low.

Organizational Withdrawal

In order to investigate the moderating role of perceived control on the politics perceptions-withdrawal relationship, three hierarchical moderated regression

Table 5. Hierarchical Regression Results Examining the Moderating Effects of Perceived Control on the Relationships Between Organizational Politics Perceptions and Job Satisfaction and Organizational Withdrawal (Study 2)

Variables	β	R^2	ΔR^2	F (Step)	df
Job Satisfaction Composite					
Politics Perceptions (A)	−.55	.31	.31	140.99***	1,344
Perceived Control (B)	.29	.38	.07	39.22***	1,343
A × B	.12	.40	.02	7.85**	1,342
Organizational Withdrawal Measures					
Negative Evaluation of Employment Decision					
Politics Perceptions (A)	.25	.06	.06	23.02***	1,341
Perceived Control (B)	−.22	.11	.05	17.55***	1,340
A × B	−.10	.12	.01	3.83*	1,339
Absences					
Politics Perceptions (A)	.07	.01	.01	1.77	1,341
Perceived Control (B)	−.05	.01	.00	<1	1,340
A × B	.13	.03	.02	5.73*	,339

* $p < .05$
** $p < .01$
*** $p < .001$

analyses were conducted. The interaction failed to account for a statistically significant increment in the variance explained for intentions to leave the current job. However, as can be seen in Table 5, the interaction accounted for a significant increment in the variance explained for negative evaluation of the decision to take the current job, as well as absences.

Under conditions of low perceived control, higher levels of organizational politics were related significantly to more negative evaluation of the decision to take the current job ($r = .34$; $p < .001$; $n = 166$). However, under conditions of high perceived control, politics perceptions were not related significantly to negative evaluation ($r = .11$; n.s.; $n = 186$). This finding is consistent with the hypothesis that organizational politics perceptions are of greater importance for withdrawal under conditions of low perceived control. In contrast, for absences, the pattern of the interaction between perceived control and politics perceptions diverged from the predictions of the model. Under conditions of low perceived control, the correlation analyses found no significant relation between organizational politics perceptions and absences ($r = .11$; n.s.; $n = 163$), rather than the expected strong positive correlation.

Under conditions of high perceived control, higher levels of politics were associated with greater absences ($r = .27; p < .001; n = 184$).

DISCUSSION

The results of the present research provided support across two studies for the moderating influence of perceived control on the organizational politics perceptions-work outcomes relationships. Organizational politics perceptions were believed to represent a source of stress in the work environment that might lead to increased job anxiety, organizational withdrawal, and decreased job satisfaction. However, the nature of this relationship between politics and anxiety was believed to differ as a function of whether politics was perceived or interpreted as a threat or an opportunity. Furthermore, the interpretation of politics as a threat or an opportunity was believed to result from the degree of control over the nature of politics perceived in the work environment. Greater perceived control over how politics are played out should lead to organizational politics being viewed positively, as an opportunity to pursue to one's advantage, thus leading to favorable outcomes, or at least less negative ones. Little perceived control should transform politics into more of a threat, or something to fear, thus leading to unfavorable outcomes.

Study 1 focused on perceived control as a moderator of the organizational politics perceptions-job anxiety relationship, and the results provided support with a significant interaction effect in the predicted direction. However, there were several limitations of Study 1. A weakness concerns the largely self-report measurement of variables. Particularly in light of some of the constructs measured, common method variance could prove problematic. In order to minimize this potential difficulty somewhat, an alternative and more objective, operationalization of the perceived control construct was used (i.e., supervisory status). Thus, while self-report responses were gathered, the question asked concerning control or supervisory status was about a variable likely to be objectively reported.

Additionally, although a valid case was built for its use, there remains some question concerning the appropriateness of the operationalization of the moderator variable. While a logical and defensible rationale was developed, drawing from prior theory and research, it remained an empirical question whether supervisory status represented an appropriate operationalization of the perceived control construct.

Study 2 was conducted to attempt to constructively replicate and extend Study 1 results using different measures of some constructs, examining relationships with different variables, and addressing some of the limitations of Study 1.

The results of Study 2 demonstrated mixed support for the moderated consequences of organizational politics perceptions on job satisfaction and organization withdrawal as suggested in the model. As proposed, perceived control moderated the politics perceptions-job satisfaction relationship. Under low levels of perceived control, the relationship between perceived politics and satisfaction was significantly stronger and more negative. That is, employees reporting that they lacked control in the organization were more likely to be dissatisfied when they also perceived that decisions were influenced by politics. However, under conditions of high perceived control, the politics-satisfaction relationship was weaker but still negative.

The results for organizational withdrawal were differentially significant. When the degree of perceived control was low, employees who reported perceiving political behavior in their organization were also more likely to regret having decided to take their current job. The outcome of reported intentions to leave the organization was not moderated by perceived control.

A third measure of withdrawal, reported number of sick days (absenteeism), was moderated by perceived control, but not as predicted by the model. Surprisingly, when the degree of perceived control was high, the politics perceptions-withdrawal relationship was positive and significantly stronger. Employees who perceived politics and reported having greater control in the organization tended to report more absences. Under conditions of low perceived control, the relationship remained positive but was not significant.

This odd finding for the politics perceptions-absenteeism relationship may be a simple indication of the amount of discretion employees with greater perceived control have relative to others in the organization. As we know from our own experiences in organizations, more absences are tolerated or allotted for those with influence, and these same individuals may also observe the political nature of the work more frequently. These same employees also may have to endure other stressors at work, potentially inducing higher absenteeism; stressors other than low perceived control and more perceived politics were not examined in this study. Caution needs to be exercised in the interpretation of these results, due to the manner in which absences were measured. Rather than an objective index of subsequent absenteeism, this variable was measured as a self-report of the number of sick days utilized over the past year. Future studies investigating the interrelationship of control, politics, and other stressors in the workplace and their relationship to satisfaction, withdrawal, and other outcomes are necessary before these issues can be considered fully clarified.

Study 2 has several limitations, perhaps the most serious being a high reliance on self-report measurement of variables. Particularly in light of some of the constructs measured, common method variance may be a concern to some. However, this study went beyond the examination of simple bivariate relationships and applied multivariate analysis, utilizing hierarchical multiple regression analysis. Although this does not completely eliminate the possibility that obtained relationships are capitalizing on "artifactual" variance, it is more likely to reflect valid variance and is arguably an improvement over past research. Future research on perceptions of organizational politics should consider employing multiple methods; perceptual and attitudinal measures employed could be complemented with behavioral, demographic, or other more observable measures.

Two additional limitations of both Studies 1 and 2 should be noted. One is that the intermediate linkages in the proposed conceptualization were not actually tested. It was proposed that perceived control moderates relationships between politics perceptions and work outcomes because perceptions of control transform the interpretation of organizational politics into either a threat or an opportunity. Future research needs to test that linkage directly. At present, it can only be inferred that is what drives the moderating influence of perceived control.

Another general concern relates to the measures of organizational politics perceptions used in the two studies. Although a systematic effort was made to develop concise yet construct-valid measures of politics that have acceptable reliability, there is much work needed in the future to consider the multidimensional nature of organizational politics perceptions, and to develop an instrument sufficiently broad to capture the different dimensions (see Kacmar and Ferris, 1991). Furthermore, while slightly different measures of organizational politics perceptions were used in the two studies, they are conceptually quite similar and have been found to correlate substantially (r = .60) in some of our other research.

FUTURE DIRECTIONS

The conceptual model and research results reported in this paper have interesting implications both for prior work and in proposing directions for future theory and research. Ferris et al. (1989) suggested three possible reactions to perceptions of organizational politics: (1) withdraw from the organization; (2) ignore politics; or (3) become involved in politics. Whereas these notions appear interesting, in retrospect, we would suggest they are a bit too simplistic. We have proposed in this paper that these work outcomes are not directly derived from organizational politics perceptions, but, instead, are outcomes differentially associated with the degree of perceived control one has over politics, and thus, the interpretation of politics as a threat or an opportunity.

Individuals who fall into the first category will reduce their exposure to, or self-select themselves out of, an organization in which a perceived political environment is prevalent if they have no intention of becoming involved in the political arena and believe they have no control over politics. This type of reaction could lead to absenteeism or to turnover. Individuals who fall in the second category could increase their job involvement in an effort to block out the political games occurring around them if they don't believe they can control them. Such individuals may believe they will succeed not by playing the game of politics, but by becoming indispensable to the organization by performing effectively. Thus, they attempt to restore control in a different way. Finally, individuals who choose to become involved in organizational politics do so because they believe they can control it, and thus interpret it not as a threat, but as an opportunity. Such individuals are likely to increase the already political nature of the work environment, possibly leading to an increase in the dissatisfaction of individuals in the other two categories who do not enjoy organizational politics.

A second question the current study introduces also deals with the reactions of employees to a perceived political environment. More specifically, does the perception of a political environment stimulate more political behavior because individuals feel they can get ahead (and get away with it) by playing? Would employees who would normally have not played begin to indulge once the costs and benefits of political behaviors are understood? According to Madison et al. (1980), CEOs in their sample reported a high incidence of organizational politics, and more so at higher levels. These CEOs suggested that personality traits of top managers, concerning people who engaged in politics and thus set an example for and encouraged such behavior in others, might explain this. As previously mentioned, some research has reported that political behaviors are more likely to be observed as uncertainty in the situation increases. However, could perceptions of organizational politics also cause more political behaviors? Ferris et al. (1989), and Ferris, Fedor, Chachere, and Pondy (1989) proposed that this could take place through new employee socialization, social learning of particular behaviors and values, and mentoring.

This issue of how "objective" politics is perceived by others and, thus, the subjectively experienced meaning of organizational politics perceptions bears more careful examination in future theory and research. One might necessarily assume that what are regarded "objectively" as political behaviors necessarily will be regarded as such and lead to negative perceptions and consequences for bystanders. Wayne, Kacmar, Rubenstein, and Ferris (1990) investigated how subordinate influence attempts directed at a supervisor would affect coworkers in the work group. Interestingly, the ingratiation efforts of the subordinate (directed toward the supervisor) affected coworkers favorably, rather than unfavorably as one might expect, leading to increased satisfaction. While these results might appear counterintuitive, upon closer examination they make more sense. As Wayne et al. pointed out, observing a subordinate exhibiting influence tactics toward a supervisor may well be construed as manipulative by onlookers, but the term "manipulative" has two quite different definitions as mentioned by Owen (1986). In citing Webster's Seventh dictionary, he presented the following two definitions: (1) "manage or utilize skillfully"; (2) "to control or play upon as artful, unfair, or insidious means."

Thus, it appears that individuals might react positively or negatively depending on which definition they employ. Furthermore, the triggering mechanism that might determine which definition of manipulation one adopts might be the particular type of influence tactics employed by the actor.

All political behaviors are neither equally effective nor similarly perceived. It has been found that ingratiation types of political behaviors tend to be positively associated with performance ratings given by supervisors (Ferris et al., in press; Kipnis and Vanderveer, 1971; and Wayne and Ferris, 1990), likely operating on such outcomes through their effects on affect or liking. Whereas, other types of political behaviors that emphasize entitlements (i.e., claiming responsibility for positive events) can be risky because the manipulator may be perceived as egotistical, thus leading to a negative impression and negative affect (Tedeschi and Melburg, 1984). In fact, such tactics have been found to be inversely related to performance ratings (Ferris et al., in press; Wayne & Ferris, 1990).

Further refinements in the analysis of organizational politics perceptions, their meaning, and their underlying dynamics have been examined here. It was argued that whether politics perceptions lead to negative or positive outcomes is a function of whether they are perceived as a threat (i.e., to fear and be intimidated by), or as an opportunity on which to capitalize. Essentially, this perspective would suggest that individuals cognitively evaluate such situations in terms of whether the particular work environmental features or activities (i.e., in this case, politics) are personally detrimental or personally beneficial, and subsequent affect and behavior follow from the initial cognitive evaluation. This is exactly the point that James and James (1989) made recently, in their analysis of work environment perceptions.

Regardless of the limitations noted, the results of the present research warrant consideration and are interesting because they extend previous research, provide empirical tests of a new theoretical model, and, thus, contribute to the body of research on organizational politics. Specifically, these results help to explain potential consequences of perceiving the work environment as political in nature. Further research is needed to develop a more informed understanding of organizational politics perceptions, with respect to antecedents, consequences, and potential moderators. The notion of perceived control as a moderator could profit from more precise articulation of the role it plays in politics perceptions-outcomes relationships. One potentially fruitful direction for future work in this area is to explore the notion that control may well be vested to greater degrees in some job types or occupational categories than others, as noted by Adelman (1987). The theory, research, and practical implications of this notion would seem to be potentially quite interesting.

ACKNOWLEDGMENTS

The authors would like to thank Terry A. Beehr and Terence R. Mitchell for their helpful comments on an earlier draft of this article. Portions of this paper were presented at the Fifth Annual Conference of the Society of Industrial and Organizational Psychology, Inc., Miami Beach, 1990.

REFERENCES

Adelmann, P. K. 1987. "Occupational Complexity, Control, and Personal Income: Their Relation to Personal Well-Being in Men and Women." *Journal of Applied Psychology* 72: 529-537.

Allen, R. W., D. L. Madison, L. W. Porter, P. A. Renwick, and B. T. Mayes. 1979. "Organizational Politics: Tactics and Characteristics of its Actors." *California Management Review* 22: 77-83.

Averill, J. R. 1973. "Personal Control Over Aversive Stimuli and its Relationship to Stress." *Psychological Bulletin* 80: 286-303.

Bacharach, S. B., and E. J. Lawler. 1980. *Power and Politics in Organizations*. San Francisco: Jossey-Bass.

Baron, R. A. 1986. "Self-Presentation in Job Interviews: When there can be 'Too Much of a Good Thing.'" *Journal of Applied Social Psychology* 16: 16-28.

Beehr, T. A., and R. S. Bhagat. 1985. *Human Stress and Cognition in Organizations: An Integrated Perspective*. New York: Wiley.

Beehr, T. A., and R. S. Schuler. 1982. "Stress in Organizations." Pp. 390-419 in *Personnel Management*, edited by K. M. Rowland and G. R. Ferris. Boston: Allyn and Bacon.

Burns, T. 1961. "Micropolitics: Mechanisms of Institutional Change." *Administrative Science Quarterly* 6: 257-281.

Buros, O. K. (Ed.) 1978. *The Eighth Mental Measurements Yearbook*. Hyde Park, NJ: Gryphon Press.

Caplan, R. D., S. Cobb, J. R. P. French, Jr., R. U. Harrison, and S. R. Pinneau, Jr. 1975. *Job Demands and Worker Health*. Washington, DC: NIOSH Research Report.

Cohen, J., and P. Cohen, 1983. *Applied Multiple Regression/Correlation Analysis for the Behavioral Sciences* (Second edition). Hillsdale, NJ: Lawrence Erlbaum.

Dreher, G. F., T.W. Dougherty, and W. Whitely, 1988. *Influence Tactics and Salary Attainment: A Study of Sex-Based Salary Differentials*. Paper presented at the Academy of Management, 48th Annual National Meeting, Anaheim, CA.

Dutton, J. E., and S. E Jackson, 1987. "The Categorization of Strategic Issues by Decision Makers and its Links to Organizational Action." *Academy of Management Review* 12: 76-90.

Fandt, P. M., and G. R. Ferris. 1990. "The Management of Information and Impressions: When Employees Behave Opportunistically." *Organizational Behavior and Human Decision Processes* 45: 140 158.

Farrell, D. 1983. "Exit, Voice, Loyalty, and Neglect as Responses to Job Dissatisfaction: A Multidimensional Scaling Study." *Academy of Management Journal* 26: 596-607.

Farrell, D., and J. C. Petersen. 1982. "Patterns of Political Behavior in Organizations." *Academy of Management Review* 7: 403-412.

Ferris, G. R., D. B. Fedor, J. G. Chachere, and L. R. Pondy. 1989. "Myths and Politics in Organizational Contexts." *Group & Organization Studies* 14: 88-103.

Ferris, G. R., and T. A. Judge. 1991. "Personnel/Human Resources Management: A Political Influence Perspective." *Journal of Management* 17: 447-488.

Ferris, G. R., T. A. Judge, K. M. Rowland, and D. E. Fitzgibbons. (in press). "Subordinate Influence and the Performance Evaluation Process: Test of a Model." *Organizational Behavior and Human Decision Processes.*

Ferris, G. R., and K. M. Kacmar. 1992. "Perceptions of Organizational Politics. *Journal of Management* 18: 93-116.

Ferris, G. R., and T. R. Mitchell. 1987. "The Components of Social Influence and Their Importance for Human Resources Research." Pp. 103-128 in *Research in Personnel and Human Resources Management,* Vol. 5, edited by K. M. Rowland and G. R. Ferris. Greenwich, CT: JAI Press.

Ferris, G. R., and J. F. Porac. 1984. "Goal Setting as Impression Management." *Journal of Psychology* 117: 33-36.

Ferris, G. R., G. S. Russ, and P. M. Fandt. 1989. "Politics in Organizations." Pp. 143-170 in *Impression Management in the Organization,* edited by R. A. Giacalone & P. Rosenfeld. Hillsdale, NJ: Lawrence Erlbaum.

Fiske, S. T., and S. E. Taylor. 1984. *Social Cognition.* Reading, MA: Addison-Wesley.

Frost, P. J. 1987. "Power, Politics, and Influence." Pp. 503-548 in *Handbook of Organizational Communication,* edited by F. M. Jablin, L. L. Putnam, K H. Roberts, and L. W. Porter. Newbury Park, CA: Sage Publications.

Gandz, J., and V. V. Murray, 1980. "The Experience of Workplace Politics." *Academy of Management Journal* 23: 237-251.

Ganster, D. C. 1989. "Worker Control and Well-Being: A Review of Research in the Workplace." Pp. 3-23 in *Job Control and Worker Health,* edited by S. L. Sauter, J. J. Hurrell, Jr., and C. L. Cooper London: Wiley.

Ganster, D. C., and M. R. Fusilier, 1989. "Control in the Workplace." Pp. 235-280 in *International Review of Industrial and Organizational Psychology* edited by C. L. Cooper and I. Robertson. New York: Wiley.

Gardner, W. L., and M. J. Martinko, 1988. "Impression Management in Organizations." *Journal of Management* 14: 321-338.

Gilmore, D. C., and G. R. Ferris. 1989. "The Effects of Applicant Impression Management Tactics on Interviewer Judgments." *Journal of Management* 15: 557-564.

Gould, S., and L. E. Penley. 1984. "Career Strategies and Salary Progression: A Study of Their Relationships in a Municipal Bureaucracy." *Organizational Behavior and Human Performance* 34: 244-265.

Greenberger, D. B., and S. Strasser. 1991. "The Role of Situational and Dispositional Factors in the Enhancement of Personal Control in Organizations." Pp. 111-145 in *Research in Organizational Behavior,* Vol. 13, edited by L. L. Cummings and B. M. Staw. Greenwich, CT: JAI Press.

Hirschman, A. 1970. *Exit, Voice, and Loyalty: Responses to Decline in Firms, Organizations, and States.* Cambridge, MA: Harvard University Press.

Hurrell, J. J., and M. A. McGlaney. 1988. *Exposure to Job Stress: A New Psychometric Instrument.* Cincinnati: National Institute of Occupational Safety and Health.

Jackson, S. E., and J. E. Dutton. 1988. "Discerning Threats and Opportunities. *Administrative Science Quarterly* 33: 370-387.

James, L. A., and L. R. James. 1989. "Integrating Work Environment Perceptions: Explorations into the Measurement of Meaning." *Journal of Applied Psychology* 74: 739-751.

James, L. R., and A. P. Jones. 1974. "Organizational Climate: A Review of Theory and Research." *Psychological Bulletin* 81: 1096-1112.

Kacmar, K. M., and G. R. Ferris. 1991. "Perceptions of Organizational Politics (POPS): Development and Construct Validation." *Educational and Psychological Measurement* 51: 193-205.

Karasek, R. A., Jr. 1979. Job Demands, Job Decision Latitude, and Mental Strain: Implications for Job Redesign." *Administrative Science Quarterly* 24: 285-308.

Kipnis, D., and S. M. Schmidt. 1988. "Upward-Influence Styles: Relationship With Performance Evaluations, Salary, and Stress. *Administrative Science Quarterly* 33: 528-542.

Kipnis, D., S. M. Schmidt, and I. Wilkinson. 1980. "Intraorganizational Influence Tactics: Exploration in Getting One's Way. *Journal of Applied Psychology 65:* 440-452.

Kipnis, D., and R. Vanderveer. 1971. Ingratiation and the Use of Power. *Journal of Personality and Social Psychology* 17: 280-286.

Lerner, A. 1980. Orientation to Ambiguity." In *Uncertainty: Behavioral and Social Dimensions.* edited by S. Fiddle. New York: Praeger.

Lewin, K. 1936. *Principles of Topological Psychology.* New York: McGraw-Hill.

Liden, R. C., and T. R. Mitchell. 1988. "Ingratiatory Behaviors in Organizational Settings." *Academy of Management Review* 13: 572-587.

Madison, D. L., R. W. Allen, L. W. Porter, P. A. Renwick, and B. T. Mayes. 1980. "Organizational Politics: An Exploration of Managers' Perceptions." *Human Relations* 33: 79-100.

Marx, K. (1844) 1963. *Karl Marx: Early writings.* Translated and edited by T. B. Bottomore. London: C. A. Watts.

Matteson, M. T., and J. M. Ivancevich. 1987. *Controlling Work Stress.* San Francisco: Jossey-Bass.

Mayes, B. T., and R. W. Allen. 1977. "Toward a Definition of Organizational Politics." *Academy of Management Review* 2: 672-678.

Mayes, B. T., and D. C. Ganster. 1988. Exit and Voice: A Test of Hypotheses Based on Fight/ Flight Responses to Job Stress." *Journal of Organizational Behavior* 9: 199-216.

McCrae, R. M. 1984. "Situational Determinants of Coping Responses: Loss, Threat, and Challenge." *Journal of Personality and Social Psychology* 46: 919-928.

McGrath, J. E. 1976. "Stress and Behavior in Organizations." Pp. 1351-1395 in *Handbook of Industrial and Organizational Psychology,* edited by M. D. Dunnette. Chicago: Rand McNally.

Mintzberg, H. 1983. *Power In and Around Organizations.* Englewood Cliffs, NJ: Prentice-Hall.

Mirvis, P., and E. E. Lawler. 1977. "Measuring the Financial Impact of Employee Attitudes." *Journal of Applied Psychology* 62: 1-8.

Naylor, J. C., R. D. Pritchard, and D. R. Ilgen. 1980. *A Theory of Behavior in Organizations.* New York: Academic.

Owen, H. 1986. "Leadership Indirection." *Transforming Leadership,* edited by J. D. Adams. Alexandria, VA: Miles River Press.

Pedhauzer, E. J. 1982. *Multiple Regression in Behavioral Research.* New York: Holt, Rinehart, & Winston.

Pettigrew, A. 1973. *The Politics of Organizational Decision Making.* London: Tavistock.

Pfeffer, J. 1981. *Power in Organizations.* Boston: Pitman.

Pfeffer, J., G. R. Salancik, and H. Leblebici. 1976. "The Effect of Uncertainty on the Use of Social Influence in Organizational Decision Makin." *Administrative Science Quarterly* 21: 227-245.

Porter, L. W. 1976. *Organizations as Political Animals.* Presidential Address, Division of Industrial-Organizational Psychology, 84th Annual Meeting of the American Psychological Association, Washington, DC.

Porter, L. W., R. W. Allen, and H. L. Angle. 1981. The Politics of Upward Influence in Organizations." Pp. 109-149 in *Research in Organizational Behavior* Vol. 3, edited by L. L. Cummings and B. M. Staw. Greenwich, CT: JAI Press.

Riley, P. 1983. A Structurationist Account of Political Culture. *Administrative Science Quarterly* 28: 414-437.

Schein, V. E. 1977. "Individual Power and Political Behaviors in Organizations: An Inadequately Explored Reality." *Academy of Management Review* 2: 64-72.

Schmitt, N., M. J. Colligan, and M. Fitzgerald. 1980. "Unexplained Physical Symptoms in Eight Organizations: Individual and Organizational Analyses." *Journal of Occupational Psychology* 53: 305-317.

Schneider, B. 1987. "The People Make the Place." *Personal Psychology* 40: 437-453.

Schneider, B. 1975. "Organizational Climates: An Essay." *Personnel Psychology* 28: 447-479.

Schuler, R. S. 1980. "Definition and Conceptualization of Stress in Organizations." *Organizational Behavior and Human Performance* 25: 184-215.

Schuler, R. S., and S. E. Jackson. 1986. "Managing Stress through PHRM Practices: An Uncertainty Interpretation." Pp. 183-224 in *Research in Personnel and Human Resources Management*, Vol. 4, edited by K. M. Rowland and G. R. Ferris. Greenwich, CT: JAI Press.

Seashore, S. E., E. E. Lawler, P. Mirvis and C. Camman. 1982. *Observing and Measuring Organizational Change: A Guide to Field Practice*. New York: Wiley.

Spielberger, C., R. L. Gorsuch, R. Lushene, P. R. Vagg, and G. A. Jacobs. 1983. *Manual for the State-Trait Anxiety Inventory*. Palo Alto, CA: Consulting Psychologists Press.

Sutton, R., and R. L. Kahn. 1986. "Prediction, Understanding, and Control as Antidotes to Organizational Stress." In *Handbook of Organizational Behavior*, edited by J. Lorsch. Englewood Cliffs, NJ: Prentice-Hall.

Tedeschi, J. T., and V. Melburg. 1984. "Impression Management and Influence in the Organization." In *Research in the Sociology of Organizations*, Vol. 3, edited by S. B. Bacharach and E. J. Lawler. Greenwich, CT: JAI Press.

Tetrick, L. E., and J. M. LaRocco. 1987. "Understanding, Prediction, and Control as Moderators of the Relationship Between Perceived Stress, Satisfaction, and Psychological Well-Being." *Journal of Applied Psychology* 72: 538-543.

Thompson, S. C. 1981. "Will it Hurt Less if I can Control it? A Complex Answer to a Simple Question." *Psychological Bulletin* 90: 89-101.

Tushman, M. 1977. "A Political Approach to Organizations: A Review and Rationale. *Academy of Management Review* 2: 206-216.

Vrendenburgh, D. J. and J. G. Maurer. 1984. "A Process Framework for Organizational Politics." *Human Relations* 37: 47-66.

Wayne, S. J., and G. R. Ferris. 1990. "Influence Tactics, Affect, and Exchange Quality in Supervisor-Subordinate Interactions: A Laboratory Experiment and Field Study." *Journal of Applied Psychology* 75: 487-499.

Wayne, S. J., K. M. Kacmar, D. Rubenstein, and G. R. Ferris. 1990. *Subordinate Upward Influence Effects on Coworker Responses*. Manuscript submitted for publication.

Weber, M. (1924) 1947. *The Theory of Social and Economic Organization*. Translated and edited by A. H. Henderson and T. Parsons. Glencoe, Ill.: Free Press.

Wood, R. E., and T. R. Mitchell. 1981. "Manager Behavior in a Social Context: The Impact of Impression Management on Attribution and Disciplinary Actions." *Organizational Behavior and Human Performance* 28: 356-378.

HOW SENTIMENTS ORGANIZE INTERACTION

Robert K. Shelly

ABSTRACT

The question how sentiments organize interaction in task groups is examined. First, sentiment structures in groups are described. Then, the simple principle that we prefer to give chances to perform and positive evaluations to those we like is enunciated. This is sufficient, together with ideas from the Expectation States research program, to describe how sentiment structures organize interaction. Two research questions are addressed by an experiment. Sentiments organize interaction in the experiment, but more weakly than task ability or formal authority. A discussion considers how to test the ideas in additional situations.

Interpersonal sentiments organize social interaction. This principle of affective social structure has been recognized since Moreno's *Who Shall Survive* (1953) and Homans' *The Human Group* (1950). Most studies have examined how sentiments order frequency of contact with others in a group. Precisely how sentiment structures organize other aspects of interaction is less well

Advances in Group Processes, Volume 10, pages 113-132.
Copyright © 1993 by JAI Press Inc.
All rights of reproduction in any form reserved.
ISBN: 1-55938-280-5

understood. By contrast, status generalization research programs provide rich explanatory environments for how social differences organize interaction. This is especially true of work in Expectation States Theory, carried out by Berger and his collaborators (cf. Berger, 1992).

This paper specifies how interpersonal sentiments organize interaction by clarifying concepts, specifying group situations where the process occurs, and identifying basic propositions about how the process works. An approach suggested by Fararo and Skvoretz (1986) in their work on state organizing processes is taken. In particular, I apply the ideas developed by Berger and others in Expectation States Theory to show how sentiment structures have state organizing properties. Relevant research is scattered throughout the literature of sociology and psychology. For instance, work on cohesion is examined to specify how intergroup processes affect interaction within groups. Principles from research traditions in sociology and psychology are identified to develop an understanding of how sentiment organizes interaction.

The paper is divided into several sections. First, we identify concepts to be explored. The definitions that result from this process are employed throughout the discussion. Second, we specify situations in which sentiments organize interaction. For instance, research on cohesion and solidarity enriches our understanding of how the strength of group bonds organizes interaction. Third, we describe how we think intragroup sentiment patterns organize interaction by providing differential opportunities for actors to perform and differential rewards for performances. Finally, we present evidence from a research program to support the conjectures about how sentiments organize interaction.

TERMS

Research in both sociology and psychology has examined relationships between sentiment and interaction. Terminology in studies that examine sentiment patterns and interpersonal organization is confusing. Terms are often used with one meaning in one set of studies and another meaning in a different set of studies. A short list follows of terms and definitions that are employed in this paper, to avoid confusion. Interpersonal terms are distinguished from terms that have groups as their principal referents. The definitions are consistent with usage in the research literature.

Affect is an attitude state held by an individual toward some object. In the traditional attitude literature this is the feeling state experienced by an individual toward an object. We wish to use the term to refer to an emotional reaction on the part of an actor toward an object. This reaction may be positive or negative, of varying strength, and may or may not result in direct action. The important point is that an actor experiences the emotion that predisposes him or her to respond to an object in some way. Objects may be persons, ideas, or inanimate.

A *sentiment* is an affective state in which the object is a person. Sentiments may be positive or negative, and the object may be either the person experiencing the sentiment or some other actor. By definition, sentiments vary in strength. The objective in defining sentiments separately from affect is to distinguish emotions with an actor as target from those with a nonactor as target. We feel sentimental toward a person, but not an inanimate object.

Liking is a low-strength positive sentiment toward another actor. An actor who is liked by a focal person is the object of positive emotional reaction. It is possible to develop hypotheses regarding such relations. For instance, liking another actor is likely to engender positive responses to objects linked to him or her. These may be other actors or inanimate objects.

Disliking is a low-strength negative sentiment toward another actor. An actor who is disliked is the object of negative emotional reaction. Again, it is possible to relate disliking to other social responses. A disliked other may be attended to less often than one who is well liked, for instance.

Love and hate relations are beyond the scope of this discussion. Research evidence and theoretical discussions suggest these sentiments are qualitatively different from liking and disliking. While both sets of concepts focus on sentiments, intensity, complexity of emotional response, and range of expression vary. I wish to restrict the current discussion to the conceptually simpler problems of liking and disliking.

The *structure of sentiments* in a group of actors is the distribution of sentiment relations over actors in the group. Such a structure might specify liking between some actors and disliking between other actors in the group. Love and hate relations might exist between still others. A structure of sentiments may be specified by one way or reciprocated sentiment. All actors need not be connected to all others by the same sentiment relations. The group is assumed to be connected and closed; that is, it is possible to reach all members from any arbitrarily chosen one. A similar definition of connection is described by Granovetter (1973) and elaborated by Freeman (1992). A modern highway map, with thicker and thinner lines of varying color, is a good pictorial analogy for the structure of sentiments in a group; colors and thicknesses of lines represent sentiment relations in the group.

The focus of this paper is the structure of sentiments and how they organize interaction. Findings from research on liking and disliking governs much of the search for principles to guide theorizing. Issues raised in cohesion research and its relationship to sentiment structures are examined first.

COHESION AND SENTIMENT STRUCTURES

Two properties of groups must also be examined. *Cohesion* is often used to describe the strength of bonds holding a group together. The term refers to a group property that varies from one group to another. One definition of cohesion employs the structure of sentiments in a group. Cohesive groups are characterized as having dense positive sentiment structures. Noncohesive groups have sparse positive sentiment structures or dense negative structures. Other criteria exist as well. Cohesive groups often have high ingroup identification and high outgroup hostility.

Solidarity is also often used to define how closely tied together members of a group might be. The important difference between solidarity and cohesion is that solidarity often refers to how members of the group are tied to the group. While cohesion is an intragroup property that varies across groups, solidarity is a property of groups that varies in both intragroup and intergroup relationships between actors.

Confusion about the role of sentiment structures in defining scope conditions results from the early research on cohesion carried out by Festinger and his colleagues (Festinger, Schachter, and Back, 1951). This research identified cohesion as a pattern of liking for other members of the group and showed that well-liked others were talked to more than less-liked others. The conceptual confusion resulting from linking cohesion and liking is still widespread (Burt, 1987; Evans and Jarvis, 1980; Murdock, 1989). However cohesion is defined, members of cohesive groups show positive sentiments for other group members. They are more likely to interact more with one another, conform to group norms, and be more satisfied with their experiences in the group than are members of noncohesive groups.

Some research on intergroup effects of cohesion has been carried out. Sherif's Robbers Cave experiment describes conditions under which cohesive bonds between members of groups structure interaction within and between groups (Sherif, Harvey, and White, 1955). This study showed that members of cohesive groups have positive perceptions of ingroup individuals and negative perceptions of outgroup individuals and act on these perceptions. Similar findings regarding intergroup effects of cohesion have been demonstrated in the study of athletic teams (Mizruchi, 1991). None of the studies of cohesion have examined effects of intragroup variation in sentiment structures on the organization of behavior within groups.

We focus on how intragroup social organization is affected by intragroup sentiment structures, and leave for following work theoretical developments linking intergroup effects on intragroup social organization, effects of intragroup structures on intergroup organization, and effects of intergroup structures on intergroup organization.

SITUATIONS

Most research on interaction in groups focuses on task-oriented groups. In such groups, actors share common orientations toward completing a valued task. The research on liking, sentiment relations, cohesion, and solidarity is broader in its scope and often includes groups without a task orientation. It is sufficient for members of a group to be interested in continued interaction with no goal specified other than association.

Decision-making groups have a common orientation toward some outcome, but are not necessarily task-focused. A group considering a decision about recreational activity is not a task group in the traditional sense, but a collective orientation to the outcome is maintained by members of the group until a decision is reached. Groups with nontask goals may slip into and out of decision-making activities. Such groups are of interest while making decisions, but not when engaged in purely associative activities.

BASIC IDEAS

We begin with the assumption that a group of actors must reach some decision, and then assume that a structure of sentiment bonds exists within the group such that at least one member likes one other member of the group more than some other member. The discussion begins with the two-actor case. The three-actor case is considered after ideas are developed for the simpler case. Members of the group are assumed not to be differentiated by any other state-organizing property of social structure. We take advantage of developments in the Expectation States research program to detail how we think sentiments organize interaction.

Berger and Conner (1974) describe how power and prestige orders emerge in initially undifferentiated groups. Recent work by Fisek, Berger, and Norman (1991) and Balkwell (1991a) has refined this discussion. It is possible to describe how actors form expectations about one another's relative skills and abilities and to make predictions about measures of interaction in initially undifferentiated and initially differentiated groups. The formal theory developed in this work specifies equations that predict interaction behavior in groups. These equations become important in determining the pattern of differentiation in the observable power and prestige order of a group. Relative frequencies of interaction behaviors reflect observable power and prestige orders in groups that are the realization of social organization produced by state-organizing processes.

The full fundamental sequence of behavior (Berger and Conner, 1974, p. 90) is described first. Interaction in a group begins at a start point. This start is followed by an action opportunity. An action opportunity is a pause in the speech of another person, a question, or a nonverbal indication that someone should speak. The action opportunity is followed by a performance output. In a task group this may be a proposal for solution of the task. In such a decision-making group as is being discussed, the proposal might be "Let's go see the movie at the Bijou." This performance is followed by another action opportunity (often implicit). The action opportunity is followed by an evaluation that may be either positive or negative. A positive evaluation is one in which the actor is praised for his or her contribution, or members agree with the suggestion of the actor. A negative evaluation is one in which the actor is criticized for his or her actions, or members disagree with the suggestions of the actor and offer alternative suggestions. This is the end point of the fundamental sequence of behavior.

A full fundamental sequence of behavior may leave out certain steps. For instance, a minimal fundamental sequence has a start point, performance output, and end point. Action opportunities and evaluations may be ignored by actors in a group.

Berger and Conner (1974, p. 98ff) suggest differences in power and prestige orders of homogeneous groups emerge when actors' ideas or contributions are differentially evaluated by others in the group. Good evaluations are those which elevate the status of an actor and his or her standing in the group, while. bad evaluations reduce the status of the actor. The power and prestige order reflects these evaluations over time. Actors who make good contributions are given more chances to perform and actors who make bad contributions are given fewer chances to perform.

The process is similar in heterogeneous groups, except that some actor(s) start with an advantage. Members who possess high status characteristics are given more opportunities to perform, are more likely to receive positive evaluations, and are more likely to influence the decisions of the group. Members who possess low status characteristics are less likely to be given opportunities to perform, are more likely to receive negative evaluations, and are less likely to influence decisions of the group (Berger and Conner, 1974, p. 101).

Sentiment structures organize interaction in much the same way status characteristics organize interaction in heterogeneous groups. Sentiment structures provide a pattern of organization that determines who will be given opportunities to perform in a group, whose contributions will be more likely to receive positive rewards, and who will be more influential in a group. How this process works in several different cases will be described. First, several propositions linking sentiment structures to the emergence of power and prestige orders in groups are presented:

Proposition I.   Given actors in a decision making group with a sentiment structure as the only salient social organization in the group:
(1) Actors give chances to perform and positive evaluations to those they like. Actors accept influence from those they like.
(2) Actors deny chances to perform and give negative evaluations to those they dislike. Actors reject influence from those they dislike.

To link the sentiment structure to the formation of expectations for behavior and the emergence of differentiated power and prestige orders, a second proposition, which links behavioral inequalities with expectations, is necessary. Berger and Conner (1974), Fisek, Berger and Norman (1991), and Balkwell (1991a) identify a similar statement as the key to emergent organization of power and prestige orders.

Proposition II. Behavioral inequalities based on sentiment structures lead to formation of expectations for task behavior. These inequalities in performances, opportunities to perform, and influence cause a structure of expectations for task behavior. These expectations are reflected in the observable power and prestige order of the group.

TWO PERSON GROUPS

The first case is a situation in which one actor likes another in the group, but this is not reciprocated. In this situation, A likes B, but B does not like A. B may find A repugnant, B may be indifferent to A, or B may be unacquainted with A. In such a situation, A will be more likely to give B positive evaluations for any performances B makes, and be more likely to provide B with opportunities to make contributions. This willingness to give positive evaluations to those we like and to provide more chances for them to perform is sufficient to organize interaction. This organization produces differentiation in the observable power and prestige order of the group. Under these circumstances, it is to be expected that A's liking for B will eventually lead to a situation where A will hold high expectations for B's behavior.

If we presume that B comes to like A in the above example, then we can complete specification of a power and prestige order for the group. Under this circumstance, B is likely to give A positive evaluations, and opportunities to perform. B will come to have high expectations for A in the same way A came to hold high expectations for B. Since A and B have high expectations for each other, the dyad will form a structure of interaction with equal levels of activity.

How might B come to like A? Two possibilities exist. Ridgeway and Johnson (1990) describe a process where the status structure of a group influences the emergence of liking structures in a group. Briefly, they suggest that receipt of positive rewards produces feelings of positive affect toward the source of the reward. Over time this leads to the emergence of a positive socioemotional bond being formed between the source of the reward and the actor receiving the reward. It may also be the case that B "discovers" some reason to like A. Research on liking relations shows that similarity, attraction, and common attitudes produce liking in social relationships (See Shelly, 1988 for a review of this work.). Thus, B may come to like A as a result of interaction or because of factors external to the situation.

For the second case, consider the situation where one actor dislikes another actor in the group, but the negative sentiment is not reciprocated. Suppose A dislikes B and B neither likes nor dislikes A, but they must work together on a common task. A is less likely to give B positive evaluations or action opportunities. This is sufficient to differentiate A and B in the power and prestige order of the group. A will come to hold low expectations for B because he or/she has given B fewer chances to perform and is critical of B's contributions. If B dislikes A, a process similar to that described above for liking will lead to the formation of low expectations by B for A's behavior. Since A and B have formed low expectations for each other, we would expect them to interact with one another as equals.

B may come to dislike A if A has given B low evaluations because an actor will experience negative socioemotional reactions to another actor who provides negative reward (Ridgeway and Johnson, 1990). B also may dislike A if the two actors are dissimilar to one another, if B finds A unattractive, or if they have conflicting attitudes (Shelly, 1988).

The third case is one in which A and B reciprocate their liking relation for one another. A likes B and B likes A, OR A dislikes B and B dislikes A. In this instance A and B will come to hold similar expectations for each other. If A and B like one another, the power and prestige order will be one where both have high expectations for one another. If A and B dislike one another, the power and prestige order will be one where both have low expectations for one another. In both instances, completion of the expectation structure is determined by the sentiment structure of the group.

The fourth case to be considered is one in which A and B exhibit a differential pattern of liking for one another. In this case, A likes B but B dislikes A. In such a situation, A will be more likely to give B opportunities to perform, give B more positive evaluations, and be more likely to accept influence from B because he or she likes B. B, on the other hand, is less likely to give A opportunities to perform, less likely to give positive evaluations, and less likely to accept influence from A.

This pattern of behavior will lead to the formation of unequal expectations for behavior. A will come to think that he or she is less capable at the task than B. B will come to see A as less capable as well. A will have low expectations for self and high expectations for other in the situation. B will have high expectations for self and low expectations for other.

The formation of the structure of liking relations to produce differentiated patterns of interaction is less straightforward than earlier cases. Actors enter this situation with patterns of liking and disliking already specified. Such a circumstance is not difficult to imagine. Employees may be assigned to work teams without regard to their liking for other members of the team. Emergence of liking and disliking patterns in such a group is not as easy to speculate about. Ridgeway and Johnson (1990) posit that differential structures of liking emerge from patterns of reward actions in a group. It is possible that a pattern of differential liking and disliking will emerge in a group as actors discover information about each other.

LARGER GROUPS

We now consider how sentiment structures organize interaction in groups with more than two actors. The extension is straightforward. Work by Berger and his colleagues suggests that actors form expectations for the behavior of others by a pairwise process. That is, actor A first forms expectations for actor B by interacting with him or her. Actor A forms expectations for actor C by interacting with him or her. This continues until all possible pairs in a group have formed expectations for one anothers performance (Berger, et al., 1974; Fisek, et al. 1991; Balkwell, 1991a; Skvoretz, 1988).

This sequence of pairwise interaction, formation of expectations, and patterning of who talks when, how often they speak, and how their contributions are evaluated in the group is the process that governs the emergence of an observable power and prestige order in a group that is structured only by a pattern of liking relations. If we think of three actors, A, B, and C, and the possible liking relations between them, it is possible to make a prediction about the power and prestige structure that emerges in an interaction sequence between them. A simple proposition states this relationship.

Proposition III. Given three or more actors in a group with a structure of sentiments between the actors:
(1) An actor will first form an organized interaction pair with another actor in the group based on their liking for one another.
(2) The focal actor will then form organized interaction pairs with all other members of the group in turn.
(3) These interaction pairs result in the observable power and prestige order of the group.

For the first case, suppose A likes B, B does not like A, and that B likes C. Such a pattern leads to the formation of a power and prestige order in which A has high expectations for self and low expectations for B. B has high expectations for self relative to C, but low expectations for self relative to A. To complete the expectation structure of the group, assume that a transitivity principle governs relations in incompletely specified liking structures. Under such an assumption, A would have high expectations for self and low expectations for C. A rank order of performance expectations can be derived from an incompletely specified structure of sentiment relations. In this case, I expect $A > B = C$ is the appropriate order.

Principles other than transitivity could be advanced to complete a structure of sentiments. I could choose a rule stating that if two actors do not dislike one another and one of them likes the other, then the emergent sentiment relation will be based on the emergent power and prestige structure between the two actors. This interaction rule produces a dynamic that does not lead to stable expectation structures until expectations for performance have been formed, followed by emergence of new sentiment structures, followed by formation of new, stable expectations for performance for all members of the group.

The emergence of structures of expectations in situations where actors dislike one another is analogous to the liking situation. If A dislikes B, and B dislikes C in a group, then we expect that A will come to hold low expectations for B and that B will hold low expectations for C. Structure completion rules in such a situation are more complex than in the earlier example. This pattern is an example of the interaction rule "my enemy's enemy is my friend." If A and C adopt such a rule, they may come to have high expectations for one another. Substantive content of sentiments rather than structural properties may govern structure completion processes in this instance.

It is possible to specify other sentiment structures. For instance, A and B may like one another but dislike C. Other structures include one in which all three actors like one another, or one in which all three actors dislike one another. Each case provides an instance where an emergent power and prestige order can be hypothesized based on the sentiment structure. If all actors in a group like one another, then we hypothesize they would form high expectations for self and for all others in the group. If three actors dislike one another, we expect actors would form expectation structures in which they have high expectations for self and low expectations for all others. If A and B like one another, but dislike C, then we expect they would have high expectations for self and for each other, but low expectations for C.

Finally, we repeat an important distinction. Cohesion and solidarity organize interaction within a group by providing actors with opportunities to act and persuade others. These properties vary between groups and do not define structures within a group. We assert that they do not organize interaction within the group in the same way as sentiment structures. It is possible that organized patterns of interaction vary with the degree of group cohesion. However, the organizational impact of cohesion and solidarity is between groups, not within them as is the case for sentiment structures.

The theoretical specifications of the Expectation States research program have been able to link diffuse status characteristics and specific status characteristics to the formation of expectations for behavior. The particular developments of interest are those that graph the structure of cognitions of the actor and predict expectations from the graph (Berger, Fisek, Norman, and Zelditch, 1977; Balkwell, 1991a). In this formulation, the path between the status characteristic and performance of the task varies in length with the relevance of the status characteristic to the task.

The strength of the organizing power of a particular status characteristic is an inverse function of the length of the cognitive path linking it to the task outcome. A diffuse status characteristic requiring a lengthy path to be linked to a task outcome is theoretically a weaker organizer of performance expectations than is an ability directly linked to the task outcome. An experiment comparing diffuse status characteristics as organizers of interaction with specific, task relevant characteristics should show stronger organizing effects for task-relevant status characteristics than for diffuse status characteristics. Observable power and prestige orders in such groups should be more strongly differentiated than those in which diffuse characteristics are salient.

The same principle applies to the formation of expectations based on sentiment structures. Sentiments are weak organizers of interaction. This is a consequence of the way sentiments organize interaction. We have suggested that sentiments organize interaction by providing differential opportunities to talk and by providing differential chances to receive positive evaluations and avoid negative evaluations of performances. Such chances to perform and receive evaluations organize interaction indirectly, through the formation of expectations based on performances. This indirect organization is similar to the organization of interaction that occurs when diffuse status characteristics are salient.

Task abilities are theoretically the strongest organizers of expectations, because cognitive chains linking them to task outcomes are shortest. Diffuse status characteristics and formal role structures, with longer chains of cognition linking them to outcomes, are weaker organizers of interaction. We speculate that sentiment structures are no stronger as organizers of interaction than diffuse status characteristics.

Indeed, we expect sentiment structures are weaker organizers of interaction than diffuse status characteristics. We expect that high-status people will be good at tasks. We have no such expectations for our friends until performance attempts are made. We are then more willing to give them additional opportunities to perform and positive evaluations of performance attempts they have made. Eventually, we form expectations for behavior based on these performances and our evaluations of them. This is a more circuitous route to differentiated performance expectations than any of the others examined here.

If the above ideas are supported empirically, then it will be possible to take advantage of other features of Expectation States theories to develop additional ideas about the relationship between sentiments and interaction. Balkwell

(1991b) has reviewed research on the combining principle of Expectation States theory. He has shown that how diffuse and specific status characteristics organize behavior in several experiments is governed by a combining rule, rather than some other cognitive principle. The combining rule says actors use all cognitive information about status characteristics of others to form expectations for behavior.

If sentiments organize interaction in decision-making groups, and if they are salient, then combining principles may be used to predict behavior in groups where sentiment and status structures are consistent with one another and in groups where such structures are inconsistent with one another. Preliminary results in research examing this question look encouraging.

To recapitulate, we suggest that sentiment structures organize interaction in groups. This organizational process is similar to the process identified by Joseph Berger and his collaborators in the Expectation States research program. Actors differentially distribute opportunities to perform, and positive and negative evaluations, based on the sentiment structure in a group. Well-liked others are given more chances to perform and are more likely to receive positive evaluations. Disliked others are given fewer chances to perform and are more likely to receive negative evaluations of performances attempts. This, in turn, leads to formation of performance expectations that result in behavioral differences in the observable power and prestige order of the group. Well-liked others become the high status members of the group and disliked others become the low status members of the group.

AN EXPERIMENT

An experiment to test these ideas is part of a larger research program in face-to-face decision-making groups in which several different types of social structure are manipulated. Task ability, formal status patterns, and sentiment structures, as well as combinations of structures, may be salient simultaneously. The experimental situation allows creation of complex social structures that may be congruent or incongruent. It also permits several dependent measures of group structures that emerge. Three-person groups are the simplest ones in which to test these ideas.

Face-to-face interaction, besides being more natural than more controlled settings, allows for many measures of power and prestige orders of groups. We report only the set of measures based on how much each person talked in the discussion for three conditions of the experiment.

Paid volunteer subjects were recruited from basic social science classes at Ohio University. A recruitment questionnaire asked for name, age, phone number, gender, and the names of friends signing up together. This information was used to construct groups and schedule experimental sessions.[1]

Individual differences in propensity to act can affect observable power and prestige orders of groups (Willard and Strodtbeck, 1972; Conner, 1985). To control for this potential confound, a scale was used to screen subjects (Schutz, 1966; 1957). This scale asked volunteers to rate how they usually behave in groups ranging from complete domination to complete submissiveness. Participants were assigned to groups to match their scale scores.

The task ability manipulation employed the National Aeronautic and Space Administration "Lost on the Moon" survival exercise. The task asks volunteers to rank fifteen items for their value in surviving on the moon after the hypothetical crash of a lunar lander. NASA has developed a best solution to this task (NTL Institute for Applied Behavioral Science, 1969). Volunteers complete the task for the first time as part of the recruiting questionnaire. Subjects rank items twice more during the experiment. The second ranking occurs during group discussion, and the third is a part of a post-discussion questionnaire.

Task ability is assessed by comparing pretest answers to the NASA solution. Task scores are correlation coefficients between subjects' rankings and the ranking provided by NASA.[2] Highly skilled people became candidates for the expert position in the Expertise Condition. Subjects with intermediate skill become candidates for all other positions.

Opinion items in the questionnaire were the basis for creating sentiment structures in groups. Group members in the Sentiment Condition were told they had many opinions in common with some other member of the group. They also were told this high level of agreement would help them to get along well with that person. This technique has shown success as a mechanism to create liking (Byrne, 1971; Banikiates and Niemeyer, 1981; Broome, 1983; Cramer et.al., 1985; Byrne et.al., 1986). Person A was told that they agreed with persons B and C, while B and C were told they agreed with A.

Following selection for an experimental session, participants were matched with others for a particular session on age, race, and gender.[3] If ability was the basis for differentiation, we used skill level to select members for a group. Assignment was done such that members were unlikely to know others in their groups. Subjects who met the criteria were assigned to groups in various conditions of the experiment. This assignment process randomized subjects across conditions of the experiment, except for the person who was the expert in the ability condition of the experiment. This condition was labeled the Expertise Condition.

A Control Condition in which no structures were salient served as baseline for comparisons of the organizing effects of various structures. Before the group discussion, participants in the Expertise Condition were told of skill differences between themselves and others. Participants in the Sentiment Condition were told that others agreed or disagreed with them about opinion items. Groups were told about only those structures salient in their experimental session.

Following instructions, members discussed the NASA task for twenty minutes with a goal of a consensus solution. The group received a deck of cards with each item in the list on a separate card, and arranged these cards in the order they thought correct. We recorded the discussion on videotape for later analysis.[4]

After the discussion, participants completed another questionnaire, were interviewed about their impressions of the experiment, and debriefed. Participants accepted manipulations (for instance, of liking), engaged in active discussion, and did not become suspicious of deceptive features of the situation.

We report an analysis of interaction data for three conditions of this experiment. The data were analyzed, using four categories for coding the interaction. For each group, contributions to the task, suggestions for procedural rules, and summaries were coded as performance outputs. Pauses in interaction, questions to others in the group, or questions to the group as a whole were coded as action opportunities. Statements that evaluated the contributions of others were coded as either positive or negative reward actions. Correlations near .90 were obtained between two coders who worked independently. Similar coding schemes have been proposed by Bales (1950), Berger and Conner (1974), and others.

The sum of each of these types of acts for each person in a group is that person's Basic Initiation Rate. Performance outputs are the most common behavior, followed in frequency by action opportunities, positive reward actions, and negative reward actions. Negative reward actions were very rare, with only one or two per group on average. Measures comparing interaction rates across persons in a group were calculated by taking proportions of the number of acts in a group for which any one person was responsible.

RESULTS

Two research questions are examined. First, we want to determine whether or not sentiment structures organize observable power and prestige orders of groups. Second, we want to determine if the relative strength of the organizing principles that have been suggested is reasonable. In both cases, the appropriate test is based on observing differences in number of acts for each actor, the relative proportion of acts performed by an actor in the group, and the extent to which groups organize themselves in the ways hypothesized.

Table 1 contains data on the Basic Initiation Rate. This rate is expected to be higher in a comparison if the actor has high status relative to another; that is, high status actors are expected to talk more than low status actors in a group.

Differentiation in the Basic Initiation Rate is measured several different ways. Table 1 shows two variations of a simple measure of differentiation in which B's and C's mean Basic Initiation Rate is compared to A's mean Basic Initiation Rate.

Table 1.   Directly Observed
Measures of Participation

A.   Basic initiation rates

Condition	BIR by A	Mean of B & C of A
Control	94.0	94.8%
Sentiment	97.9	82.5%
Expertise	132.6	68.2%

B.   Mean basic initiation rate by position and condition

Position	A	B	C
Condition			
Control *	94.0	91.2	87.0
Sentiment *	97.9	78.9	82.6
Expertise **	132.6	104.5	76.5

* The differences between means for A and B and C are not signficant.
** The differences between means for A and B are not significant; the differences between A and C are signficant ($p < .001$).

In Panel A of Table 1 we see that, on average, B and C offer contributions to the group at about the same rate as A in the Control Condition of the experiment. The extent of differentiation in the Sentiment condition of the experiment shows that B and C contribute approximately four acts for every five contributed by A. The extent of differentiation in the Expertise condition shows that actor A contributes much more than the average B or C. A contributes approximately three acts for every two contributed by B and C. Note also that A talks about thirty percent more in the Expertise condition than in the other conditions.

Panel B of Table 1 contains a summary of how often A is the most talkative person in the group as compared to B and C, by providing the mean Basic Initiation Rate for each actor in the three person group. Our interest is in whether or not A talked the most in comparison to B and/or C. The data show the same effect as that observed in Panel A. A talks more than B and C when in a group with an imposed social structure. The effect is more pronounced when task ability is salient than when sentiment structures are salient. A planned comparisons analysis of variance showed only the position effect between A and C in the Expertise condition to be significant. We suspect this outcome is due to high variability in this interaction data.

It is also possible to answer the research questions by examining the rank order in the interaction profile for each actor in the group. For instance, A may not talk the most in the group, but may not talk the least either. In such a circumstance A may be exhibiting some evidence of the organizing power

Table 2.   Derived Measures of Participation

A.   Rank position of A in the observed power and Prestige order

	High	Medium	Low
Condition			
Control	4	2	4
Sentiment	4	4	2
Expertise	7	2	1

B.   Means of basic initiation rate Proportions

Position	A	B	C
Condition			
Control *	.328	.342	.330
Sentiment **	.374	.285	.341
Expertise ***	.434	.290	.276

   * Differences between means for A, B, and C are not significant.
  ** Difference between means for A and B is significant (p < .05); the difference
     between means for A and C is not significant.
*** Differences between means for A and B and C are significant (p < .001).
     Differences between means for Person A are significant for the Expertise/Control
     comparison but not the Expertise/Sentiment or Control/Sentiment comparisons.

of an imposed social structure. It is instructive to examine what proportion of the total activity in a group is accounted for by each actor to reduce the effects of variability in amount of talking.

Table 2 presents results from these more complex treatments of the data. Panel A of Table 2 provides another demonstration of differentiation effects. Here, the relative part played by each member of the group is portrayed by his or her rank in the group interaction profile. The table presents a simple frequency count of how often person A talked the most, how often they were the second most frequent talker, and finally, how often they talked the least. If an imposed structure is a perfect organizer, A should always talk the most. This is clearly not the case for any of the experimentally imposed structures. Sentiment structures organize interaction slightly more than the random pattern seen in the Control condition, but less than the stronger organization observed in the Expertise condition. This effect, while consistent with the hypothesis, is not significant.

Panel B of Table 2 presents the mean proportion of total acts for each position in the group. For this table, each actor is assigned a value based on his or her proportion of acts in the group. A mean value is then found for each position (A, B, or C) in each condition of the experiment. Cell means range from a low of .285 to a high of .434. Position in the Control Condition has no impact on which actor speaks the most. Each position contributes approximately one-third of the effort to solve the task. In the Sentiment Condition, A begins to dominate the discussion, but the effect is not strong. The effect is strong in the Expertise Condition. The position effect is significant for some of these comparisons. Again, the effect of sentiment structures as organizers is less than the effect of expertise structures.

The pattern is consistent with results from other three-person discussion group studies (Bales, 1950). All groups organize themselves so interaction is differentially distributed in reaching common goals. In three-person groups, the most talkative person accounts for about forty percent of the acts, the next most talkative, about one-third, and the least talkative, about one-fourth.

When considered with the results from Table 1 and from Panel A, these results show that sentiment organizes interaction in the way hypothesized, and is a weak organizer, as suggested by the second hypothesis.

DISCUSSION

We have proposed that sentiment structures organize interaction in the same way status structures organize interaction in groups. The ideas employed to advance this theoretical discussion are those developed in the Expectation States Research program to explain how status structures organize interaction. We are not suggesting that sentiment structures are status structures; rather we suggest that they are state-organizing structures (see Fararo and Skvoretz, 1986). The organizing process is based on the simple principle that we are more likely to give opportunities to perform to those we like and are more likely to evaluate positively their contributions.

From this principle, and from the ideas from Expectation States theories, it is possible to describe how sentiment structures organize interaction. The organization process is described for sentiment structures that map patterns of differential liking within groups. Cohesion and solidarity do not provide differential internal group structures and, hence, do not lead to predictions of differential participation on the part of group members. Local structures may be entirely composed of reciprocated liking relations. They also may be composed of combinations of liking and disliking relations. Neutral, or affectless, relations with others may also be part of local structures.

The differential distribution of chances to perform and of evaluations that result from sentiment structures governs emergence of observable power and prestige orders of groups. This process leads to assignment of high states of performance expectations to those we like and low states of performance expectations to those we dislike. States of performance expectations are assigned to neutral others based on the structure of relations with others in the group. In groups with disliked others, neutrals should be assigned high performance expectations. In groups with liked others, neutrals should be assigned low performance expectations.

The experiment that is described here provides some evidence that ideas about how sentiments organize interaction are reasonable. First, groups in which sentiment structures are imposed organize interaction so well-liked actors are given more chances to perform. Second, the observable power and prestige order in the imposed sentiment condition is consistent with the hypothesis that sentiments are weaker organizers than status structures, as predicted from the social process thought to be at work.

Some of the comparisons are not significant, however. While the patterns in the data are consistent with the hypotheses, the small number of groups in each condition and the high level of variability in the data result in some comparisons being statistically nonsignificant. A larger number of groups in the data set would provide clearer answers to the questions raised by these comparisons.

It is also possible that a better understanding of how sentiment organizes interaction depends on instantiations of sentiment structures not pursued here. For instance, we did not try to manipulate disliking in the experiment reported here. Others who have tried to do this (Webster, 1980, and personal communication; Byrne, 1971) indicate that it is especially difficult with student subject populations in typical experimental settings. Students are strongly predisposed not to dislike one another. Second, a stronger test of these ideas will result from specification of other sentiment structures than the one presented here. Unreciprocated liking among three actors and reciprocated liking between two actors in a three-person group are examples of structures that might be explored in further tests.

ACKNOWLEDGMENT

Research reported here was conducted at Ohio University. It was supported in part by grants from Ohio University Research Challenge Funds and from DAAL03-86-D-0001 from the Naval Training Systems Center. The views, opinions, and findings contained in this report are those of the author and should not be construed as an official Department of the Army position, policy, or decision. Research assistants who have worked on the project include Cynthia Blackburn, Scott Gordon, Brady Hamilton, Gretchen Holderman, and William Miller at Ohio University, and Paul T. Munroe and Max Nelson-Kilger at Stanford University. The work has benefited from discussions with Joseph Berger and Murray Webster. Any errors which remain are my responsibility.

NOTES

1. We use this information to construct groups so that individuals are not likely to know one another. The Ohio University Institutional Review Board for Human Subjects Research approved these procedures.

2. Spearman's rho is a good measure to use here because it is based on a difference score, produces comparable results for comparable rankings, and has metric properties.

3. Gender has been shown to be a status organizing characteristic. Effects appear in mixed-sex groups (Meeker and Weitzel-O'Neil, 1977; Ridgeway, 1988). We know of no evidence to suggest that it will consistently affect interaction in homogeneous groups.

4. The camera sat in plain view of the subjects in the experiment. The videotape provides a record of interaction that may be repetitively analyzed by the research team, so that high levels of reliability in coding of interaction may be attained.

REFERENCES

Bales, Robert F. 1950. *Interaction Process Analysis: A Method for the Study of Small Groups.* Cambridge, MA: Addison Wesley.

Balkwell, James W. 1991a. "From Expectations to Behavior: Animproved Postulate for Expectation States Theory." *American Sociological Review* 56:355-369.

Balkwell, James W. 1991b. "Status Characteristics and Social Interaction: An Assessment of Theoretical Variants." Pp. 135-176 in *Advances in Group Processes,* Volume 8, edited by Edward J. Lawler, Barry Markovsky, Cecilia Ridgeway, and Henry A. Walker. Greenwich, CT: JAI Press.

Banikiates, Paul G. and Greg J. Neimeyer. 1981. "Construct Importance and Rating Similarity as Determinants of Interpersonal Attraction." *British Journal of Social Psychology* 20:259-263.

Berger, Joseph. 1992. "Expectations, Theory, and Group Processes." *Social Psychology Quarterly* 55:3-11.

Berger, Joseph and Thomas L. Conner. 1974. "Performance Expectations and Behavior in Small Groups: A Revised Formulation." In *Expectation States Theory: A Theoretical Research Program,* edited by Joseph Berger, Thomas L. Conner, and M. Hamit Fisek. Cambridge, MA: Winthrop.

Berger, Joseph, M. Hamit Fisek, Robert Z. Norman, and Morris Zelditch, Jr. 1977. *Status Characteristics and Social Interaction.* New York: Elsevier.

Berger, Joseph, Morris Zelditch, Jr., and Dana P. Eyre. 1989. "Theoretical structures and the Micro/Macro Problem." Pp. 11-32 in *Sociological Theories in Progress,* edited by Joseph Berger, Morris Zelditch, Jr., and Bo Anderson. Newbury Park, CA: Sage.

Broome, Benjamin J. 1983. "The Attraction Paradigm Revisited: Responses to Dissimilar Others." *Human Communication Research* 10:137-151.

Burt, Ronald S. 1987. "Social Contagion and Innovation: Cohesion versus Structural Equivalence." *American Journal of Sociology* 92: 1287-1335.

Byrne, Donn. 1971. *The Attraction Paradigm.* New York: Academic.

Byrne, Donn, Gerald L. Clore, and George Smeaton. 1986. "The Attraction Hypothesis: Do Similar Attitudes Affect Anything?" *Journal of Personality and Social Psychology* 51:1167-1170.

Conner, Thomas L. 1985. "Response Latencies, Performance Expectations, and Interaction Patterns." Pp. 189-214 in *Status, Rewards, and Influence,* edited by Joseph Berger and Morris Zelditch, Jr. San Francisco: Jossey-Bass.

Cramer, Robert Ervin, Robert Frank Weiss, Michele K. Steigleder, and Susan Siclari Balling. 1985. "Attraction in Context: Acquisition and Blocking of Person-Directed Action." *Journal of Personality and Social Psychology* 49:1221-1230.

Evans, Nancy J. and Paul A. Jarvis. 1980. "Group Cohesion: A Review and Reevaluation." *Small Group Behavior* 11: 359-370.

Fararo, Thomas J. and John J. Skvoretz. 1986. "E-State structuralism." *American Sociological Review* 51:591-602.

Festinger, Leon, Stanley Schachter, and Kurt Back. 1950. *Social Processes in Informal Groups.* New York: Harper.

Fisek, M. Hamit, Joseph Berger, and Robert Z. Norman. 1991. "Participation in Heterogeneous and Homogeneous Groups: A Theoretical Integration." *American Journal of Sociology* 97: 114-142.

Freeman, Linton C. 1992. "The Sociological Concept of Group: A Test of Two Models." *American Journal of Sociology* 98: 152-166.

Granovetter, Mark. 1973. "The Strength of Weak Ties." *American Journal of Sociology* 78: 1360-1380.

Homans, George C. 1950. *The Human Group.* New York: Harcourt Brace.

Meeker, Barbara F. and Patricia A. Weitzel-O'Neill. 1977. "Sex Roles and Interpersonal Behavior in Task-Oriented Groups." *American Sociological Review* 43:91-105.

Moreno, Jacob L. 1953. *Who Shall Survive?* Beacon, NY: Beacon House.

Mizruchi, Mark S. 1991. "Urgency, Motivation, and Group Performance: The Effect of Prior Success on Current Success Among Professional Basketball Teams." *Social Psychology Quarterly* 54: 181-189.

Murdock, Peter E. 1989. "Defining Group Cohesiveness: A Legacy of Confusion?" *Small Group Behavior* 20:37-49.

NTL Institute for Applied Behavioral Science. 1969. *Ten Exercises for Trainers.* Washington, DC.

Ridgeway, Cecilia. 1988. "Gender Differences in Task Groups: A Status and Legitimacy Account." Pp. 188-206 in *Status Generalization,* edited by Murray A. Webster, Jr. and Martha Foschi. Stanford, CA: Stanford University Press.

Ridgeway, Cecilia and Cathryn Johnson. 1990. "What is the Relationship Between Socioemotional Behavior and Status in Task Groups?" *American Journal of Sociology* 95:1189-1212.

Schutz, William C. 1966. *The Interpersonal Underworld.* Palo Alto, CA: Science and Behavior Books. Also 1957 as Firo: A Three Dimensional Theory of Interpersonal Behavior.

Shelly, Robert K. 1988. "Social Differentiation and Social Integration." Pp. 366-378 in *Status Generalization,* edited by Murray Webster, Jr. and Martha Foschi. Stanford, CA: Stanford University Press.

Sherif, Muzafir, O. J. Harvey, and B. J. White. 1955. "Status in Experimentally Produced Groups." *American Journal of Sociology* 60: 370-379.

Skvoretz, John. 1988. "Models of Participation in Status Differentiated Groups." *Social Psychology Quarterly* 51: 43-57.

Webster, Murray A. Jr. 1980. "Integrating social processes." Paper SES7907574. Washington, DC: National Science Foundation.

Willard, Don and Fred L. Strodtbeck. 1972. "Latency of Verbal Response and Participation in Small Groups." *Sociometry* 35: 161-75.

THE INFLUENCE OF GROUP PROCESS ON PSEUDOSCIENTIFIC BELIEF: "KNOWLEDGE INDUSTRIES" AND THE LEGITIMATION OF THREATENED WORLDVIEWS

Raymond A. Eve and Francis B. Harrold

ABSTRACT

We show in this article how scholarly considerations of the origins of pseudoscientific belief have tended to be based on a view of such beliefs as arising out of individual factors such as ignorance, superstition, or psychopathology.

We will suggest, on the contrary, that pseudoscientific belief often originates in normal group processes. We will show how common errors in human reasoning are often influenced by emergent norms within one's peer groups.

We will also show how many pseudoscientific beliefs are frequently learned normally from mass media. Finally, and most importantly, we will show that far from being the result of individual ignorance or personality distortion, much pseudoscientific belief actually arises normally out of a struggle among cultural

Advances in Group Processes, Volume 10, pages 133-162.
ISBN: 1-55938-280-5

traditionalists, modernists, and postmodernists, and that this is a group struggle for control of the means of cultural reproduction. The struggle for the control of the means of cultural reproduction involves competing phenomenological generations and attempts at the legitimation of a particular world view among several such generations by claiming unique scientific creditworthiness.

INTRODUCTION

It may strike many as strange that in modern Western society, surrounded by "high technology" and scientific advancement, many individuals cling to pseudoscientific beliefs that withstand any amount of confrontation with empirical data or reasoned scientific argument. Research on this phenomenon tends to depict adherence to pseudoscientific belief as arising from a lack of education or from personal psychopathology. After a brief review of previous views on the matter, we will demonstrate that far from being a matter of individual ignorance or psychopathology, the existence of widespread pseudoscientific support for certain ideas grows out of normal group processes at both the "macro" and "micro" levels of social interaction.

Science and its practitioners occupy a position of high regard and great influence in American life (National Science Board, 1989). Scientific research is funded by billions of tax dollars, and scientists, from biomedical researchers to geologists to sociologists, are widely consulted regarding important questions of public policy. However, at the same time, many Americans hold beliefs quite incompatible with the methods and findings of scientists—denying, for example, that human evolution ever occurred, or believing that extrasensory perception is a well-established phenomenon, or accepting claims that thousands of children are sacrificed annually by Satanists. These and many other related beliefs are often termed pseudoscientific; they are marked, not by the outright rejection of science, but by the claim of scientific authority for propositions which are not supported by scientific examination.

Until recently, social scientists have tended to attribute a person's pseudoscientific beliefs to either ignorance or stupidity, or to some pathology of the personality (such as authoritarianism or religious fanaticism). There are strong reasons to reject such explanations as totally adequate. In order to begin an investigation into the matter, however, it is necessary to arrive at some clear understanding of what is denoted by the term "pseudoscience." Let us turn our attention now to that task.

DEFINING PSEUDOSCIENCE

What do we mean here by "pseudoscience?" Definitions of the term vary (see, for example, Schadewald, 1983; Hines, 1988; and Radner and Radner, 1982).

There are, however, certain common definitional elements. A common theme is that pseudoscientific claims are based on procedures and assertions that claim the status of scientific validity, but that do not warrant such validity upon close inspection, because proper scientific procedure was not actually followed in the course of developing them. For example, some such claims involve hypotheses that are not testable or falsifiable. Other pseudoscientific claims uncritically accept myths and legends as reliable evidence concerning past events. Pseudoscientific beliefs are often held as sacrosanct, and, therefore immune to modification in light of competent criticism or new scientific advances. Pseudoscientists often conduct "research by exegesis"—by interpreting and criticizing existing scientific methods and findings rather than conducting their own scientific investigations. Pseudoscientific research is usually published either by small special-interest presses or by mass-market trade publishers. Consequently, it is seldom subjected to peer review .

Commonly encountered forms of pseudoscience include "UFOlogy" (such as Erich von Däniken's books, alleging earthly visits by ancient astronauts); Immanuel Velikovsky's tales of "cosmic collisions;" parapsychology; and even certain fads in mental health therapies (Hines, 1988; Radner and Radner, 1982; Stiebing, 1984; and Cazeau and Scott, 1979).[1] A recent and "white-hot" example of pseudoscience is the allegedly scientific study methods used by religious cult investigators to detect supposedly widespread Satanic cult activities in the U.S. These same investigators further participate in pseudoscience by advancing the associated diagnosis of "ritual abuse survivor" syndrome (Richardson, Best, and Bromley, 1991). We have also, elsewhere, characterized "scientific creationism" as yet another example of pseudoscience, and have discussed at length why we consider it so (Eve and Harrold, 1991).

Before further analysis of the importance of group influence on pseudoscientific belief, we must first acknowledge that there is controversy over both the notion and the definition of our dependent (or exogenous, if one prefers) variable, that is, "pseudoscience." Some scholars, such as Truzzi, have argued against even employing the term (Truzzi, 1979). Its use is said to be unnecessarily pejorative, and inadvisable because a number of formerly unorthodox claims have, in time, turned out to be perfectly legitimate science. An example would be the theory of plate tectonics in geology. Truzzi would prefer to call similar unorthodox claims "parascience" or "protoscience." While granting that the boundary between science and pseudoscience may sometimes be uncertain, we reject the position that *absolutely nothing* can be termed pseudoscientific in nature. Plate tectonics was eventually accepted because it generated testable hypotheses that were repeatedly confirmed using normal scientific methods. In contrast, the "scientific creationism" promoted by some religious groups *cannot* be considered a science because its literature contents itself almost entirely with trying to pick holes in evolutionary science (e.g., Gish, 1985). No new science is offered in place of the mainstream consensus in the geology, archaeology, physics, paleontology, chemistry, etc., which creationists reject. In the *Statement of Belief* of the Creation Research Society (a prominent group which publishes a scientific creationist journal) we are told that

> The Bible is the written Word of God, and because we believe it to be inspired thruout [sic], all of its assertions are historically and scientifically true in all of the original autographs.

It hardly seems necessary to point out that a "scientific investigation" whose propositions are stated at the outset to be inherently not falsifiable can hardly be termed a true science, "proto-" or otherwise. Thus, we maintain that the term "pseudoscience" can indeed be justifiably applied in some cases.

WHAT IS TO BE EXPLAINED?
THE DISTRIBUTION OF PSEUDOSCIENCE

In order to explain the distribution of a dependent variable, one must first be sure one knows what pattern it is that one wishes to explain. Therefore, this section describes some of the more interesting pseudoscientific beliefs commonly encountered among the U. S. public.

One of the best-documented such topics is creationism. Nearly half of the American public rejects the scientific consensus on humankind's evolutionary origins. In a 1991 Gallup Poll, 47 percent of respondents agreed with a "strict-creationism" statement, specifically one that "God created man pretty much in his present form at one time within the past 10,000 years" (Sheler and Schroff, 1991). As might be expected, agreement decreased with increasing education, but was still notable. For example, 25 percent of college graduates indicated agreement. This latter figure is consonant with a national study of high-school biology teachers (Eve and Dunn, 1990), where 25 percent agreed with the same statement, and with a study of three samples of college students (in Texas, California, and Connecticut), where the figures were 28 percent, 19 percent, and 19 percent, respectively (Harrold and Eve, 1987). Furthermore, many of the college students in the latter study indicated agreement with a prime tenet of "scientific creationism." Some 42 percent, 25 percent, and 30 percent of the three samples, respectively, agreed that "There is a good deal of scientific evidence against evolution and in favor of the Bible's account of creation." Such data reinforce the impression gained from well-publicized struggles in courts and school boards that many Americans reject evolution, often on putatively scientific grounds (Eve and Harrold, 1991).

A whole range of other claims which lack scientific support are accepted by many Americans. We will not attempt a systematic survey of them, but will note Gallup Poll results indicating the following percentages of public acceptance of the reality or efficacy of: ESP (51 percent); precognition (37 percent); astrology (29 percent); Sasquatch or "Bigfoot" (13 percent); and the Loch Ness Monster (13 percent) (Gallup, 1979). In a similar vein, among our previously cited three groups of college students, 32 to 36 percent (depending on the sample) agreed that UFOs are alien spacecraft, and 28 percent to 37 percent agreed that the "Lost Continent" of Atlantis actually was home to a great civilization before sinking beneath the sea (Harrold and Eve, 1987).

Some information is also available regarding another main focus of this chapter, Satanic cult fears. Let us first clarify the pseudoscientific element of these fears. Our topic is *not* belief in Satan. In most definitions, Satan is a supernatural entity, and thus not amenable to scientific study. What *is* amenable to scientific study is the claim that members of Satanic cult groups commit thousands of crimes annually, including kidnapping and murder, in the course of cult rituals. It may well be that belief in Satan facilitates acceptance of this claim, since such belief allows for an evil being who actually directs the alleged crimes. A Gallup poll found that 39 percent of the public accepted the existence of "devils" (Gallup 1979). However, one could believe in Satan and *still* reject cult crime claims as unsubstantiated—or, alternatively, one could reject Satan's existence and nonetheless believe that Satanic cult fanatics were wreaking havoc on society. We are unaware of national survey data on acceptance of Satanic crime claims, but in a recent study of approximately 500 university students from three institutions in a large Southwestern metropolitan area, approximately two-thirds of the students felt that "Satanism is a real and immediate threat to the well-being of American families," "Satanism is extensive in U. S. society," and that "more police protection from Satanist activities is needed." Nearly half felt that "Satanism is pervasive in our public schools" (Roy, Eve, and Shupe, 1992). If the figures are this high for university students, it seems safe to conclude that the prevalence figures in the general population must be even higher.

Thus, it is clear that there is no shortage of pseudoscientific beliefs among the public to be explained. Many earlier works have suggested that these prevalence levels are the result of ignorance, superstition or personal psychopathology. We will argue instead that much of such beliefs can be traced to the rational acquisition of information where the source has frequently included pronouncements that there is scientific evidence to support these opinions.

PERSPECTIVES ON THE ETIOLOGY OF PSEUDOSCIENTIFIC BELIEFS

Serious attempts by social scientists to explicate the origins of pseudoscientific beliefs are surprisingly recent. The earliest work along these lines actually was conducted by clinical psychologists. For example, Carl Jung involved himself intensely for part of his career in the study of "spiritualism" (seances, table-turning, mediums, etc.), a very popular topic in Europe and America early in this century. Like most other early explanations, Jung's resorted to the idea of primitive impulses and sometimes even individual psychopathology as the appropriate mechanisms for a subject's belief in spiritualism. Jung most frequently attributed such beliefs to the Freudian psychological defense mechanism of "projection," that is, the notion that deep-seated subconscious conflicts led individuals to see a resolution of their problems in acceptance of the pseudoscientific phenomena as real (Jung 1977). Other explanations for pseudoscientific belief have also tended to emphasize individual psychology, including such factors as ignorance and fanaticism (e.g., Cloud, 1977), common errors in perception and memory (Hines, 1988), or downward socioeconomic mobility (Cavanaugh, 1983:193). However, over time there has been a trend in the direction of explanations for pseudoscientific beliefs that places greater emphasis on the importance of group pressure and group dynamics in accounting for the acceptance of such beliefs.

Singer and Benassi (1981) have suggested a useful typology of the sources of pseudoscientific beliefs. Specifically, they suggest that there are at least four main sources of such beliefs: (1) cognitive biases, or common errors in human reasoning, (2) media coverage of science and science issues, (3) poor or erroneous science education practices, and, finally, (4) sociocultural factors.

Cognitive Biases

Human beings seem predisposed in a number of ways to make systematic errors in processing information and reaching conclusions. For example, there is the "confirmatory incident." "My mother says that she just *knew* I was going to call" (implying, of course, that some form of ESP was involved). Such incidents, which are consistent with a preferred explanation, are highly salient and likely to be remembered, while those that are not consistent may be conveniently forgotten. Another example of a cognitive bias is "accidental shaping," found in nonhuman species as well as our own. Pigeons reinforced randomly in a Skinner box develop bizarre behaviors, like hopping on one leg and bobbing their heads, in the apparent (pseudoscientific?) misapprehension that these actions are the ones that deliver food to the tray. As Herbert Spencer once said, "Contiguity is more powerful than the faculty of reason."

A detailed discussion of cognitive biases as a source of pseudoscientific belief properly belongs to the domain of learning psychologists and is therefore beyond the scope of this chapter. Such a discussion, in most cases, would have little to do with the effect of group influence on the development of the belief. However, there is one area even within the current typological category where group influence can be readily seen to be important.

In the 1950s, Solomon Asch performed his famous experiments illustrating the influence of group pressure on perceptual judgments (Asch 1955). Asch had naive subjects enter a room with seven or eight other apparently naive subjects. Each member of the group was then asked to look at three black lines on a card and tell which was equal in length to a standard line shown simultaneously on another card. The judgment should have been simple; the wrong lines were clearly either longer or shorter than the criterion line. Each subject was to call out which line matched the criterion line. However, the other subjects were actually confederates of the experimenter. There were twelve trials in all, and during the first two, each individual called out a correct choice, thus creating the impression that the judgments were easy to make correctly. However, on the next trial, all but the subject would choose an "erroneous" line. When faced with this group pressure, and even with unambiguous sensory data available to them, a third of the subjects agreed with the erroneous judgment of the majority. When the subjects who had gone along with the majority were debriefed after the experiment, most indicated that they covertly recognized that the majority decision was in error, but had gone along anyway out of fear of the social consequences of violating the emergent norm established by the others.

Some of the pseudoscientific responses one finds in the general population are likely to have been influenced in this same way, especially when such opinions are expressed in a group. While it might be hard for some individuals to accept that all of the rocks of the Grand Canyon could have been formed as a result of the biblical Deluge, as advocated by scientific creationists (e.g., Morris, 1974), they are nevertheless unlikely to raise doubts if they assume that others in the congregation do espouse such beliefs. This is especially true if these others constitute one of the individual's primary reference groups.

However, in the above example, we are assuming that individuals are *consciously* altering their publicly given opinions in acquiescence to a group norm. This would not go very far toward explaining why they might give such an answer on an anonymous questionnaire, where the power of the group would seem to be absent.

Nonetheless, there is an interesting way in which the effects of the group may still be present even in such a situation. This is apparent in another classic experiment in social psychology by Muzafer Sherif and O. J. Harvey (Sherif and Harvey, 1952). They conducted an experiment to test for the existence of a phenomenon that they termed "the autokinetic effect." Subjects were introduced into a darkened room where they saw a point of light on a screen. They were then asked how much the light was moving. While the light was not actually moving, many subjects expressed the opinion that it was. Sherif and Harvey explained this by pointing to "the removal of situational anchorages" (i.e., the darkened room gives no perceptual clues against which to "anchor" the position of the point of light). From this experiment and in subsequent work (see Sherif and Sherif 1969), Sherif drew several conclusions, among them:

1. The more uncertain the situation, the greater the scale
   on which judgmental reactions will array themselves.
2. The more uncertain the situation, the greater the
   impulse toward convergence of opinion when in a group situation.

It is the second point that is of particular interest to us here. An example of this phenomenon at work may well be seen in some reports of UFO witnesses. Typically, we find that such observers' initial estimates of the size, distance, and speed of an object in the sky are widely divergent, probably because of the lack of situational anchorages. Looking high in the sky removes objects on the surface of the earth as reference points. If one grossly misjudges the size of an object as larger than it actually is (e.g., mistaking a nearby weather balloon for a very large distant object), then one will greatly overestimate the magnitude and speed of any apparent direction change. UFO investigators have noticed a tendency for initially variable estimates of size, speed, and distance of a sighting to converge as witnesses discuss with each other what each believed he saw. It seems likely that this is a direct analogy of the "autokinetic effect." Similarly, recent depictions of alien beings by self-proclaimed UFO abductees show a remarkable similarity among descriptions. A high percentage of abductees describe their captors as creatures each of which has a small body, large head, and two enormous eyes and lacks a protruding nose. Apparently, separate abductees really are seeing similar aliens, or, alternatively, a group norm has emerged that is powerful enough to shape different individuals' perceptions of what they "saw."

However, a caveat is in order at this point. We have implied that the uncertainty of a situation increases the probability of conformity to a group norm (including, for example, the acceptance of the pseudoscientific beliefs found among those in one's reference groups). Such a proposition is in keeping, certainly, with the results of the studies by Sherif and by Asch, but it should be pointed out that even those authors never really identify the respondent's subjective reasoning that leads to the adoption of the group beliefs. Certainly, we will not be able to do so here either. There exists, therefore, a need for new research projects to confirm or reject the proposition that situational uncertainty leads to increased vulnerability to pseudoscientific beliefs. If the proposition can be supported, then there is also a need for future studies to identify the specific mechanisms by which this occurs. Perhaps, the matter is as simple as the old Marxist dictum that "external threat creates internal cohesion"—even if it is cohesion to a specific abstract definition of the situation.

Mass Media and Pseudoscientific Belief

By often providing erroneous images of the scientific method and scientific findings, and sensationalistically reporting pseudoscientific claims, the media also contribute to acceptance of such beliefs. In the survey of college students cited above (Harrold and Eve, 1987), from 26 to 36 percent of the three samples accepted the proposition that humans and dinosaurs lived at the same time. While some of this acceptance can be attributed to some students' religious commitment to a recent creation, others apparently uncritically accept Hollywood images of Raquel Welch in a fur bikini being chased about by a *Tyrannosaurus rex.* Similarly, many contemporary television programs (variously labelled as "documentaries," "reality programs," or "infotainment") carry breathless accounts or "reenactments" of flying saucer encounters and abductions, as well as the detection of mysterious monsters (such as Bigfoot and the yeti) and a seemingly endless series of expeditions in search of the remains of Noah's Ark on Mount Ararat in Turkey. One popular magazine with a science theme, *Omni,* often gives favorable coverage to "fringe" science claims. Even major news organizations sometimes trumpet unconventional claims, while ignoring or downplaying their eventual refutation. Some years ago, for instance, one of the authors saw rather excited coverage on a network evening news program of a videotape claimed to document the tracking of an airplane by a UFO off of the New Zealand coast. Only many months later did he see the claim refuted (the "UFO" was apparently reflected light from a fishing fleet at night)—on the PBS science program *Nova,* whose viewership is relatively small. As far as the author could ascertain, the network news did not cover the claim's refutation.

Pseudoscientific beliefs are also promoted by the mass media because of the way in which they typically present controversial issues. A value deeply embedded in American culture is that of letting everyone "have his say" before an issue is decided. Enshrined in the Constitution's free speech and due process provisions, and institutionalized in broadcasting's "equal-time" regulations for political candidates, this estimable value results in the media's generally presenting controversies in terms of two viewpoints (and spokesmen), each given an equal opportunity to make his case. This practice neatly dovetails, by the way, with the perceived need among journalists to depict conflict in order to make a story interesting. Thus, a scientific controversy may be presented as little more than a personal dispute between two prominent scientists. It is an easy step from belief that each side should get "equal time" to an unconscious decision that since each side gets equal time, they must be at least nearly equal in scientific merit and number of supporters.

Creationists have taken advantage of this value in their struggles to insert scientific creationism into public school curricula (Eve and Harrold, 1991). They first argue that they are promoting science, not religion. They thus seek for their views the imprimatur of scientific authority, which is highly valued in American society, if not broadly understood (Toumey, 1991). Then, they charge that evolutionists are being unfair and closed-mindedly stifling dissent by not letting each side have equal time in the science classroom. "Let the children hear both sides, and then make up their own minds," they say. This frames the issue in such a way that their opponents appear to many to be censors of free speech.

Before leaving the topic of media influence, it is necessary to mention the "alternative media" which have been developed to give the appearance of scientific legitimacy to what remain, in actuality, fringe claims. Examples would be publishing houses set up especially to promote creationism or UFOlogy. It has been widely noted by social theoreticians that Western society in general, and U. S. society in particular, seem to be entering a new stage of economic development termed "postindustrialism." The postindustrial society is said to be one that has completed the stage of developing heavy industry and essential consumer goods. After this, a society seems to begin to devote an ever larger segment of its economy to the sale of services (by hairstylists, accountants, fashion consultants) and to the production and management of information (computer programmers, systems analysts, artificial intelligence and "expert systems" designers, etc.). An increasing percentage of the society's "capital" is represented by the rather intangible entity of information. Whereas, in industrial society, the problem faced by the would-be successful individual is how to make his product appear more valuable than those of his competitors, under postindustrial circumstances the problem becomes how to make, not his "widgets," but his information, more valuable than a competitor's.

Michael Cavanaugh (1983) has suggested that such a situation is likely to lead to the rise of "knowledge industries." These are units of the larger economy devoted to producing information in a competitive "knowledge marketplace." This, Cavanaugh suggests, is precisely the enterprise of "creation scientists." In our time, science has often been said to have replaced religion as the great legitimizer of truth claims. Under such circumstances, if claims about creationism are accepted as "scientific," the creditworthiness of the claimants is greatly enhanced.

Such acceptance of creation science has been particularly important for creationists since federal court decisions in the 1960s and 1970s struck down state laws prohibiting the teaching of evolution or mandating equal time for traditional creationism (with its frankly religious rhetoric) in public schools (Larson, 1989; and Eve and Harrold, 1991). In this context, creationist "research institutes" and think tanks have become important resources for the movement. Staffed by evangelical Protestants with Ph.D.'s in scientific and technical fields, they provide books, periodicals, videotapes, and lecturers who reassure the faithful that true science is on their side. These speakers and materials also try to persuade the uncommitted that if they can maintain an open mind, they, too, will see how flawed evolutionary theory is. Notable examples of these groups include the staunchly conservative Institute for Creation Research (ICR) in Santee, California, and the Institute for Thought and Ethics in Richardson, Texas. The latter group follows a "softer" creationist line. For example, in a supplementary text produced for biology classes (Davis and Kenyon, 1989), they avoid words like "God" and "creation" in favor of "intelligent design," which they credit with the development of life forms too complex to have been produced by "chance" evolution. The student is left to draw his or her own conclusion that the intelligent designer must have been God.

The work of such institutes circumvents the normal channels of scientific publication, usually in favor of in-house publishing organs like the ICR's Master Books. Some antievolution books also find their way into the shelves of "trade" publishers, where some pseudoscientific writers, such as Erich von Däniken, have been spectacularly successful.

The main point here is that there are many of these fringe media concerns that fit Cavanaugh's description of "knowledge industries," which fund and operate media outlets of their own. In addition to creationists, one can find similar "scientific" studies of crop circles,[2] UFO landings and abductions,[3] and so on. Each of these knowledge industries has learned to utilize mass media techniques to assert the scientific validity of its claims.

KNOWLEDGE INDUSTRIES AND SYMBOLIC INTERACTION THEORY

George Herbert Mead (1934) laid the foundations for an understanding of how people utilize the processes identified as central to the theory of symbolic interaction to arrive at a definition of their own reality and their situation within it. Essentially, Mead argued that the meaning of a thing was not an intrinsic property of the thing, but instead arose out of the interactions one has with other human beings with regard to the thing. For example, the Navaho Indians produced a "Swastika" design of black on red on their ceremonial blankets long before Hitler used the same symbol to represent the Nazi party. So which is this twisted cross, a religious symbol or one of nationalistic pride? The answer, of course, is that its only meaning derives from that assigned to it through social interaction by each of the relevant groups. Unfortunately, however, there was never anything in Mead's theory to tell us which particular definition of a symbol would arise in a given set of circumstances out of an infinitely large set of *possible* meanings that could have been assigned to the symbol.

To some degree, a partial answer to this inadequacy can be found within the recent criticisms of science arising largely out of the work of Jurgen

Habermas (1987) and the "deconstructionist" camp. These writers attempt to demonstrate that the meaning of rhetorical symbols is largely determined by the role of differential power among groups, both large and small. In other words, they argue that those occupying privileged positions in society have a disproportionate ability to establish which particular meanings for symbols will be considered "natural" or "real.[4] While this type of influence has been largely subconscious in the past (with males, Europeans, scientists, and holy men) having had more say in the meaning to be assigned to symbols than the social pariahs, we now have had the arrival on the scene of "knowledge industries" (Cavanaugh, 1983) driven by those who specifically set out to manipulate which symbols will be chosen to be the shared ones. Examples of such knowledge industries are readily seen now not only in so-called creation science, but in the "marketing" of Presidents, and the selling of Western culture to less developed nations. In other words, a knowledge industry is a special-interest group which sets about trying to legitimate a particular set of symbolic meanings as being more valid, and therefore valuable, than any competing interpretations.

The foregoing remarks constitute, of course, a set of propositions which are, as yet, largely untested. They are presented in hopes that a new generation of studies will be undertaken to assess their veracity. More specifically, it is hoped that such new studies would also focus on understanding how group dynamics operate to determine how specific shared symbols are chosen from the universe of all possible symbols that potentially could be adopted collectively. It seems likely such studies will reveal a dramatic increase in recent years in competition among knowledge industries for control of the "means of cultural reproduction" of various groups (as distinct from a purely Marxist economic struggle) (Oberschall, 1984). Such an outcome would be expected as more societies move into a postindustrial stage where the dominant form of capital is not labor, money, or "widgets," but, instead, is represented by the production of information and the control of the meanings associated with its availability. The central problem for modern economies heretofore based on material capital has been how to make one's own capital (such as gold, or its symbol, e.g., paper currency) more valuable than that of other groups. Now, however, as information becomes the dominant form of capital, a new but analogous problem arises, specifically, that of how a group makes its information more valuable than that of other groups. This is the primary task of knowledge industries, either to gain credit for their own special symbol system, or, at least, to discredit the validity of the other group's information (symbols). This is one of the major ways in which groups struggle to control the means of cultural reproduction.

Knowledge Industries at Work: Constructing a Moral Panic

Yet another example of the operation of a knowledge industry all dressed up as a science, but which is actually a pseudoscience, is the illustrative case of the recent coverage, in both the mass media and the mental health profession literature, of alleged Satanic abductions, "ritualistic" sexual abuse, and sacrifice of adults and children. This recent attention is best viewed as a case of a "moral panic" (not unlike the other witch hunts of earlier times, or the McCarthy-era "Red Scare"). Many people are clearly readily accepting claims that there is much "hard" evidence (intended to imply "scientific" evidence) to support fears of a general outburst of Satanism. We will turn our attention first to these allegations, then to a brief consideration of their legitimacy, and will finally return to our main point in this section concerning the role the media have played in creating this "feeding frenzy."

Just before Halloween of 1989, it was widely rumored that Satanic cult activities were picking up around the U. S., and that the cultists were planning the abduction (and, some said, sacrifice) of blond, blue-eyed children just before or on Halloween. Victor (1990:288) has reported that such rumors surfaced in 21 other sites across the country that year, and many more have followed, since.

Perhaps part of the hysteria can be traced back to the fact that since 1984 there have been well over 50 Satanic cult seminars taught by police "experts" at professional conferences for educators, social workers, mental health professionals (especially those working in psychiatric facilities for youth), victim advocates, probation officers, corrections officials, and clergy (Hicks, 1991).

There are two ways in which these seminars are of interest here. First, many of the "cult cops" allege that there is much empirical evidence for the alleged activities. Second, a number of police and mental health "experts" have emerged who are fully prepared to testify in court to the scientific validity of a "ritualistic abuse syndrome" among cult survivors. Both sets of claims can probably best be understood as instances of pseudoscience.

Hicks (1991) believes that there are several precipitating factors in the development of the cult crime model. One was the publication of *Michelle Remembers* (Smith and Pazder, 1980), a popular book which was one of the first of the "cult survivor accounts." The book presented a lurid tale of sexual and physical abuse of a young woman at the hands of a Satanic cult. ("Michelle" later recanted her story, but the same media which had made so much of her original expose paid little attention to her change of heart). Michelle was identified as schizophrenic (allegedly as a result of her experiences with the cult), a fact which dovetailed nicely with the 1980 decision of the American Psychiatric Association to legitimate in the Diagnostic and Statistical Manual of Mental Disorders the diagnosis of "multiple personality disorder" (or MPD) as a type of "dissociative disorder."

In recent years, attention to the possible existence of such cults has grown intense, at least in part because of massive televised coverage since 1983 of alleged child abuse cases at various day-care centers (some of which episodes were later to be said to be connected to Satanism) and to coverage of the Matomoros, Mexico cult slaying of an American college student.[5]

There have been many resultant "cult crime seminars" which typically present a four-tiered model of Satanism (Hicks, 1991). The model depicts many young people innocently drawn into the Satanic web by fantasy-role-playing games such as "Dungeons and Dragons," or by rock music.[6] The cult crime model's final tier describes well-placed, apparently responsible adults who are said to be seeking demonic empowerment through human sacrifice and ritual child abuse. The seminars presenters tell their audiences that Satanic cults are no rare phenomenon. So-called "Scientific" estimates are sometimes presented, suggesting that as many as 50,000 victims per year are being sacrificed. If one adds claims of physically and sexually abused persons to the total, some estimates have been given of up to two million victims of Satanists per year. Such large figures bear close scrutiny regarding their scientific validity and reliability.

Carlson and Larue (1989) have reported that the National Child Safety Council estimates that there are approximately 240,000 missing child reports filed each year. They have also found that, annually, more than 120,000 of these reports are solved within several weeks. Nearly all the remainder are related to child custody battles, runaway episodes, or cases of "throwaway" children. Carlson and Larue go on to report that, based on FBI Uniform Crime Report statistics for one recent year, a mere 67 cases of missing children could even *possibly* have been the result of stranger abductions (assuming they were *all* abducted by Satanists). A nearly inescapable conclusion is that Satanic alarmists have given into the temptation to grossly exaggerate their allegedly scientific figures in order to bolster their cause. Such seminars and statistics can be clearly seen to be another case of a knowledge industry busily trying to legitimate a particular social construction.

Knowledge Industries at Work: Constructing the Ritualistic Abuse Syndrome

There is also need to raise some questions about whether media practices have helped to create a new class of medical and psychiatric ("scientific") experts on the topic of "ritualistic child abuse." The appearance of such apparently legitimate experts is becoming common on television, before college social work classes, and even as expert witnesses in court cases.

The rise in the number of these experts can probably be traced back, at least in part, to the increasing acceptance by the public and the media of the widespread prevalence of family violence and abuse. Initially, training of increasing numbers of clinicians to recognize the symptoms of wife and child battering was on relatively safe scientific grounds—black eyes, broken bones, and so on are straightforward data. However, the courts increasingly confront enormous difficulties in establishing *after the fact* just what should be considered accidental injury and what is clearly abuse. One eventual outgrowth of this problem was the popularization within the medical and psychological professions of the "post-traumatic stress syndrome." With the advent of this concept, things begin to get a bit more ambiguous. Now we are no longer speaking of physical evidence collected at the time of the alleged event (pictures, semen samples, or hair or skin under the assailant's fingernails). Instead, we are asked to accept that certain subtle psychological traits can be taken as scientific evidence of a *prior* assault. Even so, this diagnosis has been sufficiently widely accepted as scientific that some court decisions have been based, at least partially, on the diagnosis that a victim evidenced a "post-traumatic stress syndrome."

This sequence of events made it natural in the eyes of some to develop more recently the diagnosis of "ritualistic abuse syndrome" (a subclassification of the more amorphous "cult survivor syndrome"). With the recent development of the "ritualistic abuse syndrome" diagnosis, we move into uncharted waters. This diagnosis has its origins in the 1980s classification by the American Psychiatric Association of "multiple personality disorder" (MPD) as one of the associative disorders. It is the increasing and uncritical acceptance of the identifiability of this alleged disorder (MPD) which lies at the heart of the identification of the even slipperier diagnosis of "ritualistic abuse syndrome." Children alleged to have been ritualistically abused by demonic cults are said to be so traumatized that their only way to deal with the experience is to employ a "dissociative mechanism." That is, they develop more than one personality so that the injured self can hide from view, and the events themselves become unconscious unresolved memories. The recently preferred treatment for such alleged victims is to try to "recover" the conflicted subconscious memory of victimization through hypnosis, and then to attempt to induce "abreaction"— the reexperiencing, under hypnosis, of the traumatic events in such a way that they can be "processed" and returned to conscious memory. Consequently, no one dares to doubt when "unusual fears, survivor guilt, indoctrinated beliefs, substance abuse, sexualization of sadistic impulses and dissociative states with sadistic overtones" are advocated as a new scientifically identified syndrome (Young et al., 1991). In addition to the suspect nature of the procedures themselves by which such a syndrome are inferred, there has yet to be a single trial in the United States that has produced any other type of hard evidence of ritualistic child abuse and sacrifice. Apparently, Satanists are infinitely more clever than other types of criminals in their avoidance of the law—drug dealers get caught, inside stock traders get caught, major Mafia chieftains do not go wholly undetected, but Satanists never get caught. There is, of course, another, and likelier, conclusion that one could reach.

The depths to which the cult crime advocates will go can be illustrated in the case of "The Devil in Mr. Ingram" (Watters, 1991) wherein we find some law-enforcement officers arguing that the normal rules of criminal evidence must be suspended in the case of Satanists because, otherwise, we would never catch one!

Knowledge Industries and Pseudoscientific Beliefs

The foregoing discussion reminds us that pseudoscientific beliefs are not only to be found among the ignorant or the mentally imbalanced. On the contrary, even the most highly educated and respected members of what Peter Medawar has termed the "New Class" (Medawar, 1973) can fail to recognize when they are crossing the boundary between legitimate science and bogus science. Members of the New Class are neither owners nor managers in the classical Marxian sense, but are instead involved in the production and management of highly technical information. However, it is obvious from the examples above that membership in the New Class does not automatically make one immune to pseudoscientific beliefs. Indeed, it seems clear that, in many instances, adherence to such beliefs is just what one or more knowledge industries intend to occur. It is clear therefore that group dynamics such as suggestibility and collective symbolization are not always the results of ignorance or psychological disability. Unfortunately, we may expect to see in the future more new "sciences" (such as creationism or Scientology) and "scientific" mental health therapies which are actually burgeoning knowledge industries attempting to legitimate bogus science as the real thing.[7]

Factors Related to Science Education

A third set of factors promoting the acceptance of pseudoscientific beliefs can be added to those involving cognitive biases and the media. One might expect that the way to avoid the pseudoscientific beliefs just described would be science education. However, there is an unfortunate amount of evidence against this supposition. For example, sociologist John Miller (1983, 1987a, 1987b) has conducted national surveys intended to assess American adults' level of "scientific literacy." Miller employed a fairly modest definition of scientific literacy. To be considered scientifically literate, an adult needed to demonstrate a basic understanding of scientific methods (e.g., the role of formulating and testing hypotheses in scientific study of a subject), and a few scientific constructs (e.g., the rudiments of concepts such as DNA). Miller (1987b) found that even by these minimal standards, only about five percent of the U.S. public could be considered scientifically literate—able, for instance, to read a newspaper story about DNA research and genetic engineering, and understand its content and implications for public policy.

Only 16 percent of Miller's subjects reported having a "clear understanding" of what DNA is, compared to 57 percent who admitted that they had "little understanding" of DNA. Some 7 percent believed astrology to be "very scientific," and another 29 percent believed it to be "sort of scientific." And no less than 41 percent agreed that "Rocket launchings and other space activities have caused changes in our weather" (Miller, 1987a). Miller has also found that only about forty-five percent of his U.S. respondents knew that the earth revolves around the sun in a period of one year (instead of a day or a month). While it is true that scientific literacy scores increased with education, still only 12 percent of college graduates were judged to be scientifically literate (Miller, 1987a). Even among Americans holding postgraduate degrees, the proportion was an unimpressive 18 percent.

Nor is there much evidence that superior science education is producing a new generation of scientifically sophisticated citizens. In a recent study by the International Association for the Assessment of Educational achievement (or IEA), science achievement levels of secondary students from thirteen developed nations were assessed for the years 1983 and 1986 (IEA, 1988; and Walsh, 1988). American students made discouragingly low scores in all areas of science tested, never finishing higher than eighth in any subject. In biology, they scored last. Some indication as to why such scores are so low may be found in the study by Eve and Dunn (1990). For example, 19 percent of the high school biology teachers they surveyed claimed that dinosaurs and people were once contemporaries, and 26 percent agreed that "some races of people are more intelligent than others." Obviously, people with little understanding of what science *is* are ill-equipped to evaluate pseudoscientific claims.

It might be objected that the data just presented document only a *lack* of proper instruction in the methods and findings of modern science. In this view, good science education, now in short supply, should remedy the problem of pseudoscientific beliefs.

However, such a remedy would be incomplete, for many such beliefs thrive, not only in a vacuum of scientific ignorance, but also *in spite of* the dissemination of scientific findings which contradict them. Granted, far too many students are ignorant of basic knowledge in science, and are unaware of how such findings are made. But we do not encounter militant claims by authors, students, parents, religious groups, and even teachers, that water is actually composed of helium atoms or that the sun really orbits the earth. Nor do we see demands that such alternative views be included in school science curricula. Creationism and some other pseudoscientific beliefs do, however, involve such resolute advocacy in the face of mainstream scientific consensus. Inadequate science education cannot wholly explain this advocacy, and we maintain that other, sociocultural factors are involved as well.

Furthermore, studies indicate that instruction in critical thinking and scientific methods and findings (even when specifically designed to offset pseudoscientific beliefs) have real but *limited* effects on groups of students (Gray, 1987; and Singer and Benassi, 1981). Interestingly and unfortunately, even these limited effects tend to fade over time. In fact, in Gray's (1987) pre- and posttest examination of the effects of a university-level "debunking" course, in spite of an initial drop in pseudoscientific beliefs among the students which took place immediately after the course, some such beliefs after a few months actually exceeded their precourse levels.

In a similar vein, as yet unpublished recent research by one of the present authors which focused on college students' geographic knowledge found no significant relationship between such knowledge and the number or recency of geography courses that students had taken, or even with their overall grade point average. The only strong correlations found were ascriptive characteristics of respondents, such as race, ethnicity, and gender. (For example, females constituted only 9 percent of the top quartile of all students scored in terms of overall geographic knowledge). We do not accept the idea, however, that these performance levels are somehow genetically determined. It is likelier that little-understood but powerful socialization processes are at work very early in the lives of individuals. We strongly suggest that such a pattern is *not* the result of mere ignorance at work. Instead, it seems likely that the etiological rules for how a thing is known to be "true" differ when a student leaves the campus. We need to know more about how both formal and informal group dynamics operate to produce different phenomenological constructions of both "Truth" and "Reality." The rules used to construct these entities seem to differ sharply across social settings, leading to what one might well refer to as "situated" beliefs—that is, beliefs held in one setting but not in others.

The implication of the foregoing observations is, of course, that we need to begin a whole series of studies focusing on early childhood influences on the origins of both scientific and pseudoscientific thinking as these are influenced by sociological forces. (To date, it can be argued that sociologists have paid very little attention to sociological forces and how they influence the microlevel *during child development*. For an exception, see Eve, 1986, regarding a study of the small group dynamics which promote the origins of normative behavior. Similarly, studies to date have paid little attention to systematic differences in employment of differing sets of etiological rules for the formation of phenomenological reality as these change across social settings or social strata.)

Sociocultural Factors

The last of the generic sources of pseudoscientific beliefs listed by Singer and Benassi is that of "sociocultural factors." Such factors, while the most truly sociological in nature, are, paradoxically, often the hardest to recognize. The situation is a bit like a studied consideration of air or gravity. Certain phenomena are so omnipresent in one's environment that their existence tends to pass unnoticed. So too, do many of those factors which most influence our cognitive epistemologies, and resultant social constructions of "reality," typically go unnoticed in spite of their nearly omnipresent influence on our world views. Let us illustrate how in at least one area, the creation-evolution debate, the dispute is framed less by questions of what constitutes good science than by the more general sociocultural factors we explicate in this section.

In 1974, the new fall school term started for some 45,000 students in Kanawha County, West Virginia. However, a large number of protesting parents had withheld their children from the schools (10,000 by school board estimates), picketed businesses and mines throughout the county, and prevented school and city buses from operating. Unwilling to cross lines of picketing parents, 3500 miners declined to go to work, in spite of orders to do so from United Mine Workers officials. After about a week, the schools were closed for a three-day period and a number of controversial books that had initiated the fracas were removed from the classroom, at least temporarily (Moffett, 1988).

However, according to Page and Clelland (1978:269) "During this so-called cooling off period, violence escalated." Random sniping and vandalism of school property were commonplace; schools were firebombed and dynamited; school buses were shot at with firearms, as were two highway patrol cars that were escorting a school bus. In addition, the County Board of Education building was dynamited.

"What," one might well ask, "had moved the citizens of Kanawha county to declare war on each other?" The answer is the inclusion into the required curriculum of a set of controversial public school textbooks. Alice Moore, a school board member and the wife of a fundamentalist minister, objected to the choice of a number of books. West Virginia law requires that local school districts choose their books from a state-approved list if they expect the state to pay for the books. As a result, a good deal of weight had been given to the mandate of the West Virginia State Board of Education that districts should select reading materials which placed emphasis on "minority and ethnic group contributions to American growth and culture and which depict and illustrate the inter-cultural character of our pluralistic society" (Moffett, 1988: 11). However, while Kanawha County contains the West Virginia capital of Charleston, it is otherwise composed of many small, rural communities. "The population is distinctly non-ethnic and Protestant. Less than 1 percent of the population is non-white and only 2.9 percent are first or second generation foreign born" (Page and Clelland, 1978). Alice Moore had been elected to the school board in 1970 as a result of her instigation, in 1969, of a campaign to remove the recently introduced topic of sex education from the county schools' health education program. In this too, the source of the conflict could be seen to be largely one of rural Appalachian values in opposition to those of a more cosmopolitan, secular sector in the county. Mrs. Moore had asserted that sex education was anti-Christian and anti-American, and violated God's law.

After reading the newly adopted Language Arts books, Moore denounced them as "filthy, trashy, disgusting, one-sidedly in favor of blacks, and unpatriotic" (Moffett 1988:14). At the next school board meeting, more than 1000 anti-textbook protestors appeared. Not to be intimidated, the board voted for adoption of the disputed books anyway. This decision was a precipitating event in what one author has termed "culture wars" (Hunter, 1991).

What was the nature of the topics in the books that had created such a volatile situation? The books contained a diverse selection of poetry, nonfiction, drama, short stories, and other fiction, mostly written by recent American authors. The protesting parents had pointed to passages in the books which they believed promoted a climate of sexual immorality, lack of patriotism, Communism, drug and alcohol use, violence, and most notably for our purposes, *belief in evolution*. While most of these topics are not particularly scientific in nature, belief in evolution versus creationism certainly *appears* to be. This seems a curious addition to the overall list. After all, people do not shoot at each other over whether cold fusion actually exists or over whether there is "dark matter" in deep space. Why then should evolution versus creationism inspire citizens to acts of such violence? We will suggest an answer here that may shed light not only on the question of creationism versus evolution, but may also suggest a powerful source of pseudoscience, in general. To draw back this particular curtain, however, requires that we first identify three Weberian "ideal type" groupings of members of modernized societies. We will term these types here "cultural traditionalists," "cultural modernists," and "cultural postmodernists."

Page and Clelland (1978) suggested the term "cultural fundamentalists" to describe an ideal type grouping of citizens who espouse traditional religious, political, and social values. They tend to believe that the "truth" of an assertion should be evaluated in terms of authority, tradition, faith, and a belief that God reveals the truth through scripture.

"Creation scientists" are particularly exemplary of this category. Page and Clelland suggest that members of this grouping are strongly opposed to situational ethics. For example, cultural fundamentalists believe that evolutionary theory *cannot be* valid. If the Bible is wrong about human origins, "How," they ask, "can it then be trusted in any other area?" And if this question were to become manifest, would not one also lose faith in the Bible as a guide for answers to the pressing social and scientific problems of our times? People might then begin arguing that good and evil are only manmade constructs. This, they further believe, would lead to "secular humanism" (seen as "placing man before God"). Secular humanism would then, they argue, lead (through human fallibility) to mistakes in moral judgment. The "predictable result" would be a society characterized by prostitution, pornography, drug abuse, homosexuality, Satan worship, and a host of other evils. While in substantial agreement with Page and Clelland, we prefer the term "cultural traditionalists," to emphasize that not all members of this group are religious fundamentalists. Research (Eve and Harrold, 1991; Toumey, 1990; and Loy, Eve, and Shupe, 1992) has shown that this mindset *cuts across* class lines and is, therefore, not just characteristic of the poor or other marginal groups. Some social cleavages are not "horizontal" ones (such as caste or class distinctions), but instead are "vertical" ones, which arise in protection of a threatened world view or lifestyle. Thus, we find many conservative Christians (not just lower-class ones) who are firmly committed to the epistemological principles Page and Clelland describe as cultural fundamentalism. Nor are such adherents largely backwoods ignoramuses, nor typically of pathological personality, nor do they all suffer from status inconsistency problems as some have suggested.[8]

The influence of cultural traditionalism has been in decline in the second half of the twentieth century in the face of urban heterogeneity, consumerism, and increased rationalization and secularization of society. This, it would appear, has begun to stimulate revitalization movements among conservative Christians, intended to restore cultural traditionalism to supremacy.

In contrast, "cultural modernists" (among whom are to be counted most scientists) are far less likely to believe that the Bible is the inerrant word of God. By no means do all of them flatly reject the validity of the Bible. For some, the Bible is still seen as divinely inspired, but, also, as open to metaphorical interpretation and understanding in the context of the ancient society which produced it. Some even question whether it *can* be completely adapted to our current times of rapid social change. For cultural modernists, truth is a hard-won result of human enterprise, not revelation directly from a deity. Cultural modernists are the grandchildren, epistemologically speaking, of the thinkers of the Enlightenment. These thinkers proposed that the truth about the physical universe (and often, but not always, about moral issues too) can be identified through rational hypothesis testing with empirical data.

According to Page and Clelland (1978:276-277), cultural modernism is largely an adaptation to affluence, rationalization (in a Marxian sense) of production, increasing scale of organizations, and a service economy. Further, cultural modernism is marked by at least some degree of ethical relativism and accepts a wide variety of life styles, and often compels the segmentation of religion from other spheres of life. It tolerates, or even advocates, not only rationality and creativity, but even moderate consummatory hedonism (in, for example, alcohol use and sexual practices).

Cultural traditionalists, in contrast, see the highest purpose of man as following God's will, guided by scripture and revelation, and practicing traditional family values. These principles are seen as imperatives, even if they require the denial of hedonism or the endurance of suffering. By contrast, cultural modernists believe that the purpose of human existence is self-actualization and the minimization of human suffering. They suggest that pursuit of such goals in modern society will often require frequent and quick man-made judgments about correct morality and behavior.

Since the beginning of this century, the conflict between these two types of belief systems has been most intense in areas experiencing a rapid transition from a rural past to an urban or suburban "high-tech" future. The new symbol system associated with cosmopolitanism (as well as its parent world view, cultural modernism) advocates a relativistic, flexible moral code, and hence inevitably threatens adherents of traditional mores, who see such rules as God-given and not open to question. This is indeed the pattern that Cavanaugh seems to have found when, in 1983, he identified creationism as particularly strong within North America in rapidly expanding metropolitan areas (Cavanaugh 1983:182).[9]

Unfortunately, the two systems of symbolic meaning under discussion here are profoundly irresolvable. Neither system is the result of gross ignorance or irrationality, but, instead, each is driven by powerful group processes. For example, with a few exceptions, most creationists' belief systems are essentially rational. In saying this, we hasten to add that "rationality" is a complex, multidimensional concept. Loosely following Cavanaugh (1983), it can argued that a belief's rationality should be assessed based on its position on not just one, but on *three* conceptual dimensions. Specifically, we need to ask whether a system is *logical*, whether it is *plausible*, and whether it is *creditable*. If we use the term "rational" to mean logical in the formal sense (e.g., if $A = B$ and $B = C$, then $A = C$) then "scientific creationism" *is* logical in the sense that it does not represent bizarre or magical thinking. "Flood geology" adherents, for example, recognize the logical requirement in a "scientific" account of earth history for a physical mechanism to explain the accumulation of the earth's sedimentary rocks. Similarly, to an average person scientific creationism is "plausible," in the sense that there is no a priori reason that it *could not* be true; flood geology has such surface plausibility. However, we reserve here the term "creditable" for claims about the nature of reality which can be demonstrated to be consistent with the actual operation of the physical laws which govern the universe. In this sense, scientific creationism is not creditable, because it simply cannot be made to correspond with the totality of modern observed data and the resultant scientific models of the structure of the natural world.

The conflict between evolutionists and creationists is actually an example of a struggle for control of the means of cultural reproduction masquerading as a scientific controversy. Let us illustrate this point. In 1959, the Soviet Union successfully launched Sputnik, the first orbiting satellite. The immediate U.S. response was a set of loud calls to improve our own science education system. To this end, the National Science Foundation commissioned a series of biology materials termed the Biological Sciences Curriculum Study (BSCS). The BSCS materials made evolution the central organizing skeleton for the study of biology. When these textbooks began appearing in the classrooms of schools whose constituents were often conservative Christians, parents feared they could no longer control the values being taught to their own children.[10] In some Southern states, they invoked existing laws forbidding the teaching of evolution, or supported laws guaranteeing equal time for creationism.

Subsequently, however, the children of rocket scientists and computer engineers began telling their parents that their teachers insisted that the world was less than 10,000 years old and created directly by God in less than a week.

What becomes clear is that parents on both sides of the issue felt they were losing the ability to socialize their own children. Thus, the creation-evolution debate is not actually a scientific debate, as is frequently supposed, but is instead primarily a symbolic manifestation of a struggle for control of the means of cultural reproduction.[11] Perhaps this concept can take us a long way towards explaining why the creation-evolution conflict has also been much in evidence in courtrooms and legislatures. It is precisely the courtrooms and legislatures, and the schools, which are the great legitimating institutions of modern Western societies where particular symbolic meanings are to be approved as appropriate for mass socialization.

It may be objected at this point that many types of pseudoscientific beliefs do not seem so clearly to reflect this type of group pressure in their origins as does the creationism versus evolution debate. We would agree with this observation. Indeed, past empirical studies have had relatively little success in predicting the concomitants of pseudoscientific beliefs that are not related to "core values" such as religion, ethnicity, or patriotism. The antecedents of beliefs in UFOs, mysterious monsters, or sunken civilizations, for example, are much more obscure than those of beliefs about the age of humankind, or whether Adam and Eve actually existed. We believe this situation gives strong inferential support to the very argument that much apparently egocentric pseudoscientific belief is actually strongly influenced by the social control of the group. In a sense, analogy can be made with evolution. Specifically, biological traits are not removed from a population by natural selection if there is no environmental pressure from predators, climate, or other factors to propel the selection process. Similarly, it would appear that many seemingly bizarre beliefs of individuals are unlikely to be selected out of that individual's repertoire unless they come under pressure from his social environment, and this is just what we find; beliefs *unrelated* to core values seem almost randomly distributed in the population. This is because no one really cares about such deviation from mainstream belief. Indeed, in such cases the deviation may be perceived by the social audience as entertaining and harmless. In contrast, the role of group pressure seems a likely explanation for the convergence of opinion on questions such as human origins. The answers to these question do threaten core values that the group uses to integrate its collective consciousness and organize its world view. Therefore, social control is invoked if opinion strays too far from the norm.

In a similar vein, the belief in increased Satanic activities discussed previously seems to be another manifestation of the conflict between cultural traditionalism and modernism. Such beliefs are apparently occasioned by subconscious fears among cultural traditionalists of the collapse of the nuclear family. In a recent study, Roy, Eve, and Shupe (1992) found strong support for the proposition that individuals most afraid of alleged Satanic cult activities would score strongest on a scale measuring cultural traditionalism and lowest on acceptance of cultural modernism. Indeed, fear of cultural modernism was found to be a much better predictor of their fears of Satanist activities than was the case for traditional social psychological scales of religious conservatism or orthodoxy. Faced with anxiety arising from the reduced parental role in contemporary American socialization, the reaction of many seems to be fear that "secular humanists," or care-givers in daycare centers, or lurking but invisible cult members are eager to exploit the family socialization vacuum now so much in evidence.[12]

Finally, in recent years, a third ideal type has emerged, which we call "postmodernist." Postmodernists reject the legitimacy of both traditionalism and modernism. They see in the former a continuation of the oppressive values of the past, but they view the latter as having led to overrationalization of production and of life in general, as well as rampant consumerism, runaway militarism, and environmental despoliation. They tend to see science as the "hit man" of big government, big military, and big business. Postmodernists have no use for either cultural traditionalists or cultural modernists.[13]

Like the other two ideal types, postmodernists tend to feel powerless in the face of the problems that confront contemporary Western society. Each group is therefore motivated to attempt to empower itself to reduce its anxiety. Like the other two groups, postmodernists attempt to legitimate their own symbol system as more empowering than any other. Cultural traditionalists attempt to answer problems of the present with formulas from the past; modernists believe that the exercise of science and technology are the road to mortal salvation; and some postmodernists have involved themselves in "New Age" strategies for empowerment. (One would indeed be truly powerful if one could heal or divine the future with a crystal, or gain knowledge through channeling with a powerful spiritual entity). Each group has, consciously or unconsciously, constructed its own set of knowledge industries intended to legitimate its own world view, information, and resultant agenda for change as more valid than any other.

The New Age seems actually to be several "protomovements" milling about in search of a set of generalized beliefs that would allow effective mobilization of their members. A good example would be the Wicca (or "white" witchcraft movement). Wiccans have rejected the legitimacy of the world view and symbol systems of Christianity as those of an excessively patriarchal religion arising out of ancient pastoral Hebrew social structure. They have replaced Christianity with a goddess-centered religion influenced by their notions of pre-Christian nature-centered religions ("*Mother* Earth") or native American religions which emphasize the need to balance natural forces.[14] Many New Age claims therefore again attempt to legitimate a new symbol system as "scientific." They seek the mantle of scientific legitimacy when they employ terms like harmonic convergence or crystal resonance. Many New Age claims have a strong quality of earlier attempts at science, such as alchemy, which can be viewed as quasiscientific technologies intended to liberate energy for empowerment of the practitioner. Not unlike creationists, New Agers seem to have a deeply ambivalent attitude towards science, on the one hand damning it for allegedly leading to a host of afflictions for mankind, but at the same time seeking the legitimacy for their own symbol systems commonly accorded only to science in recent years. We see whole new knowledge industries arise among New Agers alleging, for example, that natural foods and crystal vibrations are better for one's health than bureaucratized hospitals and CAT-scanning machines.

SUMMARY AND CONCLUSIONS

Early scholarly considerations of the origins of pseudoscientific belief tended to be based on a view of such beliefs as arising out of individual ignorance, superstition, or psychopathology (such as the Freudian defense mechanism of projection).

We have suggested instead that pseudoscientific belief often originates in normal group processes. We have seen that common errors in human reasoning, typically thought to be largely individualistic in origin, are often influenced by emergent norms within one's peer groups. We have seen that many pseudoscientific beliefs are merely learned normally from mass media, which frequently present pseudoscientific claims as if they had significant scientific support. Perhaps most importantly, we have learned that, in recent years, sophisticated knowledge industries have arisen to propagate pseudoscientific beliefs (while, ironically, they often claim the mantle of scientific legitimacy for such beliefs). We have listed a number of factors indicating that the rate of appearance of new knowledge industries will increase as we enter the postindustrial society. We have seen that Americans are, in general, woefully ignorant of science, and that this makes them easy prey for bogus scientific beliefs.

The struggle among cultural traditionalists, modernists, and postmodernists is a group struggle for control of the phenomenological generation and legitimation of a particular world view over all competing ones. We have, therefore, proposed the need for more research into the dynamics of group processes that both propel and reflect struggles for control over the means of cultural reproduction.[15]

Future research should identify existing and emerging knowledge industries, and the manner in which group processes operate within each to create the particular truth dominant within each collectively. Second, we must examine how group dynamics induce conformity to the particular knowledge of one's group. Finally, we need to examine how commitment to a group's shared symbol system leads to seeing outsiders as deviants and their world views and symbol systems as cause for alarm.

Let it be clear that we are not "taking sides" with cultural modernists, or cultural traditionalists, or with postmodernists. We believe that rather than taking sides, the job of social scientists is to analyze the distribution of differences in power between groups and other social aggregates, and then to show how these manifest themselves in both material and symbolic struggles.

Finally, we should not be understood to suggest that the current mainstream consensus in science is immutably and unassailably correct. To do this would be to fundamentally misunderstand the nature of a scientific theory. A theory is always but the most parsimonious manner in which the data can be accounted for *at the present time*. One of the hallmarks of science is the *desire* that the theory eventually be supplanted by a better account, even if this requires a whole paradigm shift within science itself.

In light of this, what we hope has been accomplished most obviously here is a clear recognition that belief in pseudoscience is *not* largely an individualistic matter, any more than belief in science is the result of purely individualistic factors.

The growth of knowledge industries that is a concomitant of postindustrial society implies a major new research agenda in the study of group processes.

NOTES

1. Indeed, the very notion of "mental health" is open to charges of an illegitimate pretension to the same scientific status as organic medicine. For a detailed discussion of this charge, see Szasz 1974.

2. "Crop circles" are flattened circles of grain which have mysteriously appeared in farmers' fields. Most of these have been found in England, most commonly in the area of Stonehenge. The circles themselves vary from a few feet across to a hundred or more feet across. They have been variously interpreted as UFO landing sites, New Age spiritual manifestations, a harbinger of the return of Christ, etc. Recent confessions indicate that at least some crop circles were the work of midnight hoaxers (Nickell and Fischer, 1992).

3. The major organization devoted to UFOs is the Mutual UFO Network (MUFON). MUFON claims to exist for the purpose of the scientific investigation of UFO phenomena. However, most of their meetings do not involve presentation of carefully collected data for the purposes of testing scientifically deduced hypotheses, but consist of anecdotal evidence presented

in a style similar to Alcoholics Anonymous "confessions." Even MUFON members seem inordinately found of comparing themselves to AA members—perhaps because of the public stigma often attached to followers of both movements.

4. It must be said, however, that the work of many deconstructionists seems to confuse the issue of whether the scientific method per se is valid with the more appropriate questions for social scientists of why certain problems are chosen for scientific study or whether truth claims that aspire to scientific status are always unaffected by differential power among groups. The answer to the latter is almost certainly that they are so influenced. However, such a finding is far from a refutation of science as a valuable way of finding out about the operation of the physical universe.

This confusion over the goals of deconstructionism seems to mirror uncertainty over "the demarcation problem." Specifically, how does one distinguish a scientific pronouncement that is merely a "construction" (such as certain German scientists' pronouncements in the 1930s that some race were superior to others biologically) from statements that reflect a hard-won understanding of physical reality? (We have offered to drop deconstructionists from the top of a tall building, since they should be able merely to deconstruct the concept of gravity as a social convention on the way down; so far, none will accept our offer.) Good examples of serious scholarly attempts to explicate the demarcation problem can be found in the work of Star (1989) on the construction of scientific typologies, and Majeski (1989) examining group pressures in military decison-making to conform to specific shared rules of information transmission.

5. In fact, the Matomoros murder cannot be considered a legitimate case of Satanism. Instead, the religion being followed seems to have been Santeria, a "black magic" religion derived from Africa by way of the Caribbean. There is no personification of the Devil in this religion. However, that makes little difference to conservative Christians in the U. S. who will see the murders as an example of Satanism. It may also be suggested that there are elements of xenophobia attached to atrocity tales surrounding the event. As Hispanic culture makes significant inroads into U. S. culture, such tales appear to increase in number and are characterized by a cautionary or openly reactionary quality regarding what can be expected to happen when "they" arrive here.

6. Former Presidential candidate Lyndon Larouche (now serving time for fraud in federal prison) has been very adroit at promoting various moral panics for personal gain. An unusual theory of AIDS transmission and allegations of Satanic activities against children have been two of his more notable successes. In one New Federalist publication (the organ of Larouche's organization) called "Is Satan in your schoolyard?" the First Episcopal Church of New York is identified as the center of the Satanist conspiracy in America and much popular music held to be Satanic. Indeed, not only rock music is so described, but even Wagner's operas—presumably because of the pagan imagery of Norse gods in "the Ring Cycle."

7. Some of these abuses of scientific legitimacy in the area of mental health are starting to be more commonly seen now in industry. An example is the widespread use of "personality inventories" which purport to scientifically fit the employee to the job. Particularly notable is the recent widespread adoption of the Meyers-Briggs Personality Type Assessment. To date, this widely accepted personality test has seen little attempt to assess its validity. What attempts have been made are uniformly unsupportive.

8. For a fuller discussion of why the characteristics just mentioned have been hypothesized by some as adequate explanations for membership in conservative social movements, and why such explanations are insufficient, see Eve and Harrold (1991:Chapter 6).

9. We predict that the same types of locales are also the most likely to experience episodes of social movement conflict where the basis is not creation versus evolution, but other worldview issues such as abortion rights, gay rights, legalization of adult erotic materials, and so on.

10. It is commonly forgotten that John Scopes lost the famous "Monkey trial" that took place in Tennessee in 1925. Although a higher court later reversed his conviction, the creationists won an important victory when biology textbooks after 1925 reduced or eliminated coverage of the touchy subject of evolution (Grabiner and Miller, 1974).

11. Horizontal (or class-based politics) typically involve struggles to control the means of material or economic production, while vertical status-group politics are based on a conflict of world-views. The conflict is over control of the means of legitimating values and symbol systems, and is therefore termed "cultural reproduction." The recent "deconstructionist" debates in academia based on the writing of such scholars as Habermas, Derrida, and Foucault, as well as the more general "political correctness" controversy in the media, can also be seen as examples of infant knowledge industries engaged in a struggle to legitimate their particular epistemologies through a struggle for the means of cultural reproduction.

12. For a detailed and well-developed discussion of the moral panic said to result from the breakdown of the family, see Richardson, Best, and Bromley (eds.) (1991).

13. One possible interpretation of postmodernism suggests that the effects of the social movements of the 1960s are not over yet. This view would suggest that those movements "deconstructed" militarism, rampant consumerism, blind patriotism, and perhaps even God. However, postmodernism seems much clearer about what it opposes than about what it favors.

The usual definition of "anomie" is a breakdown in the norms of group. However, anomie does not necessarily imply the total absence of a normative system; a second possibility is a proliferation of several such systems with no clear way of deciding which ones are to be legitimated. This latter would seem more descriptive of the postmodern world view. It also goes a long way toward suggesting why knowledge industries would arise to try to reverse the situation; but paradoxically, each such industry makes it less likely that another can actually succeed in its legitimating efforts.

14. The information presented here regarding Wicca, plus a great deal more interesting information on new age religions can be seen online in Compuserve Information Services special-interest users' group electronic bulletin board entitled "Religion" (subheading "New Age"). An interesting characteristic of Wiccans is the degree to which many of them are highly computer literate; the authors have elsewhere styled them "Space age witches."

15. In some ways our call is not wholly removed from the recent work performed by deconstructionist theorists interested in analysis of science's role in creating and maintaining "privilege" for certain social categories.

However, our call does differ in one respect. Deconstructionists often begin with existing "texts" of rhetoric and attempt by analysis of these to discover post hoc the forces that led to the particular nature of the rhetorical text. We, on the other hand, suggest that a more truly sociological approach should involve a structural analysis of the group processes that drive the struggle for the control of symbolic reproduction, and then use of the resultant knowledge to generate testable assumptions about the shape and dynamics of future intergroup conflicts.

REFERENCES

Asch, Solomon. 1955. *Social Psychology*. Englewood Cliffs, NJ: Prentice-Hall.

Carlson, Shawn and Gerald Larue. 1989. *Satanism in America*. El Cerrito, CA: Gaia Press.

Cavanaugh, Michael A. 1983. *A Sociological Account of Scientific Creationism: Science, True Science, Pseudoscience*. Ann Arbor, MI: University Microfilms International. Ph.D. dissertation, University of Pittsburgh.

Cazeau, Charles J., and Stuart D. Scott, Jr. 1979. *Exploring the Unknown: Great Mysteries Reexamined*. New York: Plenum.

Cloud, Preston. 1977. "'Scientific Creationism'—A New Inquisition Brewing?" In *A Compendium of Information on the Theory of Evolution-Creationism Controversy*. National Association of Biology Teachers.

Davis, Percival, and Dean H. Kenyon. 1989. *Of Pandas and People*. Dallas, TX: Haughton.

Eve, Raymond A. 1986. "Children's Interpersonal Tactics in Effecting Cooperation by Peers and Adults." In Part IV of *Sociological Studies of Child Development*, Vol. 1, edited by Peter Adler and Patricia Adler. Greenwich, CT: JAI Press.

Eve, Raymond A., and Dana Dunn. 1990. "Psychic Powers, Astrology, and Creationism in the Classroom." *The American Biology Teacher* 52:10-21.

Eve, Raymond A., and Francis B. Harrold. 1991. *The Creationist Movement in Modern America.* Boston: Twayne.

Gallup, George H. 1979. *The Gallup Poll: Public Opinion 1978.* Wilmington, DE: Scholarly Resources.

Gish, Duane. 1985. *Evolution: The Challenge of the Fossil Record.* El Cajon, CA: Creation-Life Publishers.

Gray, Thomas. 1987. "Educational Experience and Belief in Paranormal Phenomena." In *Cult Archaeology and Creationism,* edited by F. Harrold and R. Eve. Iowa City: University of Iowa Press.

Habermas, Jurgen. 1987. *The Theory of Communicative Action.* Translated by Thomas McCarthy. Boston: Beacon.

Harrold, Francis B., and Raymond A. Eve. 1987. "Patterns of Creationist Belief Among College Students." Pp. 68-90 in *Cult Archaeology and Creationism,* F. Harrold and R. Eve. Iowa City: University of Iowa Press.

Hicks, Robert. 1991. "Satanism and the Law." Pp. 175-190 in *The Satanism Scare,* edited by James T. Richardson, Joel Best, and David G. Bromley. New York: Aldine de Gruyter.

Hines, Terence. 1988. *Pseudoscience and the Paranormal.* Buffalo, NY: Prometheus.

Hunter, James D. 1991. *Culture Wars.* New York: Basic Books.

IEA (International Association for the Evaluation of Educational Achievement). 1988. *Science Achievement in Seventeen Countries: A Preliminary Report.* Oxford: Pergamon Press.

Jung, Carl G. 1977. *Psychology and the Occult.* Princeton, NJ: Princeton University Press.

Larson, Edward J. 1989. *Trial and Error: The American Controversy Over Creation And Evolution,* 2nd edition. New York: Oxford.

Majeski, Stephen. 1989. "A Rule Based Model of the United States Military Expenditure Decision-Making Process." *International Interactions* 15:2.

Mead, George Herbert. 1934. *Mind, Self and Society From the Standpoint of a Social Behaviorist.* Chicago: University of Chicago.

Medawar, Peter B. 1973. *The Hope of Progress.* Garden City, NJ: Anchor.

Miller, John D. 1983. "Scientific Literacy: A Conceptual and Empirical Review." *Daedalus* 112:29-48.

———. 1987a. "The Scientifically Illiterate." *American Demographics* 9(6):26-31.

———. 1987b. "Scientific Literacy in the United States." Pp. 19-40 in *Communicating Science to the Public,* edited by D. Everett and M. O'Connor. New York: Wiley.

Moffett, J. 1988. *Storm in the Mountains.* Carbondale, IL: Southern Illinois University Press.

Morris, Henry M. 1974. *Scientific Creationism.* San Diego: Creation-Life Publishers.

National Science Board. 1989. *Science and Engineering Indicators–1989.* Washington: National Science Foundation.

Nickell, Joe, and John F. Fischer. 1992. "The Crop-Circle Phenomenon: An Investigative Report." *The Skeptical Inquirer* 16(2):136-149.

Oberschall, Anthony R. 1984. "Politics and Religion: The New Christian Right in North Carolina." *Social Science News Letter* 69:20-24.

Page, Ann L., and Donald A. Clelland. 1978. "The Kanawha County Textbook Controversy: A Study of the Politics of Lifestyle Concern." *Social Forces* 57:265-281.

Radner, Daisie, and Michael Radner. 1982. *Science and Unreason.* Belmont, CA: Wadsworth.

Richardson, James T., Joel Best, and David G. Bromley, eds. 1991. *The Satanism Scare.* New York: Aldine de Gruyter.

Roy, Lon, Raymond A. Eve, and Anson B. Shupe. 1992. "The Modern Satanism Scare as a 'Moral Panic:' An Empirical Examination of the Role Played by Cultural Traditionalism." Unpublished paper.

Schadewald, Robert J. 1983a. "The Evolution of Bible-Science." Pp. 182-300 in *Scientists Confront Creationism,* edited by L. Godfrey. New York: Norton.

_____. 1983b. "Creationist Pseudoscience." *The Skeptical Inquirer.* 8:22-35.

Sheler, Jeffrey L., and Joannie M. Schrof. 1991. "The Creation." *U.S. News & World Report,* December 23, 1991:55-64.

Sherif, Muzafer, and O. J. Harvey, 1952. "A Study in Ego-Functioning: Elimination of Stable Anchorages in Individual and Group Situations." *Sociometry* 15:272-305.

Sherif, Muzafer, and Carolyn Sherif. 1969. *Social Psychology.* New York: Harper and Row.

Singer, Barry, and Victor A. Benassi. 1981. "Occult Beliefs." *American Scientist* 69:49-55.

Smith, Michelle, and Lawrence Pazder. 1980. *Michelle Remembers.* New York: Congdon & Lattes.

Star, Susan Leigh, and James R. Griesemer. 1989. "Institutional Ecology, 'Translations' and Boundary Objects: Amateurs and Professionals in Berkeley's Museum of Vertebrate Zoology." *Social Studies of Science* 19:3.

Stiebing, William H., Jr. 1984. *Ancient Astronauts, Cosmic Collisions, and Other Popular Theories About Man's Past.* Buffalo, NY: Prometheus.

Toumey, Christopher P. 1990. "Social Profiles of Anti-evolutionism in North Carolina in the 1980s." *The J. of Elisha Mitchell Scientific Society* 106(4):93-117.

_____. 1991. "Modern Creationism and Scientific Authority." *Social Studies of Science.* 21:681-699.

Truzzi, Marcello. 1979. "On the Reception of Unconventional Scientific Claims." Pp. 125-137 in *The Reception of Unconventional Science,* edited by M. Mauskopf. Boulder, CO: Westview.

Victor, Jeffrey S. 1990. "The Spread of Satanic-Cult Rumors." *The Skeptical Inquirer,* Spring, 287-290.

Walsh, John. 1988. "U.S. Science Students Near Foot of Class." *Science and Society* 239:1237.

Watters, Ethan. 1991. "The Devil in Mr. Ingram." *Mother Jones,* (July) 16.

Young, W. C., R. G. Sachs, B. B. Braun, and R. T. Watkins. 1991. "Patients Reporting Ritual Abuse in Childhood: A Clincial Syndrome." *International Journal of Child Abuse and Neglect* 14.

AN EXPECTED VALUE MODEL OF SOCIAL EXCHANGE OUTCOMES

Noah E. Friedkin

ABSTRACT

A new approach to social exchange is developed and illustrated. The approach carries forward a line of work on power structures that was initiated with French's formal theory of social power. The approach moves beyond the rank-order prediction of actors' resource outcomes that is characteristic of extant social exchange hypotheses, and provides baseline predictions of the amount of resources each actor is expected to acquire through social exchange. Under baseline assumptions, the approach provides a simple account of the literature's intriguing findings that the most centrally located actors in exchange networks do not necessarily acquire the most resources via exchange processes. The baseline predictions of the approach provide a null hypothesis against which the merits of more refined alternative hypotheses can be assessed. I illustrate how the baseline assumptions may be relaxed by introducing a formal hypothesis in which an actor's bargaining behavior is related to the actor's vulnerability to exclusion from social exchange. For the several cases that were examined, the hypothesis does a credible job of predicting actors' absolute amounts of acquired resources.

Advances in Group Processes, Volume 10, pages 163-193.
ISBN: 1-55938-280-5

INTRODUCTION

In the formal theory of social power proposed by French (1956), and in Cartwright's (1965) integration of the literature on interpersonal leadership and control, social power is defined as potential interpersonal influence and a power structure describes the pattern of opportunities for direct interpersonal influences among a set of actors. Along the lines of this approach, Friedkin (1986) suggested that predictions about a power structure's outcomes could be derived in the form of *expected values*, that is,

$$E(y_i) = \sum_{i=1}^{K} P(\mathbf{R}_i)\, y_i,$$

where $\mathbf{R}_i$ is one the K possible networks of interpersonal influence that might occur in a power structure, $P(\mathbf{R}_i)$ is the probability of $\mathbf{R}_i$, and y_i is an outcome of $\mathbf{R}_i$. The outcome of interest to Friedkin was consensus; however, he also illustrated the applicability of the approach to predicting structural outcomes (e.g., dominant coalitions) and patterns of disagreement in a network of power.

Here, I generalize Friedkin's (1986) *expected value model* of social power by allowing power structures to represent opportunities for interpersonal transactions other than interpersonal influence. In this more general approach, a power structure might indicate an opportunity for information flow, social support, or social exchange. I then develop and illustrate the application of this expected value model to networks of social exchange.

A considerable body of theoretical work and experimental findings has developed on network exchange phenomena.[1] The prevailing theoretical agenda of the recent work has been to construct a measure of point-centrality that, when applied to a network of potential exchanges, correctly predicts the resources actors acquire through negotiated agreements in these networks (Bonacich, 1987; Cook and Emerson, 1978; Cook, Emerson, Gillmore, and Yamagishi, 1983; Emerson, 1969, p.396; Markovsky, Willer, and Patton, 1988; Marsden, 1983 and 1987; and Stotle and Emerson, 1977). Substantial controversy has developed in the pursuit of this agenda (Cook, 1982, p. 188; Cook, Gillmore, and Yamagishi, 1986; Markovsky, Willer, and Patton, 1990; Willer, 1986; and Yammagishi and Cook, 1990).

For reasons that are developed in the present work, pursuit of a general structural index of social power is not likely to be fruitful. Instead, I argue that many of the problems that presently engage social exchange theorists are addressable from an expected-value approach to social exchange outcomes. First, from baseline assumptions, the approach correctly predicts actors' rank order for the resources they acquire through social exchange; in particular, the approach explains the experimental findings on certain networks in which the seemingly most central actors do not acquire the most resources. Second, the approach is not limited to rank-order predictions; it predicts the observed amount of resources acquired by actors through social exchange and allows a judgment (predicated on baseline assumptions) on whether these predictions are noteworthy. Third, the approach is sufficiently general that it accommodates different regimes of social exchange: negative exchanges, positive exchanges, multiple exchanges, mixed exchanges, and resource flows. Fourth, and finally, the approach provides an intellectual framework that (a) contributes to disentangling and clarifying certain complex theoretical issues and (b) points to areas where additional formal development of social exchange theory might be undertaken. To illustrate this last point, I introduce a formal hypothesis on the relationship of actors' bargaining behaviors and vulnerabilities to exclusion from social exchange. For the several cases that were examined, this hypothesis does a credible job of predicting actors' absolute amounts of acquired resources.[2]

EXPECTED VALUE MODEL OF SOCIAL POWER

In this section, a general description of the approach is provided, so that it may be viewed as applicable to a variety of social processes, including social influence, social exchange, social support, and information flow. There are five steps involved in deriving predictions about the expected values of a power structure's outcomes. These steps are described in tandem with a simple illustration (Table 1).

Delineation of the Power Structure

The approach starts with the delineation of a power structure as a pattern of opportunities for relational events among a set of actors. The power structure of a population is represented as either a graph or digraph with lines connecting actors whenever there is a positive probability of some type of relational event. The occurrence of a line indicates that the particular event is possible; the absence of a line indicates that the event is not possible. Whether the power structure consists of directed or undirected lines will depend on the type of relational event. In the case of social influence and information flow, the relation will be directed. In the case of social exchange, the relation may be undirected if it simply represents the occurrence of a negotiated agreement between two actors about a division of available resources.

Depending on the relation under consideration, a path of power lines such as $i \to j \to k \to l$ may or may not be meaningful. For example, if the relation is interpersonal influence, such a path will indicate an opportunity for indirect

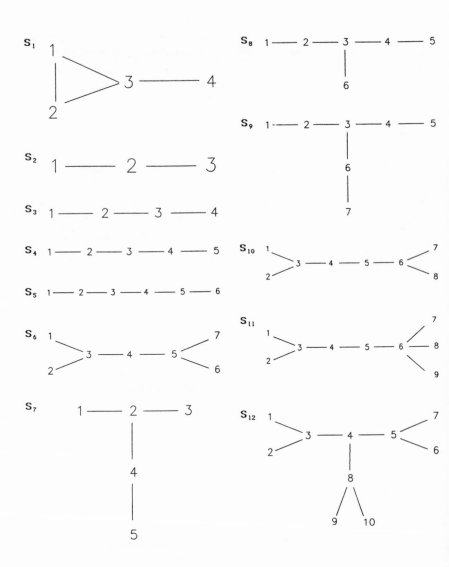

Figure 1. Networks

influence via intermediaries. However, if the relation is a negotiated agreement, the path may not imply agreement between nonadjacent actors.

Various power structures comprised of undirected lines are illustrated in Figure 1. I concentrate initially on S_1, a network consisting of four positions {1,2,3,4} and four lines {1-2,1-3,2-3,3-4}.

Delineation of the Sample Space

The sample space of the power structure consists of the different **R**-networks that might occur in the context of the power structure, where **R** is a pattern of interpersonal influence, social exchange, social support, or information flow. I refer to the sample space of a power structure as *unrestricted* if it contains all the possible **R**-networks: if the number of lines among actors in the power structure is v, then a maximum of 2^v alternative, labeled, **R**-networks are possible. Table 1 illustrates the sixteen members $\{\mathbf{R}_1, \mathbf{R}_2, \ldots, \mathbf{R}_{16}\}$ of the unrestricted sample space of $\mathbf{S}_1$.

Unrestricted sample spaces may contain a large number of members. For example, in a power structure with 25 lines, there are over 33 million **R**-networks in the unrestricted sample space. However, *restrictions* on the sample space will reduce the number of its members. The restriction may be empirical; that is, certain possibilities are never observed and so they are eliminated. The restriction may be theoretical; that is, certain possibilities cannot occur on theoretical grounds. The restriction may be experimentally imposed; that is, certain possibilities are not allowed to occur under the conditions of an experiment. Later, I show how the conventional designs of social exchange experiments entail important restrictions on the sample spaces of power structures.

Relative Frequency of **R**-Networks

Next, the probability of occurrence of each **R**-network in the sample space is determined. These probabilities can be determined in various ways:

(a) In the absence of data or theory, it may be assumed that each **R**-network is independent and equally likely.

(b) The probability of each **R**-network may be based on their observed relative frequencies in a large sample of realizations, for example, trials of an experiment on a fixed power structure.

(c) The probability of each network may be derived from a model of the emergence of **R**-networks. Or,

(d) The probability of each network may be analytically derived from information on the probabilities of line activation in the power structure.

For a simple illustration of the analytical approach, assume that the lines in power structure $\mathbf{S}_1$ are independent and that the probability of line activation is .60 for each of the four lines $\{1\text{-}2, 1\text{-}3, 2\text{-}3, 3\text{-}4\}$. It follows that the probability of occurrence of each $\mathbf{R}_i$ is the product of the probabilities for the relational events that have produced the network; that is, the probability of $\mathbf{R}_i$ is $p^u (1 - p)^{v-u}$ where u is the number of lines in $\mathbf{R}_i$, v is number of lines in the power structure,

Table 1. Illustration of the Approach for Power Structure S_1

R_i	Unrestricted Sample Space Power Lines	Network Image	$P(R_i)$	Isolate Count	Majority Coalition
	1 1 2 3				
	2 3 3 4				
R_1	0 0 0 0	⋮ ⋅ ⋅	.0256	4	0
R_2	0 0 0 1	⋮ —	.0384	2	0
R_3	0 0 1 0	⟋ ⋅	.0384	2	0
R_4	0 0 1 1	⟋—	.0576	1	1
R_5	0 1 0 0	⟍ ⋅	.0384	2	0
R_6	0 1 0 1	⟍—	.0576	1	1
R_7	0 1 1 0	> ⋅	.0576	1	1
R_8	0 1 1 1	>—	.0864	0	1
R_9	1 0 0 0	| ⋅ ⋅	.0384	2	0
R_{10}	1 0 0 1	| —	.0576	0	0
R_{11}	1 0 1 0	| ⋅ ⋅	.0576	1	1
R_{12}	1 0 1 1	⌐—	.0864	0	1
R_{13}	1 1 0 0	⌐ ⋅	.0576	1	1
R_{14}	1 1 0 1	⌐—	.0864	0	1
R_{15}	1 1 1 0	▷ ⋅	.0864	1	1
R_{16}	1 1 1 1	▷—	.1296	0	1
Expected Values				.784	.763

Notes: $P(R_i) = p^u (1 - p)^{v-u}$, where $p = .60$, u is the number of relations in R_i and v is number of relations in the power structure S_1.

The number of isolates is a count of the actors in R_i who are not related to any of the other actors. The occurrence of a majority coalition is defined as the presence of a connected subgraph in R_i that includes a majority of the actors.

and $p = .60$. Table 1 illustrates the calculations. The probability of R_1 is $.60^0(.40)^4 = .0256$, where none of the four power lines is active; the probability of R_2 is $.60(.40)^3 = .0384$, where one of the four power lines is active; and so forth.[3]

Outcomes of R-Networks

Fourth, the outcome(s) for each of the **R**-networks in the sample space is (are) determined. The outcome may be some feature of the structure of R_i: its density, diameter, connectivity category, point centralities, bundle sizes, and so forth; see Harary, Norman, and Cartwright (1965) for the definitions of these structural features. The outcome also may be derived from a process model. Friedkin (1986), who dealt with a power structure comprised of lines of possible interpersonal influence, applied a process model of opinion formation to each of the R_i. In Friedkin's application, the outcome was either 1 or 0, indicating, respectively, consensus or its absence in a particular R_i.[4]

Table 1 illustrates two structural outcomes that may be ascertained by visual inspection of the sample space of the power structure S_1. The number of isolates is a count of the actors in R_i who are not tied to any of the other actors. The occurrence of a majority coalition is defined as the presence of a connected subgraph in R_i that includes a majority of the actors in the power structure.

Expected Values

Finally, the expected values of the power structure's outcomes are computed. Th se expectations are weighted averages of the values of the outcomes for the **R**-networks, where each outcome is weighted by the probability of occurrence of the **R**-network:

$$E(y_i) = \sum_{i=1}^{K} P(\mathbf{R}_i) \, y_i , \tag{1}$$

where $P(\mathbf{R}_i)$ is the relative frequency of **R** in a sample space comprised of K networks and y_i is an outcome of $\mathbf{R}_i$. In Table 1, the expected values for the number of isolates and the occurrence of a majority coalition are .784 and .763, respectively.

An expectation is an indicator of the central tendency of the distribution of outcomes for the set of **R**-networks that may arise from the power structure; in the special case of a binary {0,1} outcome, the expectation is simply the probability of the outcome. If the sample space of the power structure contains relatively few **R**-networks, then it is possible to carry out the computation exactly; for each **R**-network in the sample space, a product is formed of the relative frequency of the **R**-network and the outcome, and these products are summed to form the expectation of the outcome. When the sample space of the power structure is large, the expectations may be estimated from a suitable sample of **R**-networks.

In terms of the expected values it generates, the model permits comparisons among actors in the same or different power structures. The predictions of the model may or may not be obvious depending on the particular application. It is noteworthy that the model will generate predictions where intuition cannot process the implications of a complexly configured power structure. Whether the model forwards counterintuitive conclusions in simple power structures is an open question with an answer that may vary with the application.

APPLICATION TO SOCIAL EXCHANGE

In this section, an expected-value model of social power is developed for a type of resource exchange relation that has been the focus of considerable experimental work during the past several decades: an interpersonal network comprised of two-party transactions where each transaction provides one actor in the transaction with a fraction of some amount of resources and the other party in the transaction with the remaining fraction. Initially, a baseline model for such networks will be developed and illustrated. Derived from elementary assumptions about exchange processes, this baseline model generates expected values for exchange outcomes against which the merits of more refined assumptions will be assessed.

Delineation of the Power Structure

The power structure is defined as a network comprised of (a) points that represent collective or individual actors, and (b) undirected lines that represent potential exchanges (Emerson, 1962, and 1972a, p. 56; and Cook and Emerson, 1978, fn 9). The presence of a line between two actors indicates that a particular exchange *may* occur, and the absence of a line indicates that a particular exchange *cannot* occur. This definition is consistent with Emerson's (1972b, p.70) definition of an exchange network as a "set of three or more actors each of whom provides opportunities for transactions with at least one other actor in the set" and with Cook, Emerson, Gillmore, and Yamagishi's (1983, p. 277) definition of an exchange relation as "a set of historically developed and utilized exchange opportunities" so that "the set of exchange relations is properly viewed as a subset of exchange opportunities."

Two simplifying assumptions are made about a power structure's actors and their potential transactions:

Assumption A_0:   The actors in a power structure are assumed to be *rational actors* who seek to maximize their net receipt of resources over any set of transaction opportunities provided to them.

Later, the assumption of rational action is replaced by operational statements describing the type of action that is assumed to occur.

Assumption A_1:   A power structure is assumed to be *stable* with respect to its configuration of potential exchange transactions.

Thus, a power structure is considered stable even with changes in the identities of the actors who occupy the different positions in the network.[5]

A power structure may be described as an $n \times n$ symmetric matrix $\mathbf{S} = [s_{ij}]$, where $s_{ij} = 1$ if there is the possibility of an exchange between actor i and actor j, and $s_{ij} = 0$, otherwise. Figure 1 illustrates some of the power structures that have been studied in experiments on social exchange.

Delineation of the Sample Space

The sample space of the power structure is comprised of the K different networks of exchange transactions $\{\mathbf{R}_1, \mathbf{R}_2,..., \mathbf{R}_K\}$ that might occur in the context of the power structure. Previously, I suggested that different sample spaces may be defined on the same power structure via restrictions on the domain of $\mathbf{R}_i$. Three types of restrictions are discussed, two of which are common in experimental work on social exchange. Regardless of the particular restriction, it is assumed that each of the $\mathbf{R}_i$ in a sample space is *maximal* with respect to the number of transactions:

Assumption A_2:   It is assumed that each $\mathbf{R}$-network in the sample space of a power structure is *maximal* in that no other feasible transaction could occur.

Rational actors (see assumption A_0) will not absent themselves from exchange opportunities.

Experimental work on social exchange has dealt mainly with *negative* exchange where actors are limited to one exchange on each trial of an experiment. Figure 2 illustrates this type of sample space for power structure S_9. While the unrestricted sample space of this structure contains $2^6 = 64$ members, under the negative exchange regime the feasible sample space is comprised only of the four $\mathbf{R}$-networks displayed in Figure 2.

Social exchange theorists have defined *positive* exchange as a situation in which an exchange with one actor is contingent upon an exchange with another actor. However, a more precise definition of positive exchange is required for an unambiguous delineation of its sample space. An illustration of a suitable definition might be that exchanges only occur among actors who reach a collective (three-party) agreement and that actors may not be a member of more than one three-party agreement; thus, a sample space may be defined in terms of the different triads with two or three lines that might arise among the actors in a power structure. The sample space of S_9, based on this regime of triadic combination, is also illustrated in Figure 2.

Experiments on social exchange have sometimes allowed actors to participate in more than one exchange without imposing any rule of triadic agreement such as entertained in the positive exchange situation (Markvosky, Willer, and Patton, 1988). These multiple exchange situations are referred to as *e*-exchange regimes: the 1-exchange regime is equivalent to negative social exchange; the 2-exchange regime allows actors to participate in *at most* two exchanges with diffferent actors; the 3-exchange regime allows actors to participate in *at most* three exchanges with different actors; and so on. The sample space of S_9 for the 2-exchange regime is illustrated in Figure 2.

A noteworthy feature of the three regimes just discussed is that their sample spaces are nonoverlapping; that is, any R_i appearing in the sample space of one of these regimes does not appear in the sample spaces of the other regimes. There are 13 different R_i in the three sample spaces, a number that does not exhaust the membership of the unrestricted sample space of S_9; thus, there is room for other regimes that would provide additional nonoverlapping sample spaces for this power structure.

Cook, Emerson, Gillmore, and Yamagishi (1983, p. 277) have suggested that a sample space might be defined that consists of a mixture of different exchange regimes; for example, the sample space of a mixed regime might encompass all three sets of R_i shown in Figure 2. They also suggest that R-networks based on mixed regimes are more common in natural settings than are pure regimes. A formal typological analysis of sample space restrictions would be useful, as would better theory about conditions that affect the sample spaces of naturally occurring power structures (Cook, 1982; Cook et al., 1983; Emerson, 1981; Markovsky, Willer, and Patton, 1988; and Yamagishi, Gillmore, and Cook, 1988).

Relative Frequency of Exchange Networks

The expected-value model requires a theoretically or empirically based determination of the probability of each of the R-networks in the sample space of a power structure. Ideally, the required probabilities would stem from a formal model of the social exchange process.[6] In experimental work, where data on a large number of trials is gathered, these probabilities can be estimated by the relative frequencies of the observed R-networks; this is illustrated later.

In the absence of probability estimates based on data or theory, a rudimentary baseline assumption may be employed:

Assumption A_3: Each R-network in the sample space of the power structure is *equally* likely.

Accordingly, the probability of a particular R-network is simply the reciprocal of the size of the sample space, that is, $P(R_i) = 1/K$.

Based on assumption A_3, Table 2 illustrates the probabilities of the R_i for power structure S_9 under the various sample space restrictions described in Figure 2. Four networks are possible under the 1-exchange regime and three are possible under the 2-exchange regime; hence, the probabilities of the R-networks are uniformly $1/4$ in the 1-exchange regime and $1/3$ in the 2-exchange

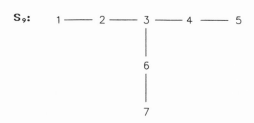

Negative (1-Exchange) Regime	Positive (Triad-Combination) Regime	Multiple (2-Exchange) Regime

Figure 2. Illustrative Sample Spaces of Power Structures S_9

regime. Six networks are possible under the triad-combination exchange regime; hence, the probability of each network is $1/6$.

Outcomes of Exchange Networks

Outcomes relevant to exchange theory include (a) the amount of resources acquired by each actor in a **R**-network, (b) the particular division of resources for each of the exchanges in a **R**-network, and (c) the structural features of a **R**-network. The capacity of an actor to acquire more resources than other

Table 2. Baseline Outcomes for Power Structure S_9

R_i	$P(R_i)$	Outcomes — Actor Resource Receipts							Outcomes — Exchange Occurrences					
		1	2	3	4	5	6	7	1,2	2,3	3,4	4,5	3,6	6,7
(a) 1-Exchange Regime														
R_1	$1/4$	0	12	12	12	12	12	12	0	1	0	1	0	1
R_2	$1/4$	12	12	0	12	12	12	12	1	0	0	1	0	1
R_3	$1/4$	12	12	12	12	12	12	0	1	0	0	1	1	0
R_4	$1/4$	12	12	12	12	0	12	12	1	0	1	0	0	1
Exp. Values		9	12	9	12	9	12	9	$3/4$	$1/4$	$1/4$	$3/4$	$1/4$	$3/4$
(b) Triad Combination Exchange Regime														
R_1	$1/6$	12	12	12	0	0	0	0	1	0	0	0	0	0
R_2	$1/6$	0	0	12	12	12	0	0	0	1	0	1	0	0
R_3	$1/6$	0	0	12	0	0	12	12	0	0	0	0	1	1
R_4	$1/6$	0	12	12	12	0	0	0	0	1	1	0	0	0
R_5	$1/6$	0	12	12	0	0	12	0	0	1	0	0	1	0
R_6	$1/6$	0	0	12	12	0	12	0	0	0	1	0	1	0
Exp. Values		2	6	12	6	2	6	2	$1/6$	$1/2$	$1/2$	$1/6$	$1/2$	$1/6$
(c) 2-Exchange Regime														
R_1	$1/3$	12	12	24	24	12	24	12	1	0	1	1	1	1
R_2	$1/3$	12	24	24	12	12	24	12	1	1	0	1	1	1
R_3	$1/3$	12	24	24	24	12	12	12	1	1	1	1	0	1
Exp. Values		12	20	24	20	12	20	12	1	$2/3$	$2/3$	1	$2/3$	1

actors in the power structure should be distinguished from the actor's negotiated resource exchange ratios; the two outcomes are not necessarily associated. For example, in e-exchange regimes where actors are allowed multiple exchanges, a strategically placed actor may acquiesce to unfavorable exchange ratios (undercutting competitors) in order to amass relatively greater amounts of resources.

A particular R_i describes a pattern of exchanges but does not stipulate the outcome of each exchange in the pattern. Ideally, the prediction of exchange outcomes is based on a refined model of the bargaining process and is evaluated against a baseline model that assumes no knowledge of the putative process. The egalitarian norm, which stipulates an even split of available resources among parties to an agreement, is the obvious candidate for a baseline approach (Cook and Emerson, 1978, p. 723; and Molm, 1981):

Assumption A_4: If a_{ij} is the amount of resources that might be divided between actors i and j, then a transaction between i and j will net each actor one-half a_{ij}.

The outcomes of this baseline assumption are illustrated in Table 2, for each of the three sample space restrictions that were described in Figure 2.

Outcomes may be transparent structural features of the **R**-network, rather than derived outcomes. For example, suppose that the probabilities of the separate exchanges are not theoretically determined and that one is concerned with predicting the most probable exchanges of a power structure (Markovsky, Willer, and Patton, 1988; and Willer and Patton, 1987). For each of a power structure's possible exchanges, the pertinent outcome is whether or not the exchange occurs in a **R**-network. Table 2 also illustrates this.

Expected Values

Our final step is computing the expected value(s) of the outcome(s). Expected values are illustrated in Table 2 where, for each sample space restriction on power structure S_9, baseline expectations are provided for (a) the amount of each actor's resources and (b) the likelihood of each of the six exchange transactions.

In power structure S_9, the relative magnitude of the resource expectation for actor 3, who appears in the most central position, depends on the particular sample space restriction. In the 1-exchange regime, actors 2, 4, and 6 are expected to acquire more resources than actor 3. The situation is reversed in the triad-combination and 2-exchange regimes.

It is *not only* an actor's *position* in a power structure that determines the actor's relative advantage (Molm, 1990; and Bacharach and Lawler 1981). Actors' resource expectations also depend on the sample space of the power structure; evidently, alternate sample spaces can switch actors' rank-order advantage (Markovsky, Willer, and Patton, 1988; and Yamagishi, Gillmore, and Cook, 1988). In addition, of course, actors' resource expectations may be affected substantially by departures from the egalitarian norm of fifty-fifty splits of resources.

THE NEGATIVE EXCHANGE REGIME

Social exchange theorists have been forced to discard conventional measures of network centrality, such as those described by Freeman (1979), and to develop new measures that are more consistent with their empirical findings. To illustrate the problem that has engaged social exchange theorists, consider again the power structure S_9 shown in Figure 2; experimental findings indicate that actors 2, 4, and 6 acquire more resources than actor 3, who appears in the most central location according to conventional measures of network centrality. Since the discovery of this structural anomaly (Cook and Emerson, 1978, pp. 726-727; and Cook et al., 1983), researchers have found similar anomalies for a variety of networks under the 1-exchange regime. The expected value model's *baseline* explanation of these structural anomalies is simply that certain "off center" actors may be relatively advantaged by virtue of a greater relative frequency of participation in exchange transactions.

Thus far, I have shown that, under baseline assumptions, the expected value model of social power correctly predicts the rank order of actors' acquired resources in S_9. Table 3 provides another illustration of actors' baseline resource expectations under the 1-exchange regime. The power structure is S_{12}, a network of ten actors that, like S_9 was studied by Cook et al. (1983). The sample space of S_{12} consists of 20 R-networks; instead of presenting images of these networks, I have listed the lines that comprise each of them. The baseline assumptions are those previously illustrated in Table 2: (1) the R_i in the sample space are equally likely and (2) the exchange agreements uniformly entail an even division of 24 resource units.

Table 3 shows that the resource expectation of actor 4, who appears most structurally central, is less than those of actors 3, 5, and 8. Again, this prediction is consistent with experimental findings. The relatively high expectations of actors 3, 5, and 8 occur because these actors are involved in *all* possible exchange networks that might arise from the power structure, while actor 4 is sometimes excluded from exchange.

Under the same set of baseline assumptions, Table 4 gives the resource expectations of actors in all the networks of Figure 1. The model's predictions concerning positions with the highest resource expectations appear consistent with available experimental evidence, with the possible exception of S_7. In a study of S_7, Markovsky, Willer, and Patton (1988, p. 227) observe the rank

Table 3. Baseline Outcomes for Power Structure S_{12}

Sample Space 1-Exchange Regime	$P(R_i)$	Actor Resource Receipts									
		1	2	3	4	5	6	7	8	9	10
R_1{4-5, 1-3, 8-10}	.05	0	0	12	12	12	0	0	12	0	12
R_2{4-5, 1-3, 8-9}	.05	12	0	12	12	12	0	0	12	12	0
R_3{4-5, 2-3, 8-10}	.05	0	12	12	12	12	0	0	12	0	12
R_4{4-5, 2-3, 8-9}	.05	0	12	12	12	12	0	0	12	12	0
R_5{5-7, 3-4, 8-10}	.05	0	0	12	12	12	0	12	12	0	12
R_6{5-7, 3-4, 8-9}	.05	0	0	12	12	12	0	12	12	12	0
R_7{5-7, 1-3, 4-8}	.05	12	0	12	12	12	0	12	12	0	0
R_8{5-7, 1-3, 8-10}	.05	12	0	12	0	12	0	12	12	0	12
R_9{5-7, 1-3, 8-9}	.05	12	0	12	0	12	0	12	12	12	0
R_{10}{5-7, 2-3, 4-8}	.05	0	12	12	12	12	0	12	12	0	0
R_{11}{5-7, 2-3, 8-10}	.05	0	12	12	0	12	0	12	12	0	12
R_{12}{5-7, 2-3, 8-9}	.05	0	12	12	0	12	0	12	12	12	0
R_{13}{5-6, 3-4, 8-10}	.05	0	0	12	12	12	12	0	12	0	12
R_{14}{5-6, 3-4, 8-9}	.05	0	0	12	12	12	12	0	12	12	0
R_{15}{5-6, 1-3, 4-8}	.05	12	0	12	12	12	12	0	12	0	0
R_{16}{5-6, 1-3, 8-10}	.05	12	0	12	0	12	12	0	12	0	12
R_{17}{5-6, 1-3, 8-9}	.05	12	0	12	0	12	12	0	12	12	0
R_{18}{5-6, 2-3, 4-8}	.05	0	12	12	12	12	12	0	12	0	0
R_{19}{5-6, 2-3, 8-10}	.05	0	12	12	0	12	12	0	12	0	12
R_{20}{5-6, 2-3, 8-9}	.05	0	12	12	0	12	12	0	12	12	0
Expected Values		4.8	4.8	12	7.2	12	4.8	4.8	12	4.8	4.8

Table 4. Baseline Predictions for Figure 1 Power Structures

Power Structure	Actors, Resource Receipt Expectations									
	1	2	3	4	5	6	7	8	9	10
S_1	8	8	12	4						
S_2	6	12	6							
S_3	6	12	12	6						
S_4	8	12	8	12	8					
S_5	6	12	9	9	12	6				
S_6	5	5	12	6	12	5	5			
S_7	4	12	4	12	8					
S_8	8	12	12	12	8	4				
S_9	9	12	9	12	9	12	9			
S_{10}	4	4	12	9	9	12	4	4		
S_{11}	4	4	12	10	9	12	3	3	3	
S_{12}	5	5	12	7	12	5	5	12	5	5

order $2 > \{4,5\} > \{1,3\}$, while the present baseline prediction is $\{2,4\} > 5 > \{1,3\}$. I return to this interesting case after introducing a more refined approach to actors' bargaining behaviors.

A REFINED MODEL

Under baseline assumptions for the 1-exchange regime, actors' resource expectations are never more than one-half of the available resources; for example, in Table 4, where actors negotiate a division of 24 units, the resource expectation of an actor cannot exceed 12 units. This upper limit holds for any distribution of $P(\mathbf{R}_i)$ under the 1-exchange regime. Hence, these baseline assumptions do not explain empirical findings that actors in certain positions of a power structure typically acquire substantially more resources than these baseline expectations. In order to explain these findings, we must relax the assumption of fifty-fifty splits in exchanges.

The success of the expected value model in predicting actors' rank-order resource outcomes is a point in its favor, and so is its manifest failure (under baseline assumptions) to account for actors who acquire substantially more resources than expected. Extant social exchange hypotheses have dealt strictly with rank-order predictions; hence, the field has not been in a position to assess and refine ideas about how actors acquire *particular amounts* of resources.

Building on the thesis that the size of actors' offers are inversely related to the relative frequency of their exclusion from social exchange, the assumption of an egalitarian split of available resources (i.e., A_4) can be replaced with a more refined approach to the bargaining process.[7] The proposed refinement, $A_{4.1}$, involves a set of assumptions dealing with (a) actors' initial offers and

(b) how actors reconcile inconsistent offers. I use the notation $A_{4.1}^{(1)}$, $A_{4.1}^{(2)}$, $A_{4.1}^{(3)}, \ldots, A_{4.1}^{(7)}$, to refer to the constituent parts of $A_{4.1}$.

The first of these constituent assumptions predicts actor i's initial offer to actor j as a function of the dependency of actor i on actor j:

Assumption $A_{4.1}^{(1)}$:   The amount of available resources initially offered by actor i to actor j is governed by an asymptotic function

$$f_{ij} = a_{ij} - b_{ij} (c_{ij})^{100d_{ij}},\qquad (2)$$

where $a_{ij} > 1$ is the amount of available resources, f_{ij} is the amount of these available resources that actor i initially offers to actor j, $0 < b_{ij} < a_{ij}$ and $0 < c_{ij} < 1$ are coefficients, and $0 \le d_{ij} \le 1$ is the actor i's dependency on actor j.

Assumption $A_{4.1}^{(2)}$:   The dependency of actor i on actor j is the probability that actor i is excluded from an exchange and that the two actors do not exchange with each other.[8]

Thus, actor i is assumed to be dependent on actor j because of the association between actor i being excluded from exchange and actor i not exchanging with actor j. Under 1-exchange regimes, this formulation of an actor's dependency simplifies to the vulnerability of the actor to exclusion (i.e., the probability that the actor is excluded from an exchange); hence, under this regime an actor makes the same initial offer to all possible transaction partners.[9] It is under multiple-exchange regimes that an actor's dependency and initial offers may vary for different transaction partners.

The curve (2) rises from $a_{ij} - b_{ij}$; that is, when actor i is least dependent on actor j ($d_{ij} = 0$), the predicted initial offer of actor i is an amount that is less than the available resources. As actor i's dependency on actor j increases (i.e., as $d_{ij} \to 1$), the predicted offer approaches the asymptote a_{ij}, which is the total available resources.

A priori values for the coefficients $\{b_{ij}, c_{ij}\}$ may be derived under two assumptions:

Assumption $A_{4.1}^{(3)}$:   Actors who are minimally dependent will offer only one unit of their available resources, and

Assumption $A_{4.1}^{(4)}$:   Actors who are maximally dependent will offer all but one unit of their available resources.

From the first assumption, $b_{ij} = a_{ij} - 1$. From the second assumption, $f_{ij} = a_{ij} - 1$ when $d_{ij} = 1$. Hence,

$$c_{ij} = \left[\frac{1}{a_{ij} - 1} \right]^{\frac{1}{100}}$$

For example, given 24 units of a resource, $c_{ij} = .969$ and some of the predicted initial offers are:

d_{ij}	0	$\frac{1}{5}$	$\frac{2}{5}$	$\frac{1}{2}$	$\frac{3}{5}$	$\frac{4}{5}$	1
f_{ij}	1	12	17	19	21	22	23

Next, we must consider what happens when two actors' offers are inconsistent; the offers of two actors, i and j, are inconsistent if their personal claims on available resources ($a_{ij} - f_{ij}$ and $a_{ij} - f_{ji}$, respectively) do not sum to the amount of the available resources:

Assumption $A_{4.1}^{(5)}$: If the sum of actors' personal claims exceed the amount of available resources, they are assumed to split the difference and settle on the average of their two offers.

Assumption $A_{4.1}^{(6)}$: If both actors want less than one-half of the available resources, they are assumed to evenly divide the available resources.

Assumption $A_{4.1}^{(7)}$: Finally, there is somewhat complex situation where one actor wants one-half the resources or more, the other wants less than one-half the resources, and the sum of their personal claims is less than the amount of available resources; in this case, it is assumed that the unclaimed portion of the available resources is allocated to the actor with the lower of the two personal claims.

Table 5 gives the predictions of the refined model under the 1-exchange regime for the power structures shown in Figure 1. The rank order of the refined model's predictions are consistent with those of the baseline model; however, in comparison to the baseline predictions, some actors' resource expectations have been considerably elevated or depressed as a consequence of imbalanced exchange ratios.

Two illustrations are provided of the steps involved in deriving predictions from the refined bargaining assumptions. The calculations are somewhat tedious; a computer program implements the approach and is available from the author upon request (Friedkin, 1991b).[10]

Table 5. Refined Predictions of Figure 1 Power Structures

Power Structure	Actor's Resource Receipt Expectations									
	1	2	3	4	5	6	7	8	9	10
S_1	5.5	5.5	20.3	0.6						
S_2	1.4	21.1	1.4							
S_3	1.4	16.6	16.6	1.4						
S_4	3.0	19.5	3.0	19.5	3.0					
S_5	1.4	19.7	5.9	5.9	19.7	1.4				
S_6	0.8	0.8	21.7	1.4	21.7	0.8				
S_7	0.6	18.7	0.6	17.0	3.0		0.8			
S_8	3.0	17.0	15.3	17.0	3.0	0.6				
S_9	4.3	18.2	4.3	18.2	4.3	18.2	4.3			
S_{10}	0.6	0.6	20.6	7.4	7.4	20.6	0.6	0.6		
S_{11}	0.6	0.6	20.3	9.2	6.7	21.4	0.4	0.4	0.4	
S_{12}	0.9	0.9	21.4	2.3	21.4	0.9	0.9	21.4	0.9	0.9

Illustration 1

Consider power structure S_9 under the 1-exchange regime. Under the baseline assumption of equally likely **R**-networks, the actor dependencies (d_{ij}) can be determined by inspection of the four networks in the sample space (see Figure 2). For instance, the dependency of actor 1 on actor 2 (d_{12}) is .25 because there is one of the four **R**-networks in which (a) actor 1 and actor 2 do not exchange with each other and (b) actor 1 is excluded from exchange. The entire matrix of dependencies is:

$$
\mathbf{D} = [d_{ij}] =
\begin{bmatrix}
* & \frac{1}{4} & * & * & * & * & * \\
0 & * & 0 & * & * & * & * \\
* & \frac{1}{4} & * & \frac{1}{4} & * & \frac{1}{4} & * \\
* & * & 0 & * & 0 & * & * \\
* & * & * & \frac{1}{4} & * & * & * \\
* & * & 0 & * & * & * & 0 \\
* & * & * & * & * & \frac{1}{4} & *
\end{bmatrix},
$$

where each asterisk indicates that no exchange between two actors is possible.

Note that the numerical entries in any row of **D** are identical (row homogeneity in **D** is a characteristic of 1-exchange regimes). Hence, an actor will make the same initial offer to all potential exchange partners; assuming that 24 resource units are at stake in all transactions, these initial offers are $\{f_{1j}, f_{3j}, f_{5j}, f_{7j}\} = 13.5$ and $\{f_{2j}, f_{4j}, f_{6j}\} = 1$ for all j.

In S_9, all exchanges occur between the two sets of actors $\{2, 4, 6\}$ and $\{1, 3, 5, 7\}$. From assumptions $A_{4.1}^{(5)}$, $A_{4.1}^{(6)}$, and $A_{4.1}^{(7)}$, all exchanges involve compromises from which actors 2, 4, and 6 receive 18 units and actors 1, 3, 5, and 7 receive 6 units. Hence, the predicted outcomes for actor i in exchanges with actor j $(i < j)$ are:

| | 1 | 2 | 3 | 4 | 3 | 6 |
	2	3	4	5	6	7
R_1	*	18.2	*	18.2	*	18.2
R_2	5.7	*	*	18.2	*	18.2
R_3	5.7	*	*	18.2	5.7	*
R_4	5.7	*	5.7	*	*	18.2

where each asterisk indicates that no exchange occurs between the two actors. Now the resource expectations of the actors may be computed; they are 4.3, 18.2, 4.3, 18.2, 4.3, 18.2, and 4.3, respectively, for the seven actors.

Illustration 2

My second illustration again concerns power structure S_9, but this time under the 2-exchange regime. Under the baseline assumption of equally likely **R**-networks, the actor dependencies (d_{ij}) can be determined by inspection of the three networks in the sample space (see Figure 2). For instance, actor 1 is not dependent on actor 2 $(d_{12} = 0)$, because in none of the three **R**-networks does the joint event occur wherein (a) actor 1 and actor 2 do not exchange with each other and (b) actor 1 is excluded from exchange. However, actor 1 is dependent on actor 3 $(d_{13} = 1/3)$ because there is one network where the two actors do not exchange with each other and where actor 1 negotiates only one of the two exchanges permited to this actor. The entire matrix of dependencies is:

$$\mathbf{D} = [d_{ij}] = \begin{bmatrix} * & 0 & * & * & * & * & * \\ 0 & * & \frac{1}{3} & * & * & * & * \\ * & 0 & * & 0 & * & 0 & * \\ * & * & \frac{1}{3} & * & 0 & * & * \\ * & * & * & 0 & * & * & * \\ * & * & \frac{1}{3} & * & * & * & 0 \\ * & * & * & * & * & 0 & * \end{bmatrix},$$

where each asterisk indicates that no exchange between two actors is possible.

Note that under the 2-exchange regime for S_9, the numerical entries in any row of **D** are not necessarily identical; thus, under this regime, an actor may make different initial offers to different potential exchange partners. Assuming that 24 resource units are at stake in all transactions, the initial offers are:

$$\mathbf{F} = [f_{ij}] = \begin{bmatrix} * & 1 & * & * & * & * & * \\ 1 & * & 16 & * & * & * & * \\ * & 1 & * & 1 & * & 1 & * \\ * & * & 16 & * & 1 & * & * \\ * & * & * & 1 & * & * & * \\ * & * & 16 & * & * & * & 1 \\ * & * & * & * & * & 1 & * \end{bmatrix}.$$

From assumptions $A_{4.1}^{(5)}$, $A_{4.1}^{(6)}$, and $A_{4.1}^{(7)}$, the predicted outcomes for actor i in exchanges with actor j $(i < j)$ are:

	1 2	2 3	3 4	4 5	3 6	6 7
R_1	12	*	19.5	12	19.5	12
R_2	12	4.5	*	12	19.5	12
R_3	12	4.5	19.5	12	*	12

where each asterisk indicates that no exchange occurs between the two actors.

The expected values for the seven actors are 12, 15, 38.9, 15, 12, 15, and 12, respectively. Under multiple-exchange regimes, an actor may acquire resources from several sources; it is an actor's net receipts that are of concern here. For example, actor 2 nets 12 units in R_1 with probability 1/3, 17 units in R_2 with probability 1/3, and 17 units in R_3 with probability 1/3; hence, the resource expectation for actor 2 is $1/3 (12) + 1/3 (17) + 1/3 (17) = 15$.

ASSUMPTIONS AND MODELS

Table 6 gives an overview of the assumptions that have been introduced and the two models that are based on them. H_0 is the baseline model which includes the assumptions (A_3) of equally likely R-networks and (A_4) of egalitarian (fifty-fifty) splits of resources in exchanges. H_1 is the refined model constructed by

Table 6. Assumptions of the Models

Models:

$H_0 = (A_1, A_2, A_3, A_{4.0}\}$ $H_0^* = (A_1, A_2, A_3^*, A_{4.0}\}$
$H_1 = (A_1, A_2, A_3, A_{4.1}\}$ $H_1^* = (A_1, A_2, A_3^*, A_{4.1}\}$

Assumptions:

A_0	Rational actors
A_1	Structural stability of power structure **S**
A_2	Maximal transaction networks (R_k, $k = 1, K$)
A_i	Probability density function for transaction networks
A_3	Equally likely R_k
A_3^*	Observed relative frequencies
A_i	Transaction outcomes
$A_{4.0}$	Egalitarian rule
$A_{4.1}$	Alternative model
$A_{4.1}^{(1)}$	Funcation for initial offers
$A_{4.1}^{(2)}$	Actor vulnerability (dependency)
$\{A_{4.1}^{(3)}, A_{4.1}^{(4)}\}$	Bounds for initial offers
$\{A_{4.1}^{(5)}, A_{4.1}^{(6)}, A_{4.1}^{(7)}\}$	Reconciling inconsistent offers

replacing assumption A_4 with the bargaining behavior assumptions $A_{4.1} = \{A_{4.1}^{(1)}, A_{4.1}^{(2)}, \ldots, A_{4.1}^{(7)}\}$. Finally, H_0^*, is obtained from H_0, and H_1^* is obtained from H_1, by replacing assumption A_3 with empirically observed relative frequencies for the R-networks; I will use these models later.

EVALUATING SELECTED ASSUMPTIONS

I now illustrate how selected components of the approach can be isolated and assessed in light of empirical data. I use data provided by Markovsky, Willer, and Patton (1988) on S_7 under the 1-exchange regime and S_9 under both the 1-exchange and 2-exchange regimes. These data were collected in experiments that rotated each subject through all of the positions of a network: in the case of S_7 (with five positions) each group of subjects participated in five such rotations and in the case of S_9 (with seven positions) each group of subjects participated in seven rotations. For each assignment of actors to positions, the experiments provided four trials; and in each trial actors sought to garner some fraction of 24 "profit points" through negotiated dyadic-level agreements. Thus, for S_7, five groups of subjects generated 100 trials, and for S_9, four groups of subjects generated 112 trials.

A_2 Maximal R-Networks

Nonmaximal networks are rare. In the 100 trials on S_7, there were three occurrences of nonmaximal transaction networks; in the 112 trials for S_9, there were four such networks under the 1-exchange regime and six such networks under the 2-exchange regime. Nonmaximal networks might be eliminated entirely by raising the value of the resources at stake or by simply requiring subjects to continue to negotiate until those exchanges that might occur have occurred.

A_3 Probability of an R-Network.

In the case of S_7 under the 1-exchange regime, the frequencies of the maximal networks are: 3 $R_1\{2\text{-}4\}$, 46 $R_2\{2\text{-}3, 4\text{-}5\}$, and 48 $R_3\{1\text{-}2, 4\text{-}5\}$, where $R_k\{i\text{-}j, \ldots\}$ indicates the line(s) in R_k. Here, the baseline assumption of a uniform probability distribution is obviously misleading.[11] In the case of S_9 under the 1-exchange regime, the frequencies of the maximal networks are: 24 $R_1\{1\text{-}2, 3\text{-}4, 6\text{-}7\}$; 23 $R_2\{2\text{-}3, 4\text{-}5, 6\text{-}7\}$; 38 $R_3\{1\text{-}2, 4\text{-}5, 6\text{-}7\}$; and 23 $R_4\{1\text{-}2, 3\text{-}6, 4\text{-}5\}$. These frequencies are not consistent with assumption of equally likely R-networks; if A_3 were true, then the probability of observing 38 instances of R_3 in 108 trials is approximately .01. In the case of S_9 under the 2-exchange regime, the frequencies of the maximal networks are: 37 $R_1\{1\text{-}2, 3\text{-}4, 4\text{-}5, 3\text{-}6, 6\text{-}7\}$; 39 $R_2\{1\text{-}2, 2\text{-}3, 4\text{-}5, 3\text{-}6, 6\text{-}7\}$; and 30 $R_3\{1\text{-}2, 2\text{-}3, 3\text{-}4, 4\text{-}5, 6\text{-}7\}$. These frequencies are more consistent with equally likely R-networks: under asssumption A_3, the probability of observing 30 instances of R_3 in 106 trials is approximately .16.

Although A_3 appears seriously flawed, currently there is no formal model to replace it. Until such a model is developed, I recommend (a) using observed relative frequencies when they are available or (b) adopting assumption A_3 with the caveat that it is a likely source of error.

A_4 Egalitarianism in Bargaining Behavior

It is recognized that egalitarian agreements do not prevail in exchange transactions. Consider the more than 1000 transactions that were generated by the three experiments conducted by Markovsky, Willer, and Patton (1988). For each of these transactions, I calculated max(PAYOFF), where max(PAYOFF) is the larger of the two amounts of resources received by the actors from the transaction. Under the egalitarian norm max(PAYOFF) = 12 for all transactions; in fact, this variable is distributed as shown in Table 7.

Table 7.   Empirical Findings on Transaction Outcomes

| | S_7 | | S_9 | | S_9 | |
| | 1-Exchange Regime | | 1-Exchange Regime | | 2-Exchange Regime | |
max(PAYOFF)	Count	%	Count	%	Count	%
12	77	40.31	54	16.67	237	44.72
13	24	12.57	34	10.49	98	18.49
14	14	7.33	19	5.86	45	8.49
15	6	3.14	9	2.78	11	2.08
16	4	2.09	5	1.54	16	3.02
17	1	0.52	10	3.09	6	1.13
18	8	4.19	8	2.47	13	2.45
19	4	2.09	10	3.09	5	0.94
20	2	1.05	18	5.56	4	0.75
21	2	1.05	15	4.63	10	1.89
22	1	0.52	25	7.72	11	2.08
23	31	16.23	117	36.11	73	13.77
24	17	8.90	0	0.00	1	0.19
Total	191		324		530	

Note: max(PAYOFF) is the larger of the two amounts of resources received by the actors from their transaction.

Table 8. Empirical Findings for Power Structures S_7 and S_9

(a) S_7 1-Exchange Regime

	Actors					Exchanges				
	1	2	3	4	5	12	23	24	36	45
PAYOFF	2	19	2	12	11	5	20	12	12	12
H_0^*	6	12	6	12	12	12	12	12	12	12
H_1	1	19	1	17	3	2	22	12	19	19
H_1^*	1	21	1	13	11	3	21	12	13	13

(b) S_9 1-Exchange Regime

	Actors							Exchanges				
	1	2	3	4	5	6	7	12	23	34	45	67
PAYOFF	5	18	3	18	5	19	4	6	19	5	18	19
H_0^*	9	12	8	12	9	12	9	12	12	12	12	12
H_1	4	18	4	18	4	18	4	6	18	6	18	6
H_1^*	5	18	3	18	5	18	5	6	20	4	18	18

(a) S_9 2-Exchange Regime

	Actors							Exchanges				
	1	2	3	4	5	6	7	12	23	34	45	67
PAYOFF	12	15	39	15	12	15	12	12	5	19	12	19
H_0^*	12	20	24	20	12	21	12	12	12	12	12	12
H_1	12	16	36	16	12	16	12	12	6	18	12	12
H_1^*	12	15	39	15	12	16	12	12	4	20	12	12

Notes: Actor PAYOFF is the observed amount of resources an actor receives on average from the transactions in R_k; exchange PAYOFF is the observed amount of resources actor i receives on average from transactions with actor j ($i < j$); in deriving the H_0^* and H_1^* predictions, assumption A_3 is replaced by the observed relative frequencies of the **R**-networks.

187

$A_{4.1}$ Dependencies in Bargaining Behavior

The available data do not describe the initial offers of actors, so that the constituent parts of assumption $A_{4.1}$ cannot be evaluated directly. Cook, Emerson, Gillmore, and Yamagishi (1983, p. 287; see also Cook and Emerson, 1978; and Molm, 1987) suggest that assumptions $A_{4.1}^{(3)}$ and $A_{4.1}^{(4)}$ may not always hold; they state, "If there are restraints on the exercise of power (e.g., equity concerns or less than fully rational negotiation), equilibrium will be reached somewhere short of... [a] maximally "exploitative" exchange ratio." Such restraints would imply that the proposed a priori solution for equation (2) may be misleading in some instances.

An indirect method of assessing $A_{4.1}$ is to examine the goodness of fit for the predicted and observed outcomes of exchange transactions. A potential methodological problem with this approach is that a lack of fit may arise from flaws in other assumptions beside those of $A_{4.1}$. Here, the problem is not serious. Assumption A_1 on the stability of the power structure is satisfied by the experimental design. Assumption A_2 on maximal R-networks is satisfied by eliminating the few networks that are nonmaximal. Assumption A_3 on the relative frequency of the R-networks is satisfied by employing observed relative frequencies.

Table 8 gives the predicted and observed outcomes for the actors and transactions in the three experiments. The fits appear excellent. We should be impressed by a close fit only if the H_1 assumptions offer substantially better predictions than baseline H_0 assumptions. Here, it appears that baseline predictions are substantially improved by introducing a more refined assumption about actors' bargaining behaviors.

DISCUSSION

As Molm (1990) has most recently emphasized, social power is manifested simultaneously as a structural potential, a process, and an outcome. Moreover, these facets of social power are manifested in different types of social relations. Thus, with respect to the structure of power, we may consider networks of potential social exchanges, interpersonal influences, or information flows; with respect to the process of power, we may consider how actors negotiate social exchange agreements, how they integrate the separate interpersonal influences upon their opinions, or how information travels through the communication channels of a network; and, with respect to the outcomes of power, we may consider the emergent distributions of material resources, total interpersonal effects, or visibility.

The present paper was motivated by the idea that the formal model of social power pursued by French (1956) and Friedkin (1986) might be made more general. The model was theoretically attractive because it simultaneously involved three dimensions of social power—structure, process, and outcome—in a coherent framework; however, the applicability of the model appeared limited to social influence relations. Thus, in extending this line of formal work on social influence to touch upon social exchange phenomena, there has been a step toward a broader theoretical framework that bridges different forms of social power and types of social relations in which these forms arise.

The present approach to social power puts new light on the prevalent hypothesis that actors' resource outcomes may be predicted from their location in a network of *potential* exchange transactions. In terms of the present theoretical framework, this hypothesis has little merit and should be replaced by efforts to explain the incidence of particular patterns of social exchanges and actors' bargaining behavior in these patterns. The basis for the skepticism is that a structure of potential exchange transactions may have very different consequences for actors' resource outcomes, depending on the intervening processes that govern the distribution of actual transactions and the content of the agreements in these transactions; see the related arguments of Friedkin (1986, pp. 114-117) and compare with Molm (1990) and Bacharach and Lawler (1981).

No previous approach to social exchange outcomes has provided a formal model that predicts the absolute values of social exchange outcomes. The present approach moves beyond the rank-order prediction of actors' resource outcomes that is characteristic of extant social exchange hypotheses, and provides baseline predictions of the amount of resources each actor is expected to acquire through social exchange. Under baseline assumptions, the approach provides a simple account of the literature's intriguing findings that the most centrally located actors in exchange networks do not necessarily acquire the most resources via exchange processes.

The baseline assumptions of the model provide a null hypothesis against which the merits of more refined alternative hypotheses can be assessed. I have illustrated how the baseline assumptions may be relaxed with the introduction of a formal hypothesis about the relationship between an actor's bargaining behavior and vulnerability to exclusion from social exchange. For the several cases that were examined, the hypothesis does a credible job of predicting actors' absolute amounts of acquired resources.

The proposed expected-value model supplies a different integration of extant ideas and findings on social exchange processes, and it provides a flexible intellectual framework in which to pursue a cumulatively refined social exchange theory. The approach disentangles three theoretical subsystems toward which more refined hypotheses may be directed. The first of these subsystems concerns social exchange regimes (i.e., sample space restrictions), the second concerns the incidence of exchange networks (i.e., the likelihoods of the separate networks in the sample space), and the third concerns actors' bargaining behavior. While social exchange theorists have recognized the existence of these subsystems, and have presented some hypotheses about subsystem interrelationships, a coherent framework has been lacking which would allow an overview of these subsystems and a disentangling of their contributions to social exchange outcomes.

An expected-value model of social exchange outcomes has been latent in the literature on social exchange and its development does not represent a dramatic new direction for the field. The approach was presaged in Emerson's (1962, p. 41) reference to French's formal theory of social power as a treatment of power toward which his own work might move. Emerson, like French, viewed power as a relation that defined opportunities for interpersonal events—exchange transactions (Emerson) or interpersonal influences (French). While neither Emerson nor French carried forward the logic of this initial formulation, Emerson (1972a, p. 56) verges on the point of departure of an expected-value approach to social power when he describes an exchange relation as giving rise to opportunities which "result in transactions with probability P_{yk}." The crucial step is the idea that power structures generate sample spaces of event or transaction networks; with this cornerstone in place, the outlines of the remaining theoretical development are straightforward.

ACKNOWLEDGMENT

I am indebted to Barry Markovsky and David Willer who generously provided data for analysis; special thanks go to David Willer for his unselfish and enthusiastic support of my approach to social exchange phenomena. This paper was presented at the Annual Meeting of the American Sociological Association, Cincinnati, August, 1991.

NOTES

1. See the following work and the literature cited therein: Cook, 1982 and 1987; Cook and Emerson, 1987; Cook, Emerson, Gillmore, and Yamagishi, 1983; Emerson, 1962, 1969, 1972a,1972b, 1976, and 1981; Marsden, 1983 and 1987; Markvosky, Willer, and Patton, 1988; Molm, 1985, 1987, and 1990; Willer, 1987; Willer and Anderson, 1981; Willer, Markovsky, and Patton, 1989; Willer and Patton, 1987; and Yamagishi, Gillmore, and Cook, 1988.

2. I draw on Markovsky, Willer, and Patton's (1988) experiments for this analysis.

3. Along these lines, for $p = .50$, R-networks are equally likely and each network has a probability of .50 $^v = 1/K$, where K is the number of R-networks in the sample space.

4. The model of consensus production employed by Friedkin (1986) has since been refined; see Friedkin (1990 and 1991), Friedkin and Cook (1990), and Friedkin and Johnsen (1990).

5. I may depart from Cook and Emerson (1978) and Cook et al. (1983) when they appear to emphasize the emergence of a stable pattern of utilized exchange opportunities. I require a stable pattern of exchange opportunities and leave open the issue of actors' commitment to one or more particular pattern(s) of these exchange opportunities.

6.   Such a formal model would take into account the development of commitments among potential exchange partners: "An actor is said to be committed to another actor in the network ot the extent that choice of current exchange partners, from among alternative partners, can be predicted from previous partnerships" (Cook and Emerson, 1978, p. 728); see also Molm (1990) for a dynamic view of social exchange networks.

7.   Here I build on the theory of bargaining provided by Cook and Emerson (1978, p. 727) in which they assume that actors will lower their offers when their offers are not accepted and on the similar thesis of Markovsky, Willer, and Patton (1988, p. 221; also see Willer, Markovsky, and Patton, 1989, pp. 324-328).

8.   Let max(deg$_i$) represent the maximum number of exchanges for actor i in any of the R-networks. Actor i is excluded from an exchange in R_k if the number the actor's exchanges in R_k is less than max(deg$_i$); let A indicate this event. Let B indicate the event that actor i and actor j do not exchange with each other in R_k. The dependency of actor i on actor j is the probability of the joint event $d_{ij} = P (A \cap B)$ in the sample space of the power structure. I settled on this measure of dependency after eliminating several plausible alternatives that performed less well.

9.   Cook et al. (1983, pp. 299-302) propose a concept of vulnerability that appears closely related to the present definition; see also Cook, Gillmore, and Yamagishi (1986) and Emerson (1978). However, in the present approach it is not strictly the number of an actor's exchange *opportunities* that determines the actor's vulnerability: the probability of an actor being excluded from exchange depends not only on the power structure, but also on (a) the sample space of the power structure and (b) the probability distribution for the R_i that comprise the sample space.

10.   This program is designed to run on an IBM PC-XT,-AT, PS/2, or compatible computer with an 8087, -287, or -387 math coprocessor (the program will not run *without* a math coprocessor). DOS 3.3 or above is required; at least 640K of memory is recommended.

11.   The similar relative frequencies of R_2 and R_3 are understandable since they are *isomorphic* with respect to their pattern of exchanges; as long as the characteristics of the actors who occupy the positions in these structures do not systematically differ by position, we would not expect any marked differences in the relative frequencies of these networks.

REFERENCES

Bacharach, Samuel B. and Edward J. Lawler. 1981. *Bargaining: Power, Tactics, and Outcomes.* San Francisco: Jossey-Bass.

Bonacich, Phillip. 1987. "Power and Centrality: A Family of Measures." *American Journal of Sociology* 92: 1170-1182.

Cartwright, Dorwin. 1965. "Influence, Leadership, and Control." Pp. 1-46 in *Handbook of Organizations,* edited by J.G. March. Chicago: Rand McNally.

Cook, Karen S. 1982. "Network Structures from an Exchange Perspective." Pp. 177-199 in *Social Structure and Network Analysis,* edited by P.V. Marsden and N. Lin. Beverly Hills, CA: Sage.

Cook, Karen S. (ed.) 1987. *Social Exchange Theory.* Newbury Park, CA: Sage.

Cook, Karen S. and Richard M. Emerson. 1978. "Power, Equity, and Commitment in Exchange Networks." *American Sociological Review* 43: 721-739.

Cook, Karen S., Richard M. Emerson, Mary R. Gillmore, and Toshio Yamagishi. 1983. "The Distribution of Power in Exchange Networks: Theory and Experimental Results." *American Journal of Sociology* 89: 275-305.

Cook, Karen S., Mary R. Gillmore, and Toshio Yamagishi. 1986. "Point and Line Vulnerability as Bases for Predicting the Distribution of Power in Exchange Networks: Reply to Willer." *American Journal of Sociology* 92: 445-448.

Emerson, Richard M. 1962. "Power-Dependence Relations." *American Sociological Review* 27: 31-40.

———. 1969. "Operant Psychology and Exchange Theory." Pp. 379-405 in *Behavioral Sociology: The Experimental Analysis of Social Process,* edited by R.L. Burgess and D. Bushell, Jr. New York: Columbia University Press.

———. 1972a. "Exchange Theory, Part I: A Psychological Basis for Social Exchange." Pp. 38-57 in *Sociological Theories in Progress,* Volume 2, edited by J. Berger, M. Zelditch, Jr., and B. Anderson. New York: Houghton Mifflin.

———. 1972b. "Exchange Theory, Part II: Exchange Relations and Network Structures." Pp. 58-87 in *Sociological Theories in Progress,* Volume 2, edited by J. Berger, M. Zelditch, Jr., and B. Anderson. New York: Houghton Mifflin.

———. 1976. "Social Exchange Theory." *Annual Review of Sociology* 2: 335-362.

———. 1981. "Social Exchange Theory." Pp. 30-65 in *Social Psychology: Sociological Perspectives,* edited by M. Rosenberg and R. Turner. New York: Academic Press.

Freeman, Linton C. 1979. "Centrality in Social Networks: Conceptual Clarification." *Social Networks* 1: 215-239.

French, J. R. P., Jr. 1956. "A Formal Theory of Social Power." *The Psychological Review* 63: 181-194.

Friedkin, Noah E. 1986. "A Formal Theory of Social Power." *Journal of Mathematical Sociology* 12: 103-126.

———. 1990. "Social Networks in Structural Equation Models." *Social Psychology Quarterly* 53: 316-328.

———. 1991a. "Theoretical Foundations for Centrality Measures." *American Journal of Sociology* 96: 1478-1504.

———. 1991b. "An Expected Value Model of Social Power: EVM System Version 1.0." Graduate School of Education, University of California, Santa Barbara.

Friedkin, Noah E. and Karen S. Cook. 1990. "Peer Group Influence." *Sociological Methods & Research* 19: 122-143.

Friedkin, Noah E. and Eugene C. Johnsen. 1990. "Social Influence and Opinions." *Journal of Mathematical Sociology* 15: 193-206.

Harary, Frank, Robert Z. Norman, and Dorwin Cartwright. 1965. *Structural Models: An Introduction to the Theory of Directed Graphs.* New York: Wiley.

Marsden, Peter V. 1983. "Restricted Access in Networks and Models of Power." *American Journal of Sociology* 88: 686-717.

———. 1987. "Elements of Interactor Dependence." Pp. 130-148 in *Social Exchange Theory,* edited by K.S. Cook. Newbury Park, CA: Sage.

Markovsky, Barry, David Willer, and Travis Patton. 1988. "Power Relationships in Exchange Networks." *American Sociological Review* 53: 220-236.

———. 1990. "Theory, Evidence, and Intuition." *American Sociological Review* 55: 300-305.

Molm, Linda. 1981. "The Conversion of Power Imbalance to Power Use." *Social Psychology Quarterly* 16: 153-166.

———. 1985. "Relative Effects of Individual Dependencies: Further Tests of the Relation between Power Imbalance and Power Use." *Social Forces* 63: 810-837.

———. 1987. "Linking Power Structure and Power Use." Pp. 101-129 in *Social Exchange Theory,* edited by K.S. Cook. Newbury Park, CA: Sage.

———. 1990. "Structure, Action, and Outcomes: The Dynamics of Power in Social Exchange." *American Sociological Review* 55: 427-447.

Stotle, John and Richard M. Emerson. 1977. "Structural Inequality: Position and Power in Network Structures." Pp. 117-138 in *Behavioral Theory in Sociology,* edited by R. Hamblin and J. Kunkel. New Brunswick, NJ: Transaction Books.

————. 1987. *Theory and Experimental Investigation of Social Structures.* New York: Gordon and Breach.

Willer, David. 1986. "Vulnerability and the Location of Power Positions: Comment on Cook, Emerson, Gillmore and Yamagishi." *American Journal of Sociology* 92: 441-444.

Willer, David and B. Anderson (eds.). 1981. *Networks, Exchange and Coercion.* New York: Elsevier/Greenwood.

Willer, David, Barry Markovsky and T. Patton. 1989. "Power Structures: Derivations and Applications of Elementary Theory." Pp. 313-353 in *Sociological Theories in Progress: New Formulations,* edited by J. Berger, M. Zelditch, Jr., and B. Anderson. Newbury Park, CA: Sage.

Willer, David and Travis Patton. 1987. "The Development of Network Exchange Theory." Pp. 199-242 in *Advances in Group Processes,* Volume 4, edited by E.J. Lawler and B. Markovsky. Greenwich, CT: JAI Press.

Yamagishi, Toshio, and Karen S. Cook. 1990. "Power Relations in Exchange Networks: A Comment on 'Network Exchange Theory'." *American Sociological Review* 55: 297-300.

Yamagishi, Toshio, Mary R. Gillmore, and Karen S. Cook. 1988. "Network Connections and the Distribution of Power in Exchange Networks." *American Journal of Sociology* 93: 833-851.

APPROACHING DISTRIBUTIVE AND PROCEDURAL JUSTICE: ARE SEPARATE ROUTES NECESSARY?

Karen A. Hegtvedt

ABSTRACT

In the last fifteen years, distributive justice issues have shared the limelight with those of procedural justice. Within social psychology, the sharing is by no means equal; emphasis on the fairness of outcomes continues to attract greater attention than concerns with the fairness of procedures. Except for the studies that specifically address the relationship between distributive and procedural justice, the domains remain relatively separate. This paper challenges the wisdom of these separate routes by highlighting parallel features of the two domains that potentially pave a way ("a single route," albeit not necessarily the only one) toward theoretical integration. The paper first examines the apparent reasons for the separation, then emphasis shifts to identification of major theoretical questions inherent in distributive and procedural justice approaches. These questions provide the basis for evaluating the two existing integrative attempts and form the foundation for proposing a more encompassing route to understanding both types of justice.

Advances in Group Processes, Volume 10, pages 195-221.
Copyright © 1993 by JAI Press Inc.
All rights of reproduction in any form reserved.
ISBN: 1-55938-280-5

INTRODUCTION

For nearly thirty years, researchers have devoted considerable attention to distributive justice concerns (see, for example, volumes by Bierhoff, Cohen, and Greenberg, 1986; Folger, 1984; Greenberg and Cohen, 1982; Lerner and Lerner, 1981; Messick and Cook, 1983; Mikula, 1980; and Vermunt and Steensma, 1990). In the last fifteen years, however, distributive justice issues have shared the limelight with those of procedural justice (see Thibaut and Walker, 1975; and Lind and Tyler, 1988). Within social psychology, the sharing is by no means equal; emphasis on the fairness of outcomes continues to attract greater attention than concerns with the fairness of procedures. And, except for the studies that specifically address the relationship between distributive and procedural justice, the domains remain relatively separate. The intention of this paper is to challenge the wisdom of these separate routes by highlighting parallel features of the two domains that potentially pave a way ("a single route," albeit not necessarily the only one) toward theoretical integration.

The paper first examines the apparent reasons for the separation and also reviews empirical findings regarding the relationship between distributive and procedural justice. Emphasis then shifts to focus on the theoretical approaches to both distributive and procedural justice issues; the brief reviews provide the means for identifying critical theoretical questions and issues that may form the foundation for integrating the two types of justice. Applying these questions and issues, we evaluate the success of the existing perspectives that address, though not necessarily integrate, both distributive and procedural justice (Folger, 1986; and Leventhal, Karuza, and Fry, 1980). We conclude with a strategy for a "single route" to distributive and procedural justice issues which recognizes the commonalities as well as the uniqueness of the two domains.

SEPARATE INSPIRATIONS BUT RELATED PHENOMENA

The contexts in which the issues of distributive and procedural justice were initially embedded account, in large part, for the separate routes that developed. Greenberg and Tyler (1987) recognize that this initial separation, furthermore, reflected the emphasis on "ends" inherent in distributive justice research and on the problem of "means" characteristic of procedural justice. Overlap between contexts emerged primarily in research examining the relationship between procedural justice and distributive justice—the relationship between means and ends.

Early approaches to distributive justice (e.g., Homans, 1961 and 1974; Adams, 1965) focused on questions of pay equity within work organizations. These perspectives designated a proportional distribution principle, which related outcomes (pay) to inputs and/or investments (e.g., effort, ability, status), as just. The perspectives, furthermore, assumed consensus on this principle. Later work expanded the focus of distributive justice in two ways. First, Walster, Berscheid, and Walster (1978) attempted to demonstrate the generality of equity to other contexts, including intimate relations. Second, a number of researchers challenged the "single rule" approach to distributive justice, arguing that other distribution principles (in particular, equality and distribution according to needs) may define justice under certain circumstances (e.g., Deutsch, 1975; Leventhal, 1976; and Reis, 1984). Although pertaining to different types of "end" distributions, this second development also highlighted contextual variation, including work, family, and peer situations.

Little attention, however, extended to the context in which discussions of procedural justice issues first emerged: the legal arena. Thibaut and Walker (1975), the first notable researchers in the area, focused primarily on legal means or strategies of conflict resolution (e.g., arbitration, mediation, and judicial review). Unlike distributive justice perspectives, such an approach assumed lack of, rather than consensus on, justice rules. More recently, Lind and Tyler (1988) have attempted to expand the purview of procedural justice to include organizational, political, and interpersonal contexts. For example, in organizations, procedural justice concerns may arise in the development of job evaluation schemes—the means by which managers assess worker performance in view of job demands. Such evaluation schemes may underlie the distribution of pay raises and thus, in effect, highlight the relationship between procedural and distributive justice.

Empirical research examines the mutual effects of individuals' perceptions of procedural and distributive justice on each other, as well as the joint effects of both types of justice on other social evaluations or concerns. The empirical investigations, furthermore, demonstrate the recently expanded breadth of procedural justice studies.

In studies of conflict resolution, the nature of the outcome (a third party's decision for or against the subject) consistently affects evaluations of procedural justice; individuals are more likely to judge the process as fair when their outcome is favorable (see Lind and Tyler, 1988). But perceived procedural fairness also positively affects perceptions of distributive justice. For example, Musante, Gilbert, and Thibaut (1983) show that group discussion over the rules of decision-making and actual control over the decision enhance evaluations of procedural justice; these evaluations, in turn, correlate highly with those of distributive justice.

That perceived procedural justice tends to enhance perceptions of distributive justice, especially under conditions of negative outcomes, is also a consistent finding in studies focusing on organizational settings (see Lind and Tyler, 1988). The effect, however, is somewhat less apparent under conditions of positive outcomes (Greenberg, 1987), perhaps owing to the simple hedonic acceptance of high outcomes. Studies have also demonstrated that procedural and distributive justice positively influence other aspects of organizational life, such as job satisfaction, evaluation of supervisors, and trust in management (e.g., Alexander and Ruderman, 1987).

Regardless of context, most of the research on the relationship between procedural and distributive justice stems primarily from theoretical perspectives on procedures, rather than those on outcomes. Such framing raises questions regarding the adequacy of the distributive justice approaches for understanding procedures, as well as the adequacy of procedural perspectives for understanding distribution issues. These questions direct attention to the purposes and tenets of distributive and procedural justice.

SEPARATE DISTRIBUTIVE AND PROCEDURAL JUSTICE ROUTES: WHAT ARE THE QUESTIONS?

Chronological examinations of the development of theoretical perspectives on distributive and procedural justice in social psychology document the separate routes and the purposes pursued by researchers.[1] The intent of the analysis is to specify the similarity of the theoretical questions and issues posed by each line of inquiry. In particular, three major questions emerge. First, how do justice principles develop? Second, what activates justice concerns in a situation? And, third, how do people respond to injustice?

Distributive Justice

The perspectives of the early distributive justice theorists (Adams, 1965; Blau, 1964; Homans, 1961, 1974) reflect the the social exchange approach to interaction. As a consequence, concern with balance between what actors bring to the exchange (inputs or investments) and what they receive from it (outcomes or rewards) provides the basis of justice. Homans' "rule of distributive justice" revolves around the expectations of exchange partners that the rewards of each will be proportional to the costs of each and that outcomes or profits (rewards minus costs) will be proportional to investments. For Adams, justice obtains with an equitable distribution in which which the outcomes of actor A in proportion to his/her inputs are equal to the proportion of actor B's outcomes to inputs.

Homans and Blau address the theoretical question of the development of justice principles by offering explanations about the source of beliefs in proportionality. Homans argues that as a result of observations of the links between investment and profits based on repeated experiences, individuals develop probabalistic expectations about investments and profits that hold in the future. In other words, actual exchange proportions become normative guidelines for future exchanges. Blau also recognizes the centrality of negotiation in the development of a proportional justice rule, but he distinguishes between the going rate (established by the rules of supply and demand for the benefits available in a relationship that set boundaries for the negotiation) and the fair rate of exchange. The fair rate represents normative standards for a just return of a given benefit stemming from *society's* need for the service and the investments needed to provide the service. The fair rate creates moral expectations that may bind together members of society.

Given the weight of these normative expectations regarding fair exchange, violations of the justice principle are likely to engender various types of reactions. The exchange theorists thus address the third question regarding reactions to injustice. Homans analyzes emotional reactions, arguing that those who receive less than expected are likely to feel angry, whereas those whose receipts are more valuable than expected may feel guilty. The extent to which an individual experiences these emotions depends in large part upon self-interest; there is likely to be a lower threshold for the experience of anger because of reward deprivation than that for guilt because of the reward benefit. Self-interest—in the form of cost/benefit analysis—also underlies Adams' formulation of behavioral and cognitive reactions to inequity.

Adams argues that when the comparison between an individual's own outcome/input ratio and that of another person results in the assessment of inequity, the person is likely to feel some sort of distress or tension. The distress varies directly with the magnitude and nature of the inequity, and motivates individuals to seek redress for the injustice they suffer. Individuals can attempt to eliminate the distress and restore equity by (1) altering their actual inputs; (2) altering their actual outcomes; (3) cognitively distorting either their inputs or outcomes; (4) cognitively distorting the other person's inputs or outcomes; (5) changing the object of the comparison from one person to another; or (6) leaving the situation. Adams suggests that individuals will choose the least costly method of equity restoration.

Although extensively investigated, most empirical studies on reactions to injustice have narrowly focused on specific types of reactions and fail to examine the choice among restoration methods (see Cook and Hegtvedt, 1983). Despite Walster, Berscheid, and Walster's (1978) more recent attempt to develop a logical series of propositions combining individual maximization, emphasis on society as the source for the equity principle, and emphasis on inequity reactions in various types of relationships, a number of implications of the exchange approaches remain untested. Most extensions of the distributive justice theories have resulted from integration with other perspectives. The development of a status-based theory of distributive justice, a shift in emphasis from exchange situations to those of allocation, and the growing attention in social psychology to cognitive issues hearalded expansion of the issues and approaches to distributive justice. Much of this work addresses the question unaddressed by the exchange formulations: the activation of justice concerns.

Berger, Zelditch, Anderson, and Cohen (1972) offered a status-value theory of distributive justice to address some unresolved issues inherent in the exchange approaches as well as to examine the activation of justice concerns. According to these theorists, a major shortcoming of the exchange approach is its emphasis solely on "local" comparisons between individuals, which omit consideration of social standards by which to evaluate one's outcomes. They argue that referential comparisons, consisting of generalized others who possess states of given characteristics associated with certain reward levels, put individual-level rewards into perspective. In a relationship, one or both actors may be unjustly rewarded in comparison to what people like themselves generally receive under similar circumstances. By introducing referential standards, Berger et al. draw attention to socially defined rewards or, in their terms, the status value of rewards that exchange theory overlooks.

Emphasis on status value leads to a somewhat different justice defintion than the exchange notion of proportionality. In the Berger et al. approach, justice exists when there is consistency between the status value of the rewards relevant characteristics and the rewards obtained. The referential standards described above represent how people have developed stereotyped conceptions of this consistency. Moreover, recent work by Jasso (1980, 1989) extends the analysis of how referential comparisons facilitate the activation of justice concerns outside the confines of dyadic relationships.[2]

A different form of movement away from emphasis on the exchange relationship emerged in the mid-1970s. The growing challenge to emphasis on a single rule of justice (equity) led to an empirical shift away from exchange to allocation situations involving the distribution of rewards to a specific circle of recipients. The multiple rules approaches to distributive justice (e.g., Deutsch, 1975; Leventhal, 1976; Leventhal, Karuza, and Fry 1980; Reis, 1984) assumes that a number of different distribution principles could define justice. In addition to the equity rule, attention focused primarily on rules of equality and of distribution according to the needs of potential recipients. The work of Leventhal, Karuza, and Fry (1980) provides the only major theoretical framework for identifying allocation preferences underlying the development of justice principles, whereas much of the empirical work notes contextual variation in justice principles and factors underlying the activation of justice concerns.

Leventhal, Karuza, and Fry (1980) propose a theory of allocation preferences that encompasses concerns with distributive justice. They assume that people prefer a distribution that helps them to achieve their goals, which may include fairness, self-interest, obedience to authority, or other pragmatic concerns. In addition, they assume that individuals assess how likely different distribution rules are to fulfill their goals. People then choose the rule or combination of rules for which the expectation of fulfilling their goal(s) is the highest.

When asked, "What distribution rule do people perceive as fair given a specific goal?" empirical results are fairly rigorous. When the situational goal is productivity, people tend to prefer equity; when the goal is group solidarity, individuals opt for equal distributions and when it is social welfare, people choose a needs-based distribution rule (Leventhal, 1976). These results emerge regardless of whether the evaluator is a potential recipient (i.e., a first-party allocator) or a nonrecipient (i.e., a third party).

Leventhal's expectancy approach, however, has not specifically guided much of the research on multiple distribution rules. Instead, the question, "What distribution rule do people prefer under certain circumstances?" has been the driving empirical force. This question, like research on the relationship between goals and distribution principles, corresponds to exchange theorists' concerns about the development of justice principles.

In the absence of a clearly articulated theoretical framework and in recognition of the conditional nature of rule preferences, findings are complicated. Cultural, situational, relational, and personal factors affect individuals' preferences for a particular distribution principle, especially in cases of first-party allocation. Generally, in impersonal task-oriented situations, American subjects who are corecipients of the allocation tend to prefer rules that seem to maximize their own rewards. For example, high task performers choose equitable distribution and low performers an equal one (see Cook and Hegtvedt, 1983). But with the existence or development of affective bonds, *materially* self-interested rule preferences are less likely to emerge.

Affective bonds may alter the nature of distribution preferences, for at at least two reasons. First, friendly relations within the group seem to signal the teamlike behavior of group members and the obligations of the allocator to the team. The role responsibility of the allocator may demand polite behavior and consideration of the feelings of others (Schwinger, 1980). Second, the affective bonds suggest more of a long-term relationship between the allocator and recipients, which suggests the additional relevance of social rewards and costs (Reis, 1981). Individuals may opt for allocations that appear more generous to others to achieve rewards such as liking and respect and to avoid costs such as dislike and ostracism. In effect, even with stronger affective bonds, allocators may still pursue their self-interest in terms of the maximization of the balance of material and social rewards, but the resulting distribution is less likely to appear materially self-interested and more likely to promote the welfare of other group members (see Hegtvedt, 1992).

Instances of third-party allocation eliminate the possibility of self-interest as a motivational basis, and draw attention to the welfare of group members. When the allocator is not a corecipient of the distribution, their decisions are more likely to encompass the impartiality (Rawls, 1971) characteristic of justice processes. Research demonstrates that third-party allocators instructed to produce a fair distribution designate ones that clearly differ from those made by corecipients, who tend to inflate their own rewards (Messick and Sentis, 1979). The distribution rule preferences of first and third parties highlight the importance of understanding how situational conditions affect the motivations underlying allocation choices. Furthermore, the contrast between interpretations of self-interested allocations and those representing justice per se parallels Lerner's (1982) concern that outcome-based justice models, derived from explicit assumptions about maximizing outcomes, fail to capture the unique qualities associated with justice.

To further understand the development of justice principles and reactions to injustice researchers have recently attempted to integrate issues of cognitive processing and social perception into distributive justice studies. In effect, such integration reveals additional issues affecting the activation of justice concerns. To determine whether equity obtains, individuals must evaluate the relevance and importance of both inputs and outcomes (Cohen, 1982) as well as how to combine multiple inputs (see, e.g., Farkas and Anderson, 1979). Typically, inputs perceived to be under a person's control are considered relevant bases for allocating rewards. Given at least two relevant inputs, some evidence suggests that people tend to overemphasize the input on which they rate highly as the basis for allocation (Cook and Yamagishi, 1983).

Perceived control in the situation is also relevant to reactions to inequity. Utne and Kidd (1980) argue that attributing an inequity to the intentional actions of another person is more likely to stimulate behaviors to restore equity rather than psychological changes in the perception of inputs and outcomes. And, if individuals perceive that a chance occurence has caused the inequity, they are less likely to be distressed by it and, consequently, are less likely to attempt to restore equity. Thus, the integration of cognitive elements into the analysis of justice and injustice processes in interaction has helped to resolve the issue regarding the choice of restoration method.

In summary, the major distributive justice perspectives and empirical work primarily address the development of distributive justice principles and reactions to injustice. Implicit or explicit in this research are assumptions about individual's motives or goals and the social context. Moreover, discussion of reward expectations, social comparisons, and attributions in the situation introduce issues of the activation of justice concerns.

Procedural Justice

Thibaut and Walker's (1975) landmark work on procedural justice addressed the first major question: the development of procedural justice rules. They examined such development by analyzing evaluations of types of conflict resolution. Drawing upon social exchange theory, they argued that people are likely to attempt to maximize their outcomes in interactions and, as a consequence, they are likely to desire control over the decisions that affect their outcomes. In some situations, however, intervention by third parties is necessary to maintain social relationships when disputes occur. Results from their studies show that third-party intervention was paricularly needed when the interests of disputants were extremely noncorrespondent. With such disputes, mediation (where an impartial third party only offers suggestions for resolution without dictating a settlement) was insufficient to resolve the conflict; in contrast, arbitration (where disputants explain their positions but a third party offers a binding settlement) was more successful.

Additional work by Thibaut and Walker refines the development of preferences for two types of arbitration. The *adversarial* procedure allows the disputants and their representatives to control the information-development process while a third party controls the decision, whereas in an *inquisitorial* procedure, a third party controls both information gathering and the decision. Results from several studies indicate that adversarial procedures, compared to inquisitorial ones, clearly enhanced subjective evaluations of the fairness of dispute resolution and, to some extent, enhanced objective evaluations of the elimination of bias and information accuracy. Also, when given the choice among five conflict-resolution methods, subjects who were unaware of what their positions would be in a dispute clearly preferred the adversarial method as a means to ensure fairness in conflict resolution.

Thibaut and Walker (1978) systematize the arguments and results of their earlier work on procedural justice in their "prescriptive" theory (see Lind and Tyler, 1988), which outlines how various procedures should achieve conflict resolution under certain conditions. An important condition involves the type of conflict at issue. They distinguish between conflicts about objective truth (cognitive conflicts) and those involving the distribution of outcomes (conflict of interests). Accuracy of the decision is the most important consequence of a procedure in the case of objective truth whereas the fairness of the decision is the most important consequence when outcomes are at issue.

Thibaut and Walker argue that the procedures used to resolve disputes over outcomes should promote a resolution that is equitable to meet a general prescription for fair distributions. Concern with an equitable distribution directs attention to consideration of individual circumstances or inputs. By vesting control over information collection and presentation in the hands of the disputants, procedural justice will thus be achieved. They note, however, that disputants may present information in a manner favorable to themselves, stemming from egocentric perceptual biases (see Ross, 1977). Recognizing the inaccuracy inherent in allowing disputants process control, they recommend inquisitional procedures for cognitive conflicts. In situations involving both types of conflict, Thibaut and Walker offer a two-stage model, beginning with low disputant process-control to resolve the cognitive conflict, followed by high disputant process-control to remedy the conflict of interests.

The initial work and theoretical framework laid out by Thibaut and Walker guided much subsequent work on procedural justice. But, like studies of allocation preference, perceptions of various dispute resolution procedures depend upon cultural and situational factors that were beyond the scope of the original theory (see Lind and Tyler, 1988). In addition, Lind and Tyler (1988) identify a serious limitation of the theory insofar as it evaluates procedures in terms of the outcomes they produce. They indicate that recent research demonstrates that procedural justice exerts a major influence on assessments of distributive justice. Moreover, as suggested by Leventhal, Karuza, and Fry (1980), fair procedures may be distinct from outcomes, and such procedures characterize situations other than those involving explicit disputes.

Another part of the Leventhal, Karuza, and Fry (1980) theory of allocation preferences revolves around procedural justice. Like their expectancy approach to distributive justice, they propose that individuals prefer the procedural rules that are most likely to fill the most important goals pursued in the allocation situation. Such rules are (1) consistency of procedures across persons and across time; (2) suppression of bias; (3) information accuracy; (4) mechanisms to correct bad decisions; (5) representativeness of participants to a decision; and (6) ethicality standards. A number of studies outside the legal context have identified these rules as fair and have examined their effect procedural justice perceptions (see Lind and Tyler, 1988).

In general, consistency across persons appears to be a strong indicator of procedural justice, whereas correctability is of less importance and more dependent upon the nature of group members' relationships (Barrett-Howard and Tyler, 1986). The importance of representativeness of participants to a decision corresponds to the desire of people to "voice" their opinions. Studies in organizations (e.g., Folger, 1977; and Kanfer, Sawyer, Earley, and Lind, 1987) and in the political arena (e.g., Tyler, Rasinski, and Spodick, 1985) indicate that voice decisions increases perceptions of fairness in the situation.

Lind and Tyler (1988) raise an important issue about the instrumental and noninstrumental functions of voice. They indicate that,

> ...voice could enhance procedural justice because voice is seen as leading to more equitable outcomes (as proposed by Thibaut & Walker, 1978) or because it provides some control over outcomes (as proposed by Brett, 1986). Alternatively, it could enhance procedural justice because it fulfills a desire to be heard and to have one's views considered, regardless of whether the expression influences the decision maker (as proposed by Tyler et al., 1985). (Lind and Tyler 1988, p. 192-93).

Some evidence demonstrates that being able to express opinions alone, without any direct consequences for the decision, affects feeling of inclusion and enhances perceptions of procedural justice (see Lind and Tyler, 1988). Similarly, providing rationale for a decision also seems to result in feelings of inclusion. In the case of procedural injustice, explanations by a decision maker regarding his/her actions enhances evaluations of procedures; emphasis on mitigating circumstances beyond the decision-maker's control is even a stronger basis for ensuring perceived fairness (Bies and Shapiro, 1988). Thus, attributions also play a role in assessments of procedural justice.

Evidence for the instrumental and noninstrumental roles of voice provides the basis for Lind and Tyler's (1988) two models of the development of procedural justice rules. First, building upon the work of Thibaut and Walker (1978) and Leventhal, Karuza, and Fry (1980), the informed self-interest model suggests that people seek control over decisions that affect their own outcomes because of interest in maximizing those outcomes. But when individuals must cooperate with others to obtain outcomes, they grow concerned with fair, rather than maximum, outcomes. Continued cooperation ensures long-term outcomes which, in effect, maximizes a person's own outcomes. To be assured of long-term gains, however, requires a person to demand fairness in decision-making procedures. Fair procedures ultimately reduce conflict and allow the relationship to continue. Thus, based on this model, people are likely to prefer procedures that directly advance their own outcomes, that provide them control over the decisions, or that reduce conflict. In addition, favorable outcomes are likely to enhance perceptions of procedural justice. Evidence demonstrating the effects of procedures independent of outcomes and evidence of the noninstrumental role of voice cast doubt on the completeness of this model.

Lind and Tyler's second approach, the group-value model, attempts to incorporate this evidence. Drawing upon group identification models (e.g., Kramer and Brewer, 1984; and Tajfel, 1978), they suggest that individuals in groups are more likely to act in ways beneficial to all group members than self-interest models would suggest. Assuming the importance of group membership, they stress the influence of the group on values and behaviors. Procedures consistent with group and individual values ensure procedural justice which, in turn, solidifies the group by regulating the group's structure and process. Presumably, individuals learn these procedures through socialization. The content of some rules may vary by culture whereas others may be consistent across cultures. For example, in all groups, people are likely to value participation in the life of their group. As a consequence, procedures allowing voice emerge as fair because they facilitate participation—regardless of outcome control. Thus the group-value model predicts the importance of the noninstrumental use of voice. The pivotal relationship between procedures and the group implies the importance of procedures over that of outcomes.

Besides predicting the nature of fair procedures, the model extends its scope to include violations of procedural justice. Given the centrality of procedures to group life, violations should engender strong feelings of anger and dislike toward whomever produces the injustice. Although evidence is consistent with many of the predictions and a recent study demonstrates the importance of noncontrol aspects of procedural justice consistent with the group-value model

(Tyler, 1989), the independence of the model from outcomes remains a shortcoming. The model fails to account for the observed decrease in evaluations of procedures following negative outcomes.

Lind and Tyler (1988) advocate accepting both models as explanations for procedural justice phenomena. Although each model is incomplete, both account for findings—even those emerging in a single study—of strong effects for outcomes and for noninstrumental voice. Reconciliation of the two models may require studies more specifically examining the underlying processes.

The work of Thibaut and Walker (1975, 1978), Leventhal, Karuza, and Fry (1980), and Lind and Tyler (1988) primarily emphasizes the nature of procedural justice rules and how they develop. Little attention directly focuses on the underlying process of activation, which may be a key to reconciling the self-interest and group-value models. Similarly, while violations of procedural rules result in lower evaluations of procedural justice, the theoretical models hardly examine the question of reactions to injustice.

Consequently, the procedural justice approaches seem to be less extensive than those of distributive justice. The indirect attention to issues of activation and reactions, however, indicates the potential parallels between approaches to distributive and procedural justice. In addition, the foregoing analyses of both procedural and distributive justice reveal the necessity of wrestling with consideration of the individual's motives or goals, perceptions, and the social context. To what extent do researchers who have attempted to combine procedural and distributive justice deal with these issues?

ATTEMPTS TO BRIDGE DISTRIBUTIVE AND PROCEDURAL JUSTICE ROUTES

Two perspectives have attempted to bridge distributive and procedural justice routes. They have done so by focusing explicitly on one section of the route (i.e., on one of the major theoretical questions), although the paving of other sections occurs by implication.

In their theory of allocation preferences, Leventhal, Karuza, and Fry (1980) offer parallel expectancy-based models for choices among both distribution and procedural rules. As described above, their models attempt to explain the development of justice principles. Although they build their perspective on previous empirical research, there have been no specific tests of each model as a judgment process. Also, given the existence of multiple goals, a preference for any one or combined set of distribution or procedure rules does not necessarily ensure justice. In combining the two models to provide a general assessment of justice, Leventhal, Karuza, and Fry state conditions for the emergence of a justice goal in the situation: (1) when role demands emphasize fairness, as for judges or mediators; (2) when other goals, for example, efficiency, productivity, control, are of minimal importance; (3) when current procedures or distributions are blatantly unfair; and (4) when the social system is pluralistic rather than monolithic, drawing attention to the interests of multiple groups.

In pursuit of justice, Leventhal, Karuza, and Fry emphasize distributive justice over procedural justice, owing to the perceptual salience of outcomes. They suggest that procedures are less salient because individuals find them routine or fail to understand them. People may attend to procedural justice concerns, however, when organizations are first formed to ensure the likelihood of fair outcomes or when people are dissatisfied with their outcomes as a means to justify changes in outcome distributions.

In their attempt to specify conditions under which fairness is a major goal, Leventhal, Karuza, and Fry have implicitly incorporated activation of justice concerns into their theory of the development of distribution and procedural rules. In their discussion of conditions under which concerns with procedural justice are likely to be prominent, they imply that their model also extends into the arena of reactions to perceived injustice—an extension not obvious from the model itself—but their integration of major theoretical questions remains skeletal. Some researchers (Lind and Tyler, 1988) question their estimation of the importance of distributive justice given recent research regarding the importance of procedures in their own right. And their specification of conditions under which individuals are likely to attend to justice issues fails to communicate the process by which they analyze the situation and respond. The analysis process underlying activation is addressed to some extent by the referent cognition model, which begins with the question of reactions to injustice rather than the development of justice rules.

Folger's (1986) referent cognition model emphasizes the role of perceptions in understanding reactions to injustice. Arguing that the core problem with equity theory is its failure to consider the way that decisions leading to a distribution are made, Folger attempts to fill this gap by focusing on referent cognitions. These cognitions refer to perceptions of the outcomes that "... people believe they 'would' have obtained...if the decision maker had used other procedures that 'should' have been implemented," (Cropanzano and Folger, 1989, p. 283). By focusing on the means of decision making, Folger tries to explain why people may react very differently to the same inequity.

Folger weaves together elements of referent cognition theory to predict reactions to inequity. The first theoretical element is the referent outcome level, which represents the "distance" between the actual outcome and the referent outcome. A high referent outcome refers to a more favorable outcome that can be imagined, whereas a low referent outcome is a state that is not much better than reality. In effect, these referent outcomes are akin to internal standards of justice based on individual past experiences (e.g., Austin, McGinn, and Susmilch, 1980; and Weick, 1966) or may stem from knowledge of referential standards (Berger, Zelditch, Anderson, and Cohen, 1972). Individuals with high referent outcomes are more likely to find inequity distressing. The second element is the perceived likelihood of amelioration, which pertains to the ability to remedy the situation in a way that decreases the severity of the consequences. A low likelihood of amelioration indicates that the situation is relatively unchangeable (and, thus, inequity consequences remain severe) and a high likelihood of amelioration suggests that remedies exist to improve the situation so that low outcomes are transient. The greater the probability of amelioration, the less likely an individual will express inequity distress. Folger, Rosenfield, Rheaume, and Martin (1983) confirm the direct effects of these elements as well as an interaction; low-referent subjects expressed less resentment than did high-referent ones when amelioration likelihood was low.

A third element of the theory concerns the extent and type of justification for the inequity. Whether or not there is a good reason for the outcome levels influences individuals' reactions to inequity. Discrepancy between actual and referent outcome levels may result in dissatisfaction but the extent of justification for the result determines whether feelings of resentment over the inequity will emerge. Concern with justification leads to examination of the procedures or "instrumentalities" leading to the distribution of outcomes. Instrumentalities function in a manner similar both to accounts, as described by Bies and Shapiro (1988) regarding procedural justice, and to attributions, as analyzed by Utne and Kidd (1980) regarding distributive justice. But, Folger specifies two types of instrumentalities: (1) actual instrumentalities, referring to procedures individuals perceive as responsible for their outcomes; and (2) referent instrumentalities, pertaining to what individuals imagine could have happened in the situation. Comparisons between these two types of instrumentalities result in assessments of the extent to which the actual instrumentalities are justified. If the imaginable procedures seem more reasonable, then people may judge the actual ones as unjustified; that judgment, especially coupled with dissatisfaction over outcomes, results in strong feelings of resentment. Results from Folger and Martin (1986) bear out these expectations, showing a significant interaction between referent level and justification when there is a single allocation. But when the situation suggests an enduring relationship between the decision-maker and similar recipients (i.e., multiple allocations by the same person), concerns with procedural justice alone affect feelings of resentment.[3]

The work of Folger and his colleagues complements prior developments in equity theory by providing greater understanding of the cognitive mechanisms that may lead two individuals to respond differently to the same inequity. In doing so, the theory ties together concerns with procedural and distributive justice and clarifies, to some extent, conditions leading to the priority of concerns with procedures over outcomes. The theory, however, hardly spells out assumptions of cognitive processing that would explain the effects of the three elements. In addition, the lack of specific assumptions regarding individual motivations inhibits understanding of how these cognitive ruminations influence the nature of behavioral reactions to injustice.

PAVING A SINGLE ROUTE FOR DISTRIBUTIVE AND PROCEDURAL JUSTICE

The approaches of Leventhal, Karuza, and Fry and Folger, although untested or incomplete, attest to the possibility of constructing a single route for the understanding of distributive and procedural justice. Central to both justice arenas are judgments or evaluations of justice—their development, their relevance or activation in a situation, and responses to evalutions indicating injustice. These commonalities, in turn, provide the groundwork for a single route for examining distributive and procedural justice.

Commonalities in the Routes Already Taken

Not surprisingly, there are both concrete and abstract similarities in the paths taken by distributive and procedural justice researchers. The concrete similarities refer to how the empirical work progessed, whereas the abstract commonalities pertain to underlying assumptions and facets of theory development. Empirical results helped to shape the latter, which are particularly useful in forging a single path.

On the concrete level, there is evidence that in both domains there were explicit attempts to broaden the scope of contexts to which issues applied. In distributive justice, this entailed movement away from organizational contexts to personal contexts, as well as to general social problems, including those with legal repercussions such as crime and sentencing. The trend was the reverse in procedural justice, with its origins in the legal arena and then its expansion to organizational, interpersonal, and political contexts.

Concomitant with the expansion of contexts was movement away from unidimensional ways to conceptualize justice, that is, in terms of equity, in the case of distributive justice, and in terms of decision control, in the case of procedural justice. The work of Leventhal (1976; and Leventhal, Karuza, and Fry, 1980) especially drew attention to the multidimensionality of justice rules. The multidimensional view, in conjunction with different contexts highlighted the conditional nature of justice and ultimately raised questions about people's motivations and perceptions in justice situations. Variation in motivations and perceptions potentially lead to differential assessments of and reactions to injustice. Indeed, an abstract understanding of motivations and perceptions is critical for formulating theoretical approaches to justice issues.

Three abstract commonalities characterize the progressive nature of the work in distributive and procedural justice. First, early researchers cast their arguments regarding the development of justice principles in terms of a social exchange approach. Homans (1961 and 1974), Adams (1965), Blau (1964), and, later, Walster, Berscheid, and Walster (1978) drew upon the notion of relationship contributions and receipts to justify the development of the equity principle. Thibaut and Walker's (1975) application of exchange principles regards their assumption that people perceive equitable outcomes as fair. Both exchange applications imply emphasis on outcomes as central to the justice process. Conceptualization of the breakdown in exchange, however, varies across the perspectives.

For Thibaut and Walker's (1975) approach, conflict of interests represents breakdown in the exchange. Intervention by a third party is their proposed means to resolve the conflict. In contrast, inequity epitomizes the breakdown in exchange in distributive justice perspectives. Rather than turning to a third party (although Walster, Berscheid, and Walster [1978] argue that this is possible), Adams (1965) proposes ways to restore equity that are within the relationship between exchange partners. He posits, furthermore, that individuals will select the least costly method.

The assumed strategy of cost minimization in selection of methods to restore injustice highlights another aspect of the social exchange roots of justice perspectives. These perspectives assume that individuals are self-interested, maximizing their outcomes. Walster, Berscheid, and Walster (1978) explicitly state that in the pursuit of self-interest, social groups develop principles for equitably administering rewards while others (Blau, 1964; and Homans, 1961 and 1974) imply self-interest in the arguments regarding the development of justice rules. Leventhal, Karuza, and Fry (1980) indicate that self-interest is one possible goal to pursue in allocation situations. Their expectancy formulation, furthermore, suggests that individuals select rules that ensure fulfillment of the most important goal—that is, they are rational actors (see Heath, 1976). Similarly, Thibaut and Walker (1975) use the self-interest assumption to justify individuals' desire for control over the decisions that affect their outcomes. Such control provides an efficient means to attain the given goal and thus may be interpreted as a rational act.

To contend that self-interested behavior gives rise to principles of distribution and procedure that people come to perceive as just does not equate self-interest with justice. Rather, as Lerner (1982) emphasizes, justice is unique from self-interest insofar as it draws attention to group welfare (see also Deutsch, 1985). Moreover, empirical work involving manipulations of contextual factors demonstrates that individuals sometimes prefer distributions that appear generous to others or procedures that do not provide control over decisions. Interpretations of the motives underlying these preferences suggest that individuals are not maximizing their material outcomes but rather enhancing group welfare. This movement toward emphasis on group welfare as a defining feature of fair distributions and fair procedures constitutes a second theoretical commonality between the justice domains.

As others (Eckhoff, 1974; and Leventhal, 1980) have noted, however, it is empirically difficult to discern fair behavior, involving the normative application of a justice principle, from quasifair behavior, stemming from the strategic use of a justice principle to obtain self-interested ends. In situations involving corecipients as decision-makers, complications arise partly from the pursuit of social rewards as well as material ones, as noted above. Use of third party allocators or decision-makers who are instructed to act fairly enhances the likelihood of just outcomes or procedures—those that promote group welfare. The assessments of both first- and third-party allocators and decision-makers, however, also depend upon their perceptions in the situation. In fact, underlying motivations may interact with interpretations of information available in the situation, such as when people make egocentric attributions for their own behavior (see, e.g., Ross, 1977).

Thus, a third theoretical commonality of the development of distributive and procedural justice approaches is their increasing emphasis on cognitive processing and social perceptions. The notion of reward expectations, derived from social comparisons, is an element of cognitive processing underlying the activation of the justice judgments predicted by distributive justice approaches. Issues of perception also characterize the research regarding the relevance and weighting of inputs and outcomes in the equity formula. Ultimately, what people perceive as just depends upon the comparisons that they invoke and how they interpret situational information; different perceptions of comparisons or information lead to differential assessments of injustice. Tempering reactions to the injustice, furthermore, are attributions regarding the source of distributive injustice. Similarly, accounts for the existence of procedural injustices affect ratings of the fairness of those procedures. Folger's (1986) referent cognition model emphasizes the importance of cognitions—even those representing hypothetical, prescriptive scenarios—in assessments and reactions to both distributive and procedural injustice.

Given the increasing attention in the 1970s to social perception (see, e.g., Schneider, Hastdorf, and Ellsworth, 1979; and Bierhoff, 1989), it is hardly surprising that justice researchers have incorporated such concerns into their work. More importantly, such analysis of perceptions provides a basis for understanding the individual's differential assessments of justice, potentially leading to conflict over what is just, and differential reactions to the same injustice. Including the cognitive components of the assessment process rounds out theories of justice, answering otherwise unresolved questions.

The parallel paths of distributive and procedural justice research, suggested by these concrete and abstract commonalities as well as by general questions regarding development of principles, activation of justice concerns, and responses to injustice, raises an additional question. Can theories of distributive and procedural justice be integrated? The attempted responses by Leventhal, Karuza, and Fry (1980) and Folger (1986) attest to beliefs in the potential usefulness of such an integration. Drawing upon the engineering of these first "integrationists," the intent of the next section is to construct the contours of a single route to distributive and procedural justice.

Paving a Single Route to Distributive and Procedural Justice

The paving of a single route to distributive and procedural justice requires integrated theorizing about each type of justice, with special attention to specification of motivations, perceptions, and comparison processes leading up to the judgments and subsequent reactions. In addition, the route must identify factors that distinguish the operation of procedural versus distributive concerns. As one possible strategy for combining distributive and procedural justice, the route sketched here begins with the question of the development of justice principles in small groups and continues with their application to assess whether justice obtains in a specific situation. The intention is that the same assumptions and scope conditions represent paving stones for responses to the three major theoretical questions. Despite the preliminary nature and skeletal form of the proposed route, its consideration should raise questions for further debate and research.

Scope Conditions

The proposed route is circumscribed by several scope conditions. First, the actors are socialized, carrying with them values of their social group and knowledge of role expectations. Second, these actors are engaged in exchange or allocation. Third, there is a minimal level of interdependence between actors such that one's outcomes are in some way tied to those of the other(s). Fourth, disputes arising in the situation pertain to interests, not "objective truths," as in Thibaut and Walker's (1978) notion of cognitive conflicts. Fifth, the justice preferences or evaluations are from the point of view of a corecipient of the exchange or allocation. And, finally, external sources do not explicitly constrain actors to be fair.

Core Assumptions

The core assumptions pertain to the critical issues of individual motivation and perceptions underlying justice assessments. Such assumptions are often implicit in justice perspectives. Explicit specification here as part of preliminary theory construction is a necessary step in deriving predictions. Empirical falsification of the derivations may ultimately cast doubt on the assumptions.

Beginning like the exchange and expectancy approaches to justice rule development, the first core assumption is that individuals are rational actors, preferring the most efficient means to their goal within a situation. The second assumption defines the goal as one of self-interest. Individuals attempt to maximize their outcomes by considering the balance of material and social rewards against similar types of costs. Although this assumption may seem incongruent with the spirit of justice per se, existing empirical evidence, as reviewed above, suggests its viability (see also Hegtvedt, 1992).

The third core assumption is that individuals tend to prefer cognitive simplicity to cognitive complexity in their information processing. It is unlikely that individuals consider all relevant information in the situation owing to its complexity, as well as to motivational and nonmotivational information processing biases. Egocentric biases, in particular, may plague assessment of one's own inputs and outcomes.

Development of Justice Principles

In order to predict the emergence of particular justice principles, individual motivations must be linked to the situational context. A fourth assumption is thus that the type of distribution or procedural rule that individuals prefer depends upon the salience of material or social rewards in the allocation or exchange situation. If the situation emphasizes material rewards (and costs), then rule preferences will explicitly reflect material self-interest (determined, for example, by position in the group or relative performance). In contrast, if the situation highlights potential social rewards (and costs), then rule preferences are more likely to correspond to whatever promotes group welfare and justice.

Two interrelated sets of factors affect the salience of material and social rewards. The first is that which Lind and Tyler (1988) attempt to capture in their group-value model: the individual's attachment to the group and acceptance of its values, concomitant with a socialized concern with group cooperation and solidarity. This general socialization background includes knowledge of the types of distributions and procedures generally considered fair in the group. For example, the ideologies of some groups tend to emphasize the needs of the individual over those of the group; accordingly, such groups are more likely to teach individualistic principles of justice, such as equity, over group-oriented ones, like equality or distribution according to needs. Group ideologies might also dictate procedural elements, such as when groups emphasize democracy or autocracy; in the former case, for example, individuals are likely to accept that participation in decisions is fair. Furthermore, in becoming socialized members of groups, individuals are likely to learn role responsibilities and obligations and to learn to discriminate between situations demanding the enactment of those role expectations.

Specific situational circumstances constitute the second set of factors relevant to the salience of material and social rewards and costs. The content of individuals' socialization, and, in particular, internalized group values, is likely to condition the meaning of the situation and their role in it. For example, values promoting group heterogeneity (e.g., in race, sex, and/or status) rather than group homogeneity may lead to different perceptions of relevant inputs in an allocation situation, and, consequently, different allotments to recipients. Consistency with the homogeneity value would lead to weighing more heavily inputs of people similar to the allocator resulting in unequal reward distributions, such as when a white, male employer pays lower wages to minorities for the same work as performed by recipients like himself.

Regardless of how specific group values interact with situational conditions, it is likely that self-interested individuals are more favorably inclined toward members of their own group owing to socialized values or immediate circumstances (see Tajfel, 1978). Thus, specific conditions that are likely to enhance attention to social rewards and costs include (1) salience of group membership; (2) group goals promoting cooperation; (3) affective bonds between individuals; (4) the expectation of a long-term relationship; (5) role demands of actors requiring impartially or recognition of their responsibility to others; and (6) the personalized nature of the setting. In addition, attributions may facilitate salience of social aspects of the situation. Knowing that the source of an actor's low inputs or other undesirable behavior is beyond his/her control may stimulate an empathetic response in the allocator. Such empathy may direct attention to the welfare of other group members and as a consequence make it less likely that that an individual's cognitive processing of information will be clouded by motivational biases.[4] To the extent that these conditions enhance concern with social rewards and costs, individuals are more likely to prefer both distribution and procedural rules that appear just.

In the development of justice principles, given that self-interested motivation rests upon outcome assessments, it is expected that individuals are more likely to be concerned initially with the creation of just distributions than with the creation of just procedures. This assumption, though consistent with Leventhal, Karuza, and Fry's (1980) argument pertaining to the salience of outcome distributions, contrasts with their suggestion that procedures are more salient when organizations are being formed. Instead, stemming from Lind and Tyler's (1988) group-value model, procedures gain attention when individuals are invested in the group. Moreover, failure to achieve a just distribution may lead to greater attention to developing fair procedures. For example, in agrarian societies, the feudal lord distributed grain to workers; as long as workers received what they considered an adequate share, they were unlikely to question the centralized allocation system. Such inattention to procedures, however, is more likely in the absence of knowledge of alternative procedures that would be more fair. Thus, development of procedural justice principles may take precident over development of distributive justice ones in the context of outcome injustice, the continuing organization of groups, and with imaginable alternative procedures.

The acceptance of a given distribution or procedure as a normative standard of justice depends not only on the preferences of individuals initially involved, but also on the repeated application of the rule and continued acceptance of its consequences. Homans (1961, 1974) captures this process by claiming, "what is becomes what ought to be." And, Berger, Zelditch, Anderson, and Cohen's (1972) conceptualization of how individuals develop expectations about how rewards "go with" levels of relevant characteristics suggests the development of normative justice rules.

Activation of Justice Concerns

The existence of specific distribution and procedural rules that are considered to be fair for given circumstances provides the basis for the comparisons activating justice concerns. Activation depends on interrelated processes of social perception and comparisons. Individuals must perceive relevant comparisons and assess information inherent in those comparisons.

Initially, when individuals assess an existing distribution or procedure as fair or unfair, they are likely to compare it with what they expected the situation to hold. The applicable normative justice distribution rule or procedure, their past experience (i.e., internal standards), their knowledge of the experience of a similar other or a group of similar others (i.e., local and referential standards), and even what they imagine should have been the case (i.e., referent cognitions) constitute potential bases for their expectations. The myriad of potential comparisons and the potential influence of cognitive biases creates a rather complex environment for justice evaluations. Törnblom (1977) has attempted a theoretical sorting of the various combinations of internal, local, and referential comparisons underlying justice judgments and subsequent reactions, but this work has generated little empirical research. More typically, empirical studies in both distributive and procedural justice attempt to limit the information available to subjects regarding potential comparisons.

While such a strategy fails to answer some theoretical questions, it is consistent with the third core assumption, that individuals tend to prefer cognitive simplicity in information processing. As a consequence of cognitive simplicity coupled with the rational strategy of minimizing costs, individuals are likely to perform only a select subset of the possible comparisons. Their comparison choices are likely to depend upon availability of information for each type of comparison and their perception of which comparison(s) will ensure achievement of self-interested outcomes. In addition, they may unknowingly bias their perception of information in order to achieve the self-interested goal, or may knowingly bias their perceptions in order to achieve consistency among comparisons that serve to achieve their goal.

For example, if situational conditions lead actors to focus more on the social rather than the material aspects of the exchange or allocation consequences (i.e., they appear concerned with just distributions or procedures) and information on all comparisons is readily available, they are more likely to emphasize the comparison of actual outcomes or procedures to those entailed by an applicable normative justice principle or implied by referential standards. These comparisons are consistent with considerations of group welfare because they represent consensually accepted rules and consistency across similar people, not unique instances. On the other hand, if people are attending primarily to their own material outcomes, they are likely to invoke the comparison that promises them the largest returns. They may, furthermore, exaggerate their previous experience with outcomes or procedures or misperceive information about others' experiences in order to achieve their goal. In effect, such individuals are using their justice evaluations strategically to ensure themselves either better outcomes or more control over outcomes.

Whether distributive or procedural concerns take precedence depends upon several features of the situation: (1) the availability of comparisons for each; (2) the stage in group organization; and (3) the existing evaluation of distributive injustice. With more information or more salient referent cognitions for outcomes, distributive justice becomes more important, whereas if the information or referent cognitions procedures are more extensive, procedural justice takes precedence. The extent to which the group perceives itself as an ongoing entity may increase attention to procedural justice over distributive justice. And, an initial evaluation of an unfair distribution is more likely to lead to an assessment of underlying procedures than an initial evaluation of procedural injustice to lead to a review of the distribution. The emphasis on distributive justice over procedural justice stems from the self-interested concern with outcomes and with the expectation that the assessment of procedures will provide a basis for determining how to respond to the distributive injustice.

Reactions to Injustice

Accompanying evaluations of distributive or procedural injustice are three interrelated types of reactions: emotional, cognitive, and behavioral. The failure of actual outcomes or procedures to fulfill expectations produces distress, as argued by equity theorists and others. The nature of the distress—its emotional components—varies, depending upon the direction and the extent to which expectations remain unfulfilled (Adams, 1965; and Homans, 1961 and 1974). The direction refers to whether the actual situation is better or worse than expected. A self-interested individual is more likely to respond to injustice that results in a worse than expected situation (unless the better situation is at the expense of others). The extent of injustice involves the "distance" between the actual and expected situation (similar to Folger's [1986] use of the term) as well as whether individuals perceive only distributive, only procedural, or both types of injustice. The greater the distance and the more detrimental the injustice, the more likely individuals will feel negative emotions (e.g., anger, indignation, dissatisfaction). To remove the unpleasant emotional state, which adds to the costs of injustice, people are likely to assess potential strategies to remedy their plight.

The assessment of justice restoration strategies requires cognitive review of the situation. Following the arguments of Utne and Kidd (1980) and Folger (1986), attributional analysis of the cause of the injustice, examination of justifications offered by the perpetrators of the injustice, and assessment of the probability of rectifying the situation (amelioration) are likely to affect the chosen strategy. As Adams (1965) suggested, rational individuals are likely to choose the least costly method of restoration, thereby serving their own interests. If the cognitive analysis results in realization that the cause of the injustice was unintentional or that the justification for it was accepted, and that the probability of rectifying the situation is low, then individuals are likely to use cognitive distortion to relieve their distress. Intentional and unjustified injustice are likely to encourage the most vehement behavioral response, but only if individuals perceive the possibility of amelioration. An exception to the above pattern of reactions may arise if individuals are focusing primarily on their own material benefits and are using justice arguments strategically; under such conditions, if the potential for amelioration is high, then actors may distort their perception of the cause of the injustice and its justification in order to demand higher outcomes or more control over procedures.

Whether procedural or distributive injustice elicits the strongest responses is likely to depend on the results of the cognitive assessment. If the injustice pertains to only one type of justice, then responses will focus only on that type. In the presence of both types of injustice, individuals are likely to concern themselves first with the procedural injustice, which, if remedied, may be an efficient means also to resolve distributive injustice. The reverse strategy—attention to distributive injustice first—does little to address procedural unfairness and, thus, would be a costly way to proceed. Even if the remedy for procedural injustice is noninstrumental voice, being able to have a say not only promulgates group values (as Lind and Tyler [1988] might argue) but also may reduce the emotional costs of injustice.

CONCLUDING REMARKS

This paper outlines a single route to distributive and procedural justice. Based on the reviews of existing theoretical perspectives in both arenas, the single route begins with the commonalities of those perspectives. It furthermore builds upon the foundations offered by Leventhal, Karuza, and Fry (1980) and Folger (1986) for integrating distributive and procedural justice concerns. The core assumptions—rationality, self-interest, and cognitive simplicity—address the critical issues of individual motivation and perceptions which implicitly or explicitly underlie any discussion of justice. Moreover, these assumptions in conjunction with others provide the basis for deriving predictions regarding each of the major theoretical questions.

As stated in the conditions necessary to predicting which justice principles will develop and, to some extent, in discerning the differential ascendency of procedural over distributive justice issues, situational factors enhancing concern for the social links among group members are of critical importance. Such links condition not only the nature of self-interest pursued in the situation, but may also affect the interpretations of inputs, outcomes, and other components necessary to the assessment of justice.

Undoubtedly, the single route to distributive and procedural justice specified above is filled with potholes and rubble. To withstand the potential traffic, the route requires greater formalization and specification of derivations. A systematic comparison of derived predictions with existing empirical work would provide a greater sense of the durability of the proposed route. And, as alluded to above, it may be necessary to construct more fully the cognitive processing which links motivations to situational factors and both to evaluations of justice. The outline of the proposed route, however, serves to highlight the parallel paths of distributive and procedural justice and to reinforce the connection between means and ends.

ACKNOWLEDGMENT

This is a revised version of a paper presented at the American Sociological Association meetings, Cincinnati. The author would like to thank the editors of this volume for their helpful comments on this paper.

NOTES

1. These skeletal reviews should familiarize the reader with major theories and research in both distributive and procedural justice. Familiarity with these issues is necessary to understanding arguments later in the paper. Readers already familiar with these theoretical domains may wish to continue reading with the section, "Attempts to Bridge Distributive and Procedural Justice Routes."

2. The work of Jasso (1980 and 1989) has been of critical importance in drawing attention to variation in the degree of injustice. Details of her formulation, however, are beyond the scope and needs of the argument presented here.

3. Cropanzano and Folger (1989) demonstrate how how different types of procedures, not simply justifications for the procedures, affect evaluations of injustice. They examined whether having voice in a decision about task performance affected subjects' feelings about a resulting inequity. Results indicated that the voice condition decreased feelings of unfair treatment; only subjects who were denied voice and who had high referent outcomes expressed resentment.

4. Nonmotivational information processing biases, however, may affect interpretation of the situation and subsequent preferences for those pursuing primarily material rewards as well as those seeking social rewards.

REFERENCES

Adams, J. Stacy. 1965. "Inequity in Social Exchange." *Advances in Experimental Social Psychology* 2:267-299.

Alexander, Sheldon and Marian Ruderman. 1987. "The Role of Procedural and Distributive Justice in Organizational Behavior." *Social Justice Research* 1:177-198.

Austin, William, N.C. McGinn, and C. Susmilch. 1980. "Internal Standards Revisited: Effects of Social Comparisons and Expectancies on Judgments of Fairness and Satisfaction." *Journal of Experimental Social Psychology* 16:426-441.

Barrett-Howard, Edith and Tom R. Tyler. 1986. "Procedural Justice as a Criterion in Allocation Decisions." *Journal of Personality and Social Psychology* 50:296-304.

Berger, Joseph, Morris Zelditch, Jr., Bo Anderson, and Bernard P. Cohen. 1972. "Structural Aspects of Distributive Justice: A Status Value Formation." Pp. 119-146 in *Sociological Theories in Progress,* edited by J. Berger, M. Zelditch, and Bo Anderson. Boston: Houghton Mifflin.

Bierhoff, Hans Werner. 1989. *Social Perception.* New York: Springer-Verlag.

Bierhoff, Hans Werner, Ronald L. Cohen, and Jerald Greenberg, eds. 1986. *Justice in Social Relations.* New York: Plenum Press.

Bies, Robert J. and Debra L. Shapiro. 1988. "Voice and Justification: Their Influence on Procedural Fairness Judgments." *Academy of Management Journal* 31:676-685.

Blau, Peter M. 1964. "Justice in Social Exchange." *Sociological Inquiry* 34:193-246.

Cohen, Ronald L. 1982. "Perceiving Justice: An Attributional Perspective." Pp. 119-160 in *Equity and Justice in Social Behavior,* edited by J. Greenberg and R.L. Cohen. New York: Academic Press.

Cook, Karen S. and Karen A. Hegtvedt. 1983. "Distributive Justice, Equity, and Equality." *Annual Review of Sociology* 9:217-241.

Cook, Karen S. and Toshio Yamagishi. 1983. "Social Determinants of Equity Judgments: The Problem of Multidimensional Input." Pp. 95-126 in *Equity Theory: Psychological and Sociological Perspectives,* edited by D.M. Messick and K. S. Cook. New York: Praeger.

Cropanzano, Russell and Robert Folger. 1980. "Referent Cognitions and Task Decision Autonomy: Beyond Equity Theory." *Journal of Personality and Social Psychology* 74:293-299.

Deutsch, Morton. 1975. "Equity, Equality, and Need: What Determines Which Value Will Be Used for Distributive Justice?" *Journal of Social Issues* 31:137-150.

_____. 1985. *Distributive Justice: A Social Psychological Perspective.* New Haven, CT: Yale University Press.

Eckhoff, Torstein. 1974. *Justice: Its Determinants in Social Interaction.* Rotterdam: Rotterdam Press.

Farkas, Arthur J. and Norman H. Anderson. 1979. "Multidimensional Input in Equity Theory." *Journal of Personality and Social Psychology* 37:879-896.

Folger, Robert. 1977. "Distributive and Procedural Justice: Combined Impact of 'Voice' and Improvement on Experienced Inequity." *Journal of Personality and Social Psychology* 35:108-119.

Folger, Robert (Ed.). 1984. *The Sense of Injustice: Social Psychological Perspectives.* New York: Plenum Press.

————. 1986. "Rethinking Equity Theory: A Referent Cognition Model." Pp. 145-163 in *Justice in Social Relations,* edited by H. Beirhoff, R. L. Cohen, and J. Greenberg. New York: Plenum.

Folger, Robert and Chris Martin. 1986. "Relative Deprivation and Referent Cognitions: Distributive and Procedural Justice Effects." *Journal of Experimental Social Psychology* 22:531-546.

Folger, Robert, D. Rosenfield, K. Rheaume, and Chris Martin. 1983. "Relative Deprivation and Referent Cognitions." *Journal of Experimental Social Psychology* 19:172-184.

Greenberg, Jerald. 1987. "Reactions to Procedural Injustice in Payment Distributions: Do the Means Justify the Ends?" *Journal of Applied Psychology* 72:55-61.

Greenberg, Jerald and Ronald L. Cohen (Eds.) 1982. *Equity and Justice in Social Behavior.* New York: Academic Press.

Greenberg, Jerald and Tom R. Tyler. 1987. "Why Procedural Justice in Organizations?" *Social Justice Research* 1:127-142.

Heath, Anthony. 1976. *Rational Choice and Social Exchange: A Critique of Exchange Theory.* New York: Cambridge University Press.

Hegtvedt, Karen A. 1992. "When is a Distribution Rule Just?" *Rationality and Society* 4:308-331.

Homans, George C. 1961, 1974. *Social Behavior: Its Elementary Forms.* New York: Harcourt, Brace and World.

Jasso, Guillermina. 1980. "A New Theory of Distributive Justice." *American Sociological Review* 45:3-32.

Jasso, Guillermina. 1989. "The Theory of Distributive-Justice Force in Human Affairs: Analyzing Three Questions." Pp. 354-387 in *Sociological Theories in Progress,* Volume 3, edited by J. Berger, M. Zelditch, Jr., and B. Anderson. Newbury Park, CA: Sage Publications.

Kanfer, Ruth, John Sawyer, P. Christopher Earley, and E. Allan Lind. 1987. "Fairness and Participation in Laboratory Procedures: Effects on Task Attitudes and Performance." *Social Justice Research* 1:235-249.

Kramer, Roderick M. and Marilyn B. Brewer. 1984. "Effects of Group Identity on Resource Use in a Simulated Commons Dilemma." *Journal of Personality and Social Psychology* 46:1044-1057.

Lerner, Melvin J. 1982. "The Justice Motive in Human Relations and the Economic Model of Man: A Radical Analysis of Facts and Fictions." In *Cooperation and Helping Behavior: Theories and Research,* edited by V. Derlega and J. Grzelak. New York: Academic Press.

Lerner, Melvin J. and Sally C. Lerner. (Eds.) *The Justice Motive in Social Behavior.* New York: Plenum Press.

Leventhal, Gerald S. 1976. "Fairness in Social Relations." Pp. 211-239 in *Contemporary Topics in Social Psychology,* edited by J. W. Thibaut, J. T. Spence, and R. C. Carson. New Jersey: General Learning Press.

————. 1980. "What Should Be Done with Equity Theory? New Approaches to the Study of Fairness in Social Relations." Pp. 27-55 in *Social Exchange Theory,* edited by K.J. Gergen, M.S. Greenberg, and R.H. Willis. New York: Wiley.

Leventhal, Gerald S., J. Karuza Jr., and W. R. Fry. 1980. "Beyond Fairness: A Theory of Allocation Preferences." Pp. 167-218 in *Justice and Social Interaction,* edited by G. Mikula. New York: Springer-Verlag.

Lind, E. Allan and Tom R. Tyler. 1988. *The Social Psychology of Procedural Justice.* New York: Plenum Press.

Messick, David M. and Karen S. Cook. (Eds.) 1983. *Equity Theory: Psychological and Sociological Perspectives.* New York: Praeger.

Messick, David M. and Kenneth P. Sentis. 1979. "Fairness and Preference." *Journal of Experimental Social Psychology* 15:416-434.

Mikula, Gerold. (Ed.) 1980. *Justice and Social Interaction.* New York: Springer-Verlag.

Musante, L., M. A. Gilbert, and J. Thibaut. 1983. "The Effects of Control on Perceived Fairness of Procedures and Outcomes." *Journal of Experimental Social Psychology* 19:223-238.

Rawls, John. 1971. *A Theory of Justice.* Boston: Harvard University Press.

Reis, Harry T. 1981. "Self Presentation and Distributive Justice." Pp. 269-291 in *Impression Management Theory and Social Psychological Theory,* edited by J.T. Tedeschi. New York: Academic Press.

———. 1984. "The Multidimensionality of Justice." Pp. 25-62 in *The Sense of Injustice,* edited by R. Folger. New York: Plenum.

Ross, Lee. 1977. "The Intuitive Psychologist and his Shortcomings: Distortions in the Attribution Process." *Advances in Experimental Social Psychology* 10:174-220.

Schneider, David J., Albert H. Hastorf, and Phoebe C. Ellsworth. 1979. *Person Perception.* New York: Addison-Wesley.

Schwinger, Thomas. 1980. "Just Allocations of Goods: Decisions Among Three Principles." Pp. 95-126 in *Justice and Social Interaction,* edited by G. Mikula. New York: Springer-Verlag.

Tajfel, Henri. 1978. *Differentiation Between Social Groups: Studies in the Social Psychology of Intergroup Relations.* New York: Academic Press.

Thibaut, John and Laurens Walker. 1975. *Procedural Justice: A Psychological Analysis.* Hillsdale, NJ: Lawrence Erlbaum.

———. 1978. "A Theory of Procedure." *California Law Review* 66:541-566.

Törnblom, Kjell. 1977. "Magnitude and Source of Compensation in Two Situations of Distributive Justice." *Acta Sociologica* 20:75-95.

Tyler, Tom. 1989. "The Psychology of Procedural Justice: A Test of the Group-Value Model." *Journal of Personality and Social Psychology* 57: 830-838.

Tyler, Tom, K. Rasinski, and N. Spodick. 1985. "The Influence of Voice on Satisfaction with Leaders: Exploring the Meaning of Process Control." *Journal of Personality and Social Psychology* 48:72-81.

Utne, Mary Kristine and Robert Kidd. 1980. "Equity and Attribution." Pp. 63-93 in *Justice and Social Interaction,* edited by G. Mikula. New York: Springer-Verlag.

Vermunt, Reil and Herman Steensma. (Eds.) 1990. *Social Justice in Human Relations.* New York: Plenum Press.

Weick, K. E. 1966. "The Concept of Equity in the Perception of Pay." *Administrative Science Quarterly* 11:414-439.

Walster, Elaine, Ellen Berscheid, and George W. Walster. 1978. *Equity: Theory and Research.* Boston: Allyn and Bacon.

AFFECT AND THE PERCEPTION OF INJUSTICE

Steven J. Scher and David R. Heise

ABSTRACT

Traditional approaches to distributive justice have seen the determination of whether or not a distribution of rewards is fair as a cognitive process, with emotion entering the process only as an outcome of a decision that the distribution was unjust. In this paper, we propose a modification of this view, namely, we propose that justice is not calculated unless the actor feels a justice-related emotion (anger or guilt). These emotions, which arise in the course of social interaction, lead to the instigation of justice deliberations. Using Affect Control Theory, we explain how the justice-related emotions could arise in situations that traditional models of justice would characterise as unjust. Thus, our theory is able to account for the existing literature on justice. We then show how our theory suggests several novel implications about situations that would be seen as unjust. Comparisons of our model to related models of justice are also discussed.

AFFECT AND THE PERCEPTION OF INJUSTICE

Any social interaction can be seen, from an individual's point of view, as a contribution of inputs, and a receipt of outputs. This is clearest in economic

Advances in Group Processes, Volume 10, pages 223-252.
Copyright © 1993 by JAI Press Inc.
All rights of reproduction in any form reserved.
ISBN: 1-55938-280-5

relationships where one party pays a second party for doing some work. More controversially, perhaps, we can view friendships and romances in a similar way. Each member in a social relationship makes contributions and receives certain benefits. In a marriage, for example, the husband may regularly "contribute" a certain amount of money, love and affection, house cleaning, and grocery shopping to the relationship. In turn, he receives from his wife additional money from her salary, her love and affection, cooking, laundry, and child care. Each member of the dyad contributes and each receives benefits. A group also can be seen from this perspective. Each group member contributes to the mutual goals of the group (even if the goal is merely continued existence), and in return, each member of the group receives certain benefits of membership.

An important corollary of this exchange perspective is that the motivation of individuals to remain in the relationship (i.e., to keep the job, to stay in the marriage, or to remain a member of the group) will depend on the extent to which they perceive their rewards as *fair* or *just*.[1] Perception of unfair rewards can lead to termination of relationships, or dissolution of groups. It also may lead to rebellions, in which control of the distribution of rewards is removed from the current allocators, in order to establish procedures for allocating rewards that are perceived as more just. Thus, the study of the causes and consequences of fairness judgments—the study of distributive justice— becomes central for understanding whether social relationships live or die, and whether economic and political structures engender satisfaction or unrest and conflict.

Of course, the importance of these types of judgments has not escaped the notice of sociologists and psychologists interested in group processes and interpersonal behavior. Research and theory in this area has progressed over three decades from relatively simple views of how people decide whether or not a distribution is just, to formal and increasingly sophisticated models of the distributive justice process.

The earliest influential theories of distributive justice were the equity theories (Adams, 1965; Homans, 1961; Walster, Berscheid, & Walster, 1973, and Walster, Walster, & Berscheid, 1978). Homans' (1961) simple formula for justice stressed the difference between the rewards (R) people receive for some input (I). Adams' (1965) more influential theory of equity proposed that a distribution would be considered just to the extent that the ratio of an individual's rewards to his or her inputs (or investments) was equal to a similar ratio for another person (i.e. $R_A/I_A = R_B/I_B$, where subscripts A and B denote two people participating in the distribution). From this simple beginning, various theorists have developed theories of equity, which have encompassed ever-broader issues, and have grown in mathematical complexity.

Jasso (1978), for example, empirically derived a logarithmic function for evaluating justice. Thus,

$$justice\ evaluation = \theta \ln \left(\frac{R_A}{R_B}\right). \tag{1}$$

In this and other specifications, R_B represents the reward level of any standard of comparison, which may or may not be an actual other person. Theta (θ) represents a "signature constant" (Jasso, 1980)—a unique weighting factor for each individual.

Although derived empirically, the logarithmic form for the justice evaluation function meets several important theoretical criteria as well (see, e.g., Jasso, 1990). Markovsky (1985) adopted this form and proposed that "injustice experience" (*IE*)—the subjective or emotional impact of an unjust situation—can be quantified by

$$IE_{AB} = \log_{JI_{AB}} \left(\frac{R_A}{R_B}\right), \tag{2}$$

where *JI* is *justice indifference,* the degree to which the person evaluating the fairness of a distribution desires justice for the particular comparison unit.[2] From this perspective, the determination of justice begins with a cognitive evaluation of the congruence of one's outcomes with the expected outcomes (i.e., the reward standard). The emotional reaction (*IE*), and therefore the likelihood of restorative action being taken, is then a function of this "congruence evaluation" and the degree to which the evaluator is indifferent toward justice in the current situation. (For other formal theories of distributive justice, see, e.g., Berger et al., 1983; Blalock & Wilken, 1979; and Harris, 1976).

These and other current theories of distributive justice all essentially derive from early equity theory[3] and invoke what Jasso (1986) has called the comparative postulate: "The mechanism by which an individual experiences the (instantaneous) sense of distributive justice involves a comparison of two quantities, the individual's actual amount or level of a good and the amount or level regarded by that individual as 'just' for him/herself" (pp. 251-252). Variations in how perceptions of justice are modeled have revolved around two major issues: who forms the basis for the comparison, and how that comparison is structured.

Identifying the Reference Standard

The initial equity theories (Adams, 1965; Walster, Berscheid, & Walster, 1973, and Walster, Walster, & Berscheid, 1978) postulated that we determine what is fair by comparing some transformation of our outcomes and inputs to the outcomes and inputs of another person in the same situation. Other theorists in distributive justice, however, have noted that there are other sources for the reference standard against which to compare one's situation. For example, Berger et al.'s (1972) "status value theory" allows that a generalized other can serve as a basis for reward expectancy (see also Berger et al., 1983). From this view, the comparison other need not be a real person, but can be some generalization of people who hold certain status characteristics. Comparison with one's own past experiences also has been suggested as a source of the reference standard. As Austin, McGinn, & Susmilch (1980) put it: "Persons, on the basis of their past experiences, form expectancies of what constitutes a satisfying *and* fair reward for various types of situations and then use these standards for evaluating current relationships" (p. 428). Thibaut and Kelley's (1959) interdependence theory (i.e. "comparison level"), and Stouffer et al.'s (1949) concept of relative deprivation are well-known examples of theories based on comparison with past outcomes.

Finding the Structure

The other debate within the literature focuses on what qualities of the reference standard and of the individual recipient should be compared, and how the information is integrated for the comparison. Formulae such as Markovsky's or Jasso's, above, describe this process. Notably, these theories describe the evaluation of the fairness of a distribution as a purely cognitive process. For example, Jasso (1980) describes the ratio of actual outcomes to expected outcomes (the "comparison ration") as "an exclusively cognitive magnitude, completely devoid of emotional content" (p. 6). Rewards and reference standards are compared, and a decision is made regarding the fairness of the rewards. "Short of claiming that actors carry out the mathematical operations, it is assumed that they will behave *as if* they do, whether consciously or not" (Markovsky, 1985, p. 827). Emotion enters the picture only as a *result* of this cognitive evaluation of justice.

In this paper, we offer a revision of the above theoretical formulations. Our view incorporates a more interactive process, with affective concomitants, and sees justice decisions as affective in origin, even if mostly cognitive in form. Theorists of distributive justice have been criticized for ignoring the "interactional character of the exchange relationship" (Deutsch, 1979, p. 30; see also Deutsch, 1983). In a review of several books on social psychological approaches to justice, Morton Deutsch (1983) characterizes the state of the justice literature: "The approach to 'justice' has been too psychological and not enough social psychological [sic]; that is, it has focused on the individual rather than upon the social interaction in which 'justice' emerges...Much of the current work ignores the relationship between conflict and justice and the process by which 'justice' is negotiated." (p. 312).

Our central thesis is that the evaluation of a rewarding act depends crucially on affective responses arising within the interaction, and that these affective responses emerge from the way the actors perceive the role-identities held by the various interactants, the definitions of the various actions, and the ways that these meanings combine in ongoing social interaction.

We begin by reviewing research showing that affect does impinge upon attributions of justice, suggesting that these judgments are affect-laden and tied to specific affective conditions and emotions. Next, we show how a general theory of social interaction would predict that these emotions could be generated in situations that typically produce judgments of injustice. Finally, we discuss various implications of this approach to distributive justice.

Affect and Justice Judgments

Traditional theories of distributive justice have discussed the emotional effects of unfair distributions as a by-product of the cognitive judgments about justice. For example, Homans (1961) writes "The more to a man's disadvantage the rule of distributive justice fails of realization, the more likely he is to display the emotional behavior we call anger. Distributive justice may, of course, fail in the other direction, to the man's advantage rather than to his disadvantage, and then he may feel guilty rather than angry..." (pp. 75-76). Other researchers have followed Homans' lead to argue that unjust distributions lead to the negative emotions of anger and guilt. The cognitive evaluation of congruency is proposed as a major element in the calculation of the emotional effects of injustice (e.g. Jasso, 1980; Markovsky, 1985). In fact, some have argued that injustice is part of the prototypical script which defines the emotion anger. For example, Lakoff (1987) describes the first event in the prototypical scenario for anger in the following manner: "There is an offending event that displeases S...*the offending event constitutes an injustice and produces anger in S*" (p. 397, emphasis added).

On the other hand, recent research has demonstrated that people's judgments about justice can be changed by their affective state. Mark & Sinclair (1992, cited in Sinclair and Mark, 1992) induced either an elated, neutral, or depressed mood in subjects who subsequently read about work situations. The payment in these stories ranged from equality, through equity, to inequitable overpayment. As predicted, pay structures that deviated further and further from equity (toward either equality or overpayment) were rated as increasingly less fair. More importantly, however, the greatest amount of variation occurred among subjects who were in a negative mood, and the least variation occurred among those in a positive mood. In other words, subjects in negative moods saw equality and overpayment as *more unfair*, relative to equity, than did subjects in positive moods. This supports the notion that subjects in a negative mood are more stringent about what they regard as fair.[4] In a second study, Mark and Sinclair (1992) found that subjects who were in good moods considered a broader range of payment for work done as fair, relative to subjects in bad moods, providing further evidence that justice judgments are linked to affect.

O'Malley and Davies (1984) also manipulated mood and measured subject's allocations of rewards to either themselves or to others. They had subjects engage in a puzzle where performance was quantifiable, and, after a fixed time period, subjects were led to believe that they had done either better or worse than another subject. They then were asked how they thought they should divide 100 raffle tickets between themselves and a partner. For subjects who performed better than the supposed other subject, mood had no effect—all subjects behaved in a fairly selfish way, taking a high percentage of the raffle tickets for themselves. However, for subjects who performed worse than the other, mood did affect the allocation strategies. In particular, subjects in a happy mood took more of the raffle tickets, and subjects in a sad mood took fewer tickets, when compared to subjects in a neutral mood.

The above studies dealt with diffuse moods rather than specific emotions; however, two studies have examined the effect of anger on attributions of blame and intentionality. Gallagher and Clore (1985) used hypnosis to induce either anger or fear in subjects, and then had them read a story about a transgression. Angry subjects were more likely to attribute blame to the transgressor than fearful subjects. Melton and Scher (1992) asked subjects to recall in detail an incident which had made them angry. Even though these subjects were most likely aware of the source of their anger, they were more likely to attribute intentionality to a transgressor they read about, than were subjects who had recalled a happy incident or those who had recalled a nonemotional incident. These studies suggest that people's moods can influence their interpretations of justice and that judgments relevant to justice can be influenced by specific emotions.

Indeed, there is a growing body of research on the effects of mood or emotional state on cognitive processes (see Isen, 1987; Schwarz, 1990; and Sinclair & Mark, 1992 for reviews). People sometimes use their current emotional state as a source of information when making judgments (see Schwarz, 1990; and Schwarz and Clore, 1988). People also use others' emotional states as information in making judgments about these others (Heise, 1989). Making inferences from how a person is feeling can serve as a useful heuristic method of arriving at decisions about complex issues.

Norbert Schwarz and Gerald Clore have demonstrated that when people are asked about their subjective well-being, those who have had their mood enhanced (e.g. by finding a dime on a copy machine [Schwarz, 1983] or by being interviewed on sunny days [Schwarz and Clore, 1983]) report being better off than those whose moods were not elevated (i.e., they did not find a dime, or they were interviewed on rainy days). The given explanation for these and related data is that the subjects were accessing their current affective state as a way to arrive at a decision about their well-being.

However, this effect only occurs when the affective information appears relevant to the judgment to be made. When the affective state was attributed to something else, it no longer supported an inference backwards to well-being, and the effect disappeared. For example, in the weather study (Schwarz & Clore, 1983), some subjects were asked at the beginning of the interview how the weather was in their town (the interviewer was presumably calling from another city). Those subjects who were phoned on a rainy day, and who were asked about the weather, subsequently reported being just as happy and satisfied with their lives as those who were phoned on sunny days. In other words, their mood, on being attributed to the weather, had no effect on their judgments about subjective well-being.[5] Consistent data were reported by Schwarz, Servay, and Kumpf (1985). They showed male cigarette smokers a fear-arousing film designed to persuade people to reduce cigarette use. Some of these subjects were given a pill (actually a placebo) and told that the pill had a side effect of arousing those who took it. Other subjects were led to believe that the pill would calm them down, and still others were told that the pill had no side effects.[6] The reasoning was that those who believed the pill was arousing them would be less likely to treat their fear as information in deciding whether to try and cut down on cigarette use, because they would attribute any arousal they felt from watching the film to the pill. On the other hand, those subjects who believed that they should be calmer because of the pill should be even *more* inclined to reduce their smoking. This is indeed what happened (although nonsignificantly so for the comparison between the "arousing pill" condition and the "no-side effects" condition).

These studies show that justice-related judgments and certain other reasoned decisions are influenced by affective states. In this paper, we go a step further and propose that the emotions associated with unjust distributions of reward—anger and guilt—may be aroused interactionally in situations where justice concerns are at issue. These emotions may then guide interactants to judgments about events being unfair, rather than vice versa. Characteristics of the interaction may lead interactants to feel angry (or guilty), and, in searching for a way to account for their predicament, they may draw conclusions regarding justice from the information offered by their emotional state. The information offered by the presence of anger is contained in the prototypical script for anger—that another acting unjustly toward us produces anger. Therefore, feeling anger in an interaction implies that one's interaction partner acted unjustly. The information offered by the presence of guilt is contained in the prototypical script for guilt (Kemper, 1990, p. 223)—that our unjust denial of what is due to others produces guilt. Thus, feeling guilt in an interaction implies we dealt with someone unjustly.

However, although these culturally-defined scripts for emotions can be accessed for information regarding justice-based decisions, this does not necessarily mean that these scripts *do* indeed describe the causes of the emotions. We must distinguish between the processes by which an emotion is aroused, and the prototypical definition of the concept held by members of a culture (Russell, 1991). Anger and guilt may arise out of normal social processes within an exchange transaction, but still provide information *as if* the emotion was caused by justice concerns.

Our proposition is this: *When people involved in a transaction feel anger or guilt, and the emotions are not ameliorated, they may decide that the transaction is unfair or unjust.* The scope is limited to transactions (exchanges) because that is where the concepts of justice or fairness are relevant, and people ordinarily would not employ these concepts to understand their emotions in other kinds of relationships. Alternative resolutions of the emotions within the transaction may also take place; anger and guilt sometimes can be reduced within transactions so as to preclude the need for managing them.

Our position relates unresolved justice-related emotions to Markovsky's (1985) concept of justice indifference. According to Markovsky's formula for injustice experience (see equation 2, above), the degree of justice indifference determines the emotional impact of a given incongruence. As indifference approaches infinity (i.e., complete indifference), injustice experience approaches zero—there is no emotional impact, no matter how large the incongruence. On the other hand, as indifference approaches its upper bound (in this case, $JI = 1$), injustice experience increases. Put another way, the more one cares about injustice, the greater the emotional impact of a given incongruence.

Our proposition also proposes a factor which influences the degree to which a particular incongruence will result in a subjective (affective) experience and justice restoring behavior. This factor is justice-related emotion. However, the route by which the consequences of injustice are modulated by affect is different from the route proposed by Markovsky for justice indifference. Specifically, Markovsky's formulation suggests that the cognitive evaluation of incongruence is made no matter what, but that the effect of that evaluation is reduced when justice indifference is high. In contrast, we propose that, if justice-related affect is absent, *no evaluation of incongruence (i.e., of injustice) will be made.* That is, increases in the intensity of the affect do not increase the effect of a given incongruence, but rather the likelihood that the degree of congruence will be 'calculated' at all.

For our proposition to be interesting, we must show that stereotypically unfair interactions can indeed produce justice-related emotions *before* participants make an actual judgment of unfairness. This would demonstrate that our proposition might be an adequate way of describing how justice judgments arise, and that "affect leads to justice-judgments" is just as plausible as "justice-judgments lead to affect." Yet our proposition still is no more than an exotic alternative to standard views unless we meet a second condition, showing that our sociology of emotion approach adds something new to the computational approach to justice. Toward this end, we will suggest how some seemingly unfair transactions might be perpetuated without judgments of unfairness, and how some seemingly benign transactions might instigate justice deliberations.

In order to meet these goals, we must first adopt a model of social interaction. We will use affect control theory (Heise, 1979; MacKinnon & Heise, 1993; Smith-Lovin & Heise, 1988) to cast the distributive justice process into social interactional terms, because the current work arose from trying to examine distributive justice in terms of that theory, and because we feel that affect control theory is the most thoroughly developed theory to account for the ever-changing meanings, actions, and feelings that make up social interaction. After an introduction to affect control theory, we will use the theory to show how social interaction could produce justice-related emotions that induce judgments of justice.

Affect Control Theory[7]

Affect control theory is so named because it posits that "the basic motivational principle...is that people construct or reconstruct events so as to maintain consistency between transient feelings and sentiments." (MacKinnon and Heise, 1993). The meanings applied to actions and actors in a social interaction, and the behaviors taken by an actor, are directed toward 'controlling' the discrepancy between an observer's pre-existing sentiments about the elements of the interaction and the transient impressions created by specific acts.

The theory uses Osgood's (1962) three dimensions of affective meaning to measure sentiments and impressions. *Evaluation* is the familiar attitudinal component of response—the feeling that something is good or nice as opposed to bad or awful. The *Potency* dimension of response measures the degree to which something is powerful or powerless. The *Activity* dimension relates to assessments of whether something seems lively and fast, or whether it seems quiet and slow. Research conducted in more than twenty nations (Osgood, May, & Miron, 1975) indicates that evaluation, potency, and activity (EPA) are universal dimensions of response to stimuli of many kinds, and that sentiments vary along these dimensions crossculturally. Individuals, behaviors, and settings all can be rated on these dimensions, and the averaged ratings index culturally defined sentiments about these social objects. Within a culture, mean EPA ratings for most stimuli are very similar for males and females and for people of different socioeconomic levels from various regions of the U.S. (Heise, 1966). Subcultures support divergent sentiments, but only for concepts that are very central to the subculture's routines (Heise, 1979: 100-102; Smith-Lovin and Douglass, forthcoming).

Predicting Transient Impressions

Interpersonal events create transient impressions of people, behaviors and settings, and these impressions can differ from the pre-existing sentiments. One paradigm for studying the impressions created by events (Gollob, 1968; Gollob & Rossman, 1973; and Heise, 1969, and 1970) is to present an event (e.g., the employee neglects the employer), then ask subjects to rate the actor (e.g., the employee) in the *context* of the event. Averaging responses across subjects yields an EPA profile assessing the impression produced by the actor engaging in the given behavior with the particular object person—an impression that usually is different from the sentiment (measured out of context) that represents what we think of the actor generally. Alternatively, subjects may be asked to rate the object person (the employer) or the behavior (neglecting), given the context of the event.

Transient impressions can be predicted quite accurately from EPA mean ratings of the actor, behavior, and object-person obtained outside an event (Gollob, 1968; Gollob & Rossman, 1973; Heise, 1969, 1970, and 1979; Heise & Smith-Lovin, 1981; MacKinnon, forthcoming; Smith-Lovin & Heise, 1982; and Smith-Lovin, 1987a, and 1987b). The predictive equations come from regression analyses in which in-context EPA ratings of event elements are regressed on out-of-context EPA ratings of the event elements (e.g., an employee, an employer, to neglect someone) plus multiplicative interaction terms. Gollob (1968) and Heise (1979, Chapter 2), note that some of the interaction terms correspond to classical notions of attitude balance.

Here, for example, is the equation based on a regression analysis across 515 events (Smith-Lovin, 1987a) for predicting the transient impression of a behavior on the evaluative dimension:

$$B_e' = 0.07A_e + 0.57B_e - 0.10B_p - 0.12B_a + 0.04O_e + 0.02A_eB_e + 0.02B_eO_e$$
$$+ 0.07B_pO_p + 0.02A_eB_p - 0.05B_eO_p - 0.03B_pO_e + 0.04B_aO_p$$
$$+ 0.02A_eB_eO_e + 0.03A_pB_pO_p + 0.02A_eB_pO_p. \qquad (3)$$

A, B, and *O* refer to the actor, behavior and object-person, respectively; the subscripts e, p, and a refer to the evaluation, potency and activity dimensions. The transient impression (the dependent variable) is denoted with a primed symbol; feelings existing before the event (the predictor variables) are denoted without primed symbols. Only significant predictors are shown.[8]

The terms in this equation can be interpreted as follows (Smith-Lovin, 1987a: 47-49). First, evaluation of the actor (A_e) is a factor determining impression of the behavior. Any behavior seems slightly better when done by a good actor or slightly worse when done by a bad actor. The coefficient of the second term (B_e) reflects a substantial degree of stability in impression formation: good acts preserve some of their goodness or their badness regardless of the circumstances of their use. The presence of the other behavior terms (B_p, B_a) can be viewed as meaning that a behavior's evaluative stability is especially great for weak and quiet acts. The identity of the interaction partner (O_e) also reflects on the behavior to a small degree, in the same manner as the identity of the actor does.

The A_eB_e term represents an actor-behavior consistency effect (Gollob, 1968). People evaluate behaviors more negatively when they are evaluatively out of character for the actor. For example, sexual abuse is seen as worse if it is carried out by a mother than if it is carried out by a deviant. Behavior-object consistency is another force affecting the evaluation of a behavior. The B_eO_e effect shows that behaviors are evaluated more positively when they are directed at evaluatively appropriate object persons (e.g., helping the deserving, or rebuking deviants). The B_pO_p effect means that powerful behaviors are evaluated more positively if they are directed toward strong people, and more negatively if directed toward weak objects.

The interaction of actor evaluation and behavior potency (A_eB_p) shows that powerful behaviors (like challenging) seem better when conducted by nice actors while weak, powerless acts (e.g., imitation, or begging) seem better when done by disvalued actors.

The B_eO_p interaction originally was identified by Gollob and Rossman (1973) in a study of actor potency. Directing good acts at a weak person (e.g., aiding a victim) or bad acts at a powerful person (criticizing a president) makes people view the action more positively. By contrast, we disdain acts of ingratiation with the powerful or of nastiness toward the weak. The behavior potency-object evaluation interaction (B_pO_e) is conceptually similar to the B_eO_p effect. Here again, the evaluation dynamics seem to reflect a "just world" assumption. Behaviors seem better if they represent gentle treatments of good objects or forceful treatments of bad objects. Conversely, a behavior seems worse if it amounts to treating a bad person in a gentle manner or a good person in a dominating manner.

The three-way interactions constitute qualifications of the two-way interactions. That is, the $A_eB_eO_e$, $A_pB_pO_p$, and $A_eB_pO_p$ interactions show that the consistency effects, B_eO_e and B_pO_p, operate most strongly when the actor is positively evaluated and powerful.

Smith-Lovin (1987a and 1987b) presented similar equations for predicting potency and activity impressions of behaviors, as well as equations for predicting *EPA* outcomes for actors, for object persons, and for the settings that are involved in events. As it turns out, the equation for predicting evaluation of an actor is very similar to the above equation for predicting evaluation of the actor's behavior, except the coefficient for A_e is much larger and the coefficient for B_e is much smaller. Impression-formation equations account for about 80 percent (or more) of the variance in transient impressions. Moreover, the equations are similar for subjects in different socioeconomic and national populations to a large degree. (See Smith-Lovin, 1987a, for a review of results from impression-change studies with U.S., Irish, and Arabic subjects; MacKinnon, forthcoming, provides results from Canadian subjects.) We therefore assume that the equations represent quite general, ubiquitous aspects of affective reactions to social events.

An important variant of impression-formation research develops equations for predicting outcome impressions when identities are combined with emotions, traits, or status-characteristics (Averett & Heise, 1987; and Heise & Thomas, 1989). These equations allow affect control theory to deal with situated identities—for example, an angry father, a wise child, a rich professor—and the equations also are the basis for theoretical modeling of emotional responses and trait attributions.

Emotions

According to affect control theory, emotions arise as an indication of how well the current situation is confirming the expected identities of interactants. Emotions are temporary affective conditions that register how events are making one seem as compared to how one is supposed to be (see also Higgins, 1987 for a similar theory of emotion). A person invokes an emotion that combines with his or her identity within a situation to generate a transient impression identical to the transient impression created by the current event. That is, the impression of an *angry father* is less good than for just *father*. So, if a father is involved in an event which makes him appear less good, he will become angry (i.e., he will make himself angry) in an effort to *feel* the impression he has created. Thereby, the person viscerally experiences how impressions created by the event relate to his or her identity. By knowing both the situational identity of an individual, and the transient impression of self created by an event, it is possible to predict the emotions that are likely. The prediction equations are obtained from the empirical equations that predict impressions of emotion-identity combinations (like angry father), given the profiles for the identity (father) and the emotion (angry). Solving the equations for emotion yields new equations that predict emotion from identity and impression.[9]

The model predicts that emotions directly correspond to how events have affected the self—an interpretation that corresponds to intuitions (e.g., events that make one look bad also make one feel bad). However, situational identities influence emotions in several ways. In the first place, people conduct themselves so as to keep transient impressions of themselves close to their identities, according to the basic axiom of affect control theory. Therefore, identity determines emotion by determining what transient impressions generally arise as an individual creates events. Additionally, the model indicates that emotions reflect how transient impressions of self *compare* to one's identity. This suggests, for example, that people experience especially good or potent or lively emotions when events make them seem more good or potent or lively than their identity warrants. Finally, an interaction effect in the equations indicates that the evaluation of one's situational identity influences the extent to which transient impressions of self translate into more extreme emotions; one consequence of this is that people with extremely negative self identities may experience chaotic emotions, or emotional lability.

The Course of Interaction

For the purposes of affect-control analyses, the basic progression in a social encounter is as follows:

First, participants comprehend social situations by recognizing people in terms of personalized identities (e.g., John, Mary), informal social stereotypes (e.g., he-man, maverick), or formal social roles (e.g., secretary, executive). The sentiments for these identities are evoked, and the sentiments serve as affective reference points for understanding each participant in the situation.

Second, participants anticipate, perform, and interpret events to confirm preexisting sentiments as much as possible. Thus, people with valued identities are expected to perform acts which create positive impressions, whereas interactants with disvalued identities are expected to behave in ways that confirm the stigma that is associated with them.

Third, expressive displays reveal how each interactant is faring in social interaction—whether the interactant seems good, potent, and lively and whether the levels of goodness, potency, and liveliness are appropriate to the interactant's identity. Positive emotion occurs when an interactant benefits from events—especially when the benefit is greater than the person's identity warrants, and negative emotion occurs when events create negative impressions of an interactant—especially when impressions are less favorable than is warranted by the interactant's identity.

Fourth, occurrences that disconfirm identities, and that are not repaired by subsequent restorative events, may cause reassessments in which identities are adjusted to fit the events that have occurred. For example, a person who is sullied by an event may move into a less valued identity or may adjust the current identity by amalgamating it with a negative mood, trait, or status characteristic.

The Affective Nature of Justice Deliberations

In this paper, we propose that an additional phenomenon can occur when people try to cope with certain emotions in the context of transactions, especially when the opportunities for restorative actions are very limited, as in the economic or political spheres. In many (if not most) interactions, people have a wide repertoire of behavioral and cognitive means to try and restore discrepancies between event-created sentiments and more fundamental sentiments. But, in some circumstances, including many exchange transactions, these options become limited. When unresolved anger or guilt is produced in these types of interactions, one possible way to deal with the emotion is to characterize the transaction as unjust, opening up the opportunity for restorative action *outside* of the transaction (e.g. retribution, reparation, or even dissolution of the relationship).

However, the emotion alone is not a sufficient cause for deciding that a transaction is unjust. We propose that justice-related emotions initiate an assessment of an individual's rewards relative to expected rewards, as modeled by traditional justice researchers. If this rationalistic assessment confirms a conclusion that the outcome was unfair, then the victim has the rhetorical ammunition to demand that the problem be resolved via extra-transactional means. However, the assessment may conclude that recent events are just and *do not* warrant extratransactional restorative events that would eliminate the emotions. In that case, the emoter is left with stressful emotions that cannot be resolved in the relationship that generated them, and the only recourse is to turn to stress management techniques (discussed, for example, by Thoits, 1990). Our model of how emotion enters justice judgments is shown in the accompanying schematic diagram (Figure 1).

We believe that more interpersonal options available in an interaction allow emotions to be resolved interactionally, and on the other hand, that constraint in transactions makes it more likely that justice deliberations will be needed. For example, at the extreme, all a worker can do is work, and any resolution of emotions generated by work depend on actions by the employer. And, at the extreme, all an employer can do is pay, so alleviation of worker emotions depends on how often (or how much) the employer pays and on the meaning of the payment for the worker. The more constrained an interaction is in this way, the less able the parties are to manage emotions interactionally, and the more likely that concerns about fairness will arise.

The crux of our argument is that justice-related emotions can arise in transactions apart from considerations of justice. We will elaborate on this matter in the next section. Here we point out that such emotions have impact because they typically are indicators of disconfirmed identities, and it is the disconfirmation of identity that actually motivates the search for a resolution. Sometimes, however, justice-related emotions will be consistent with one's identity. In situations where anger or guilt do not reflect disconfirmations of identity, we do not expect them to trigger justice deliberations. For example, viewing a mugging as a kind of transaction, the mugger could experience anger in confirmation of his mugger identity, and that anger would not lead him to deliberate about the justice of the transaction. The same emotion, however, experienced by the victim, could, indeed, initiate such deliberations.

If people are involved in a transaction
with limited opportunities for action
and
one or more of the parties experiences
justice-related emotions (anger, guilt)
and
transactional events do not resolve the emotion

then

the emoter will examine the transaction for injustice
and either

$$\begin{bmatrix} \text{the transaction will be characterized as injust} \\ \text{and} \\ \text{acts of retribution, reparation or reorganization} \\ \text{will be enabled} \end{bmatrix} \text{ or } \begin{bmatrix} \text{procedures for stress} \\ \text{management will be instigated} \end{bmatrix}$$

Figure 1.   Schematic Diagram of the Role of Emotions in Transactions

We allow that justice-related emotions that arise in a transaction also can be resolved in the transaction, as illustrated in the next section. Such a resolution is a normal outcome, and as long as the transaction diffuses the emotions that it generates, as long as the transaction allows participants to confirm their identities periodically, there is no impetus to turn to justice deliberations.

Examining the justice of a transaction involves searching for meanings of acts—for example, trying to determine the value of payments as judged by comparison with other transactions, or in terms of tradition or other criteria. As mentioned previously, affective states influence these deliberations, but we are not yet ready to model that influence. In this chapter, we simply recognize that a judgment of unfairness both legitimatizes renegotiation of the terms of the transaction and justifies side payments in the form of retributions or reparations in order to restore parties to their rightful identities.

Ideally a judgment of fairness would adjust the meanings of acts in a transaction so that in retrospect and prospect the identities of parties to the transaction are properly confirmed. That happens sometimes, but less often than seems rational because those experiencing the emotions make use of extra information that disinterested outsiders do not have. That is, a person who is angry or guilty in a transaction "knows" the transaction is unjust because they know what they are feeling, and the prototypical scripts for these emotions have injustice as the initiating condition.

The emoter often will trust his or her emotions, and therefore will not be convinced by reasonable arguments that an encounter is fair. As a result, they will not use the fairness judgment to adjust the meanings of acts. Meanwhile, though, they have to recognize that others' fairness judgments do make extratransactional solutions inappropriate. They end up with the transaction and their troubling emotions intact, and the only resolutions left to them are personal ones, like changing their self-concepts in a negative direction, or engaging in fantasy, substance abuse, or denial of what they are feeling (see Thoits, 1990). By this account, some mentally disturbed people can be seen as victims of justice.

Prototypical Transactions

We turn now to examining how justice-related emotions might arise in social interaction, and to showing how such emotions ordinarily are resolved naturally, precluding the need for justice deliberations. Our analysis derives from computer simulations of social interaction based on affect control theory's mathematical model combined with EPA ratings from a U.S. population of males (Heise and Lewis, 1988).

Traditional Factors Affecting Justice

Suppose that two people have taken the roles of employee and employer. The employee is expected to maintain a somewhat good, weak, and quiet identity in the transaction (the mean ratings of "employee" are 0.83, −0.53, −0.18 on evaluation, potency, and activity, respectively—see note 8 for details about these scales). The employer should maintain a good, very powerful, and quiet identity (0.92, 2.00, −0.68). Now the employee engages in work for the employer, and the action of working for the employer can take on a variety of affective meanings, depending on the nature of the work.

In order for our approach to be viable, we need to show that as conditions become more "unjust" (that is, as conditions become such that models of justice would predict judgments of unfairness), the justice-related emotions (anger and guilt) become more likely. If the work being done is not particularly onerous or physically demanding, neither traditional views nor intuitive beliefs would expect judgments of unfairness. Our simulations indicate that under these kinds of work conditions, the employee will not experience anger as a result of performing duties. Therefore, justice would not be considered. By changing the affective meaning of work (i.e., by experimenting with different EPA profiles), we found that anger will arise if the work is even a little unpleasant (say, −0.50 on the evaluation dimension) and if it is at least slightly active (say, 1.00 on the activity dimension). The potency of the work, which might be interpreted as its importance, is not crucial—people can get fed up with important work as well as unimportant work—so assume heuristically that the employee views the work as somewhat important (1.00 on the potency dimension).

Affect control theory predicts that an employee repetitively engaging in such work comes to see the self as slightly bad, potent, and active ($-0.76, 0.35, 0.69$). The bad impression arises because the employee is a person in a good role trying to relate to another person in a good role through unpleasant action (it would make more sense to relate to the employer through positive action, such as asking the employer about something, or advising the employer—as predicted from affect control theory). Appearing somewhat bad, potent, and active already is one step toward anger (which has an *EPA* profile of $-0.86, 0.21, 0.71$). But comparing the impression of self that the employee is creating with the profile for the employee identity reveals that, relatively speaking, the employee seems substantially less good, more potent, and more lively than should be the case. Such a disconfirmation of identity rouses the emotion of anger.

If the employer understands the work in the same way as the employee, then the employer is just as stressed by the situation as the employee. The employer, however, is able to do something about it within the rules of a transaction. Paying the employee helps bring the employee's self-impression back to where it is supposed to be and greatly dissipates the propensity to anger. Making the payment more positive by paying well can wipe out the negative impression of the employee and the negative emotion. On the other hand, failing to pay or paying poorly while continuing to demand arduous work leaves anger and primes the employee to begin deliberating over the fairness of the transaction. Within traditional models of justice, these conditions would likewise be expected to increase unfairness judgments, but we argue that it is the emotional effect which initiates this process.

Other details of the situation can influence whether anger is generated. The sequence of events is one important factor. For example, work that is only a little unpleasant has to be repeated without reward for some time before it begins to generate anger, according to simulations. The physical environment also is significant according to simulations—for example, a dispiriting setting (like working in a slum) can cause pleasant work to engender anger and thereby considerations of fairness.

ACT makes a number of nonintuitive predictions that would not be likely to be derived from traditional approaches to distributive justice. One of the most important, we feel, is the role played by the identities assigned to actors. This is the topic of the next section.

Identity and Justice Emotions

An important subset of factors affecting the arousal of justice-related emotions involves the identities of the actors. As described above, emotions are signals of the current ("transient") identity of an actor in a situation, as well as the relationship between the current identity and the preexisting ("fundamental") identity. Thus, one way to view the sequence of events we propose from situational factors (e.g., the amount of pay given, the laboriousness of the work) to emotional arousal, to cognitive deliberations about justice is to see the situational factors generating emotions related to the fundamental and transient identities and the relationship between them.

In the course of interaction, the identity a particular actor takes on can become threatened because the events in the interaction are inconsistent with that identity. One aspect of emotion is to signal such a discrepancy. Because the guiding force in social interaction, according to ACT, is to seek to reconfirm the identities of actors, the presence of emotion that is inconsistent with one's identity is an indication that some reconfirming action or cognition should take place. ACT allows for many ways for this to occur (e.g., reconceptualization of the event, additional behavior, or redefining of various identities). The new resolution suggested in this paper is the transformation of a justice-related emotion into a justice deliberation and then, if warranted, into a reparation outside the normal course of interaction.

This perspective on how justice-related emotions, and, therefore, justice deliberations, are generated leads to another implication of our view on whether or not a particular situation will be considered unjust (or, more exactly, whether or not justice will be considered at all). The way the participants assign identities for themselves and others influences the degree to which justice-related emotions are aroused. Even pleasant work will induce anger and justice deliberations in an employee who already has decided that he or she is a "dissatisfied employee," and work is more likely to engender anger if it is done for a disrespected person like a "slavedriver." On the other hand, even odious work will generate positive emotions in someone who accepts a debased identity like "slave."

Justice and Guilt

Our simulations suggest that guilt is not a very prevalent basis for justice deliberations. Affect-control simulations suggest that employees who receive lavish payments accept their fate because they are being made to look better than they should be, which generates happiness, and happiness (even when it signals disconfirmation of one's identity) is not an emotion instigating fairness deliberations. If an employee engages in a transactionally deviant act, like cheating the employer, the employee more likely will feel anger than guilt. (The anger could lead to considerations of fairness, but the deliberations might end up treating the theft as justified reparation.) An employer engaging in a deviant action, like overworking an employee, also is likely to have a bout of bad temper rather than guilt accompanying the episode.

The fact that the simulations do not suggest the arousal of guilt from overpayment should not be seen as inconsistent with our view of the justice process. In fact, there has been a persistent finding in the literature on distributive justice that people are less likely to call distributions unjust when they receive more than they expect or feel entitled to (e.g., Messick & Sentis, 1979). This so-called egocentric bias (Messick & Sentis, 1983) would be facilitated by the lack of guilt in situations where it "should" occur.

We did find that an employer accepting work from a "slave" could end up feeling ashamed, and convincing an employer that he or she is treating workers as debased beings might make the employer open to justice deliberations. On the other hand, an employer always has the option within the framework of the transaction of engaging in some small payment that alleviates the employer's guilt, even if dealing with slaves.

Implications

Throughout this paper, our argument has been that judgments about justice, and the behaviors that follow from those judgments, are instigated through the affective dynamics of social interaction. The implications of such a model are many. Anything that affects those dynamics should have a predictable effect upon instigation of justice deliberations. In this section, we will discuss several of the more noteworthy implications of this approach, focusing particularly on those implications which we feel have not been derived from traditional cognitive approaches to distributive justice.

A Situation Defined as Real

Affect control theory was developed in an explicit attempt to model the symbolic interactionist view of social interaction. By casting distributive justice as a social interaction, and modeling justice-oriented reactions in terms of this process, we can take into account the subjective nature of definitions of situations as they relate to justice phenomena.

The way that a particular situation is defined, the role-identities assigned to actors involved in a transaction, and the modification of those identities within the context of the transaction all influence whether or not a justice-related emotion arises. If either party takes on, or is cast into, certain negative identities, the transaction can produce negative emotions that can culminate in justice deliberations. For example, an employer who presents a stigmatized identity (Goffman, 1963; and Jones et al, 1984) like homosexual (EPA of −1.52, −0.79, 0.64) is likely to rouse anger in employees as they work for him or her, increasing the likelihood that injustice will be experienced. To our knowledge, no other theoretical perspective has suggested that status of the *payer* would affect the likelihood that a distribution would be seen as unjust. On the other hand, a party to a transaction who accepts a role that is both stigmatized and powerless is unlikely to develop anger and is unlikely to be discontent except by way of guilt; this is similar to the suggestion by Berger and his colleagues (Berger et al., 1972; and Berger et al., 1983) that the reward that a payee expects is derived from his or her status.

The initial identities assigned to interactants are not the only way that people can gain stigmatized identities. One important aspect of affect control theory is that behaviors, and the people who engage in them, are evaluated in the context of their current affective meaning in the eyes of the evaluator. And that meaning is constantly changing. Each act carried out by an actor can change sentiments about that actor. So, an actor can become stigmatized by engaging in a stigmatized act. Imagine a simple social interaction, involving an employer and an employee, with the employee engaged in slightly pleasant work. The interaction could have three steps:

(1)   The employee works for the employer.
(2)   The employer ignores the employee, whereupon the employee reidentifies the employer as a "stuffed shirt." And,
(3)   The employee continues the work, but now feels angry and begins considering the injustice of the transaction.

This vignette is from a simulation, but it plausibly recounts how transactions sometimes go sour as a result of acts that change the way people define their situation. Previous acts within an interaction can lead to such redefinition, and therefore can affect the initiation of justice deliberations.

Do People Typically View Unjust Distributions as Unjust?

One implication of our analysis is that there are many ways to eliminate claims of injustice. One can set up a slave society, and if everyone believes in it, it will work—a true slave understands that even the worst work is gratifying and even a small reward is cause for joy. But exploitation can take a more positive cast, too. Roughly speaking, anything that keeps people happy also keeps them from worrying about injustice. For example, an executive might be advised to put company people in nice settings, give everyone status, allocate frequent rewards even if they are small, and help everyone cover behavioral bloopers that could make others think unkindly of them. An outsider might consider the company's reward structure unjust, but most likely the possibility of injustice will never enter the employees' minds, unless something in the situation makes justice considerations salient to them (such as being asked by a social scientist whether they felt their outcomes were fair). From this perspective, we might think of affect as one factor affecting *justice indifference* (Markovsky, 1985). As discussed above, this term reflects the degree to which people care at all about justice concerns.

We do believe that the rational calculation of rewards and costs, and the comparison of one's own ratio to others', can set off deliberations about fairness. However, we believe that such calculations initiate fairness considerations by defining the meanings of costs and rewards. That is, whether one's work is pleasant or unpleasant is partly a function of comparisons with past work, with others' work, and with the extent of reward for such work. And whether one's rewards are good or bad is judged in light of comparisons with past rewards, others' rewards, and sacrifices made. If one reasons that one's work is onerous or one's rewards are poor, then engaging in that work or accepting those rewards can flame the anger that leads to injustice deliberations. Then demonstrations of inequity, inequality, need, or whatnot might convince self and others that injustice is present, and things need to be changed. We claim, though, that the process must proceed through emotion, and if the emotion gets allayed through the many interactional circumstances we have examined, then injustice will be ignored.

Related Approaches

Although we believe that our position about the role of emotion and social factors in instigating justice deliberations is relatively novel and unique, there have been related ideas in the literature. In this section, we will discuss two theories of justice which we believe have some interesting similarities to our view. First, we will briefly describe the way that *status-value theory* has been applied to distributive justice. This theory, developed by Joseph Berger and his colleagues (Berger et al., 1972; Berger et al., 1977; and Berger et al., 1983), attempts to explain how expectations for rewards could vary as a function of the "status characteristics" of the recipient. Despite a somewhat different focus, this approach has some interesting similarities with our application of affect control theory to distributive justice.

In this section, we will also discuss recent work in procedural justice, which argues (1) that judgments of the fairness of a procedure for deciding outcomes have at least as much influence on judgments of distributive fairness as information about the outcome itself (e.g, Lind, 1992), and (2) that decisions about the fairness of a procedure are based on social or interactional factors, not predominantly on factors directly related to the procedure itself (Lind, 1992; andTyler & Lind, forthcoming).

Status-value Theory

Status-value theory (Berger et al., 1977; and Humphreys & Berger, 1981) as applied to distributive justice (Berger et al., 1972; and Berger et al., 1983) argues that distributive fairness is determined by a comparison of the rewards an individual receives to the rewards that the person expects. This, basically, is the comparison postulate discussed above. However, Berger and his colleagues add an interesting dimension to this view of justice by specifying a means whereby the expected reward in a specific (local) situation can derive from the broader social context—from a referential structure that arises from considering factors outside the local situation.

The referential structure refers to the generalized ways in which "meaning is given to rewards and expectations are formed about their allocation" (Berger et al., 1972, p. 119). Status-value theory attempts to explain how the cultural meanings of interactants, rewards, and expectations are translated from this referential structure to the local situation.

The referential structure has much in common with the notion of fundamental sentiments as they are conceptualized in affect control theory. In both cases, certain information about the relationships and meaning of various elements in a particular interaction are generated from some culturally-shared structure.[10] In status-value theory, the referential structure specifies the relationships between various status characteristics of an individual and the goal-objects (rewards) that should be expected by someone who possesses those status characteristics. One aspect of this process implies that the rewards to be expected are generated based on what a *generalized other* in a similar role position would be entitled to (see also Blalock & Wilken, 1979). The comparison standard in this situation is this generalized other.

As in affect control theory, then, the degree to which a particular situation will ultimately be considered unfair is dependent upon how the perceiver assigns the identity of an individual reward recipient. If the individual perceives the recipient (which may be him/herself) as a mechanic, for example, then the generalized other to which he or she will compare is a generalized mechanic. However, the assignment of another identity would result in a different conclusion as to which generalized other to use as a comparison, and therefore, which standard of pay is fair. For example, if, instead of seeing oneself as a mechanic, a person sees her- or himself as a gas station attendant, then the person might view a lesser reward as fair.

Of course, the difference between our view and the approach taken by status-value theory relates to the process by which these differences in identity assignment affect justice evaluations. Berger and his associates (Berger et al., 1972; and Berger et al., 1983) propose that identification influences justice by affecting the choice of a standard for comparison. We have argued that identity may also influence justice judgments because identity affects whether or not justice-related emotions will be aroused, and therefore whether or not justice issues will be considered at all.

Another similarity between status-value theory and affect control theory has to do with Berger et al.'s (1972) position regarding the way that the meaning of status creates an expectancy for the local situation. They write:

Given a specific definition of the status significance of characteristics and goal objects in the local system, and specific beliefs about what one has a right to expect, the actual association of characteristics and goal objects in the local system either coincides with expectations or it does not. If it does, it follows that (1) the status values actually associated in the local system are balanced and (2) the system is 'moral' or 'just,' in the sense that it behaves in the way one has a legitimate right to expect. (p. 135)

In other words, a local system is unbalanced and unjust when there are people in it who do not have the characteristics, and do not receive the rewards, that people with that status should have. This is the same as saying that a local system is "unbalanced" to the extent that the identities of the participants are not confirmed. According to affect control theory, this will produce certain emotions. However, as we have delineated above, this does not *automatically* lead to attributions of injustice; the emotion could be resolved by other means.

Procedural Justice

Beginning with the work of Thibaut, Walker, and their colleagues (e.g., Thibaut & Walker, 1975), there has been an effort to explain how people evaluate the fairness of the procedures that are used to arrive at decisions. There are several reasons why this is relevant to questions of distributive justice. Often, the procedures used to decide upon a distribution are known by the participants in an interaction. When this is the case, the procedures used to determine the distribution of reward is an important part of the question of when distributions are considered fair. It is a question that, until recently, has been largely ignored in the distributive justice literature. However, some research has suggested that questions of whether or not a procedure for allocating rewards is fair are more important in decisions of distributive justice than any variables related to the size of the rewards allocated (see, e.g., Lind, 1992; see Tyler & Lind, forthcoming, for a review).

In conjunction with this finding about the importance of procedural factors in the judgment of the fairness of distributions, recent work by Tom Tyler and E. Allan Lind and their colleagues (see Lind, 1992; Lind & Earley, 1992; Lind & Tyler, 1988, Chapter 10; and, especially, Tyler & Lind, forthcoming) have argued quite convincingly that "people use impressions of authorities and inferences about their own place in the social scheme to generate a global impression of the fairness of an organization" (Lind, 1992, p. 17). According to this view, procedural justice judgments, even in formalized institutional and organizational contexts, are made on the same basis as more interpersonal impressions.

Tyler and Lind (forthcoming) have analyzed the factors that lead up to judgments that a procedure is fair. This analysis has led them to propose what they call a "relational model" of procedural justice. Procedures are judged to be fair based on the degree to which they give the participant standing, and the degree to which they generate trust in the authorities or procedures. *Standing* refers to the degree to which the participant feels that those who have the most importance in the procedure give the participant full status. *Trust* refers to the degree to which the authority or other important people in the procedure "appear to be trustworthy, benevolent, and unbiased" (Lind, 1992, p. 17). In other words, procedures are judged to be fair based on the degree of liking, trustworthiness, and status that the participants in the procedure (decision-makers and beneficiaries of the decision) are given. Affect control theory would make the same predictions about when judgments of injustice would be considered. Our reading of Tyler and Lind's work suggests that the procedures where trust and standing are higher are situations where the fundamental identity of the actors has a higher evaluation. The differences in perceptions of the actors when trust and standing are or are not present can be compared to the differences that appear when actors apply trait adjectives (e.g., "trustworthy" or "benevolent") to identities (e.g., the benevolent judge— see Averett & Heise, 1987). Procedures that are high in standing and/or trust would create a situation where the identities of the participants are particularly high in evaluation. In these situations, it would be unlikely for either actor to experience anger or guilt.

This is especially true because the evaluation of identities of the participants in most of Tyler & Lind's research are already fairly high. Much of their research, for instance, was conducted in courtrooms and other legal settings (e.g., court-annexed arbitration proceedings). The authority in this case is typically a judge (or an arbitrator). Judges are generally seen as somewhat good (for example, Canadian females rate judges at 1.42 on the evaluation dimension). Combining this positive identity with a trait adjective such as "benevolent" (the benevolent judge) would raise the evaluation even higher. Thus, we see our model as consistent with the theoretical and empirical work on procedural justice.

However, the two models (our model and the "relational model") are not redundant. For instance, one prediction of our model that is not explicit in Tyler and Lind's (forthcoming) relational model of justice (but which is not necessarily inconsistent with it) is that any other factors that might influence the ascribed identities of actors in the situation, and therefore would affect the likelihood of injustice-related emotions, would also affect the likelihood that justice considerations would arise. These factors might include the setting of the decision, previous behaviors of the actors, and so forth. In addition, it is possible that a sequence of negative events might push the transient impressions of participants down sufficiently that even a procedure decided on by a "benevolent judge" would evoke injustice-related emotions, and therefore, justice considerations.

Another contribution of our approach is that ACT can specify *specific* identities that have different affective meanings. That is, while Tyler and Lind suggest that procedures which raise the evaluation of participants will be viewed as fairer, our approach suggests specific types of people which will increase or decrease the likelihood that justice will be considered.

We feel, therefore, that our approach compliments that of Tyler and Lind (forthcoming). In addition to suggesting additional factors that might affect procedural justice judgements, and specific identities that might influence judgements of justice, we provide a model for how the "relational" aspects of justice judgements operate—by means of affect and emotion.

Summary and Conclusion

We have presented a theory which attempts to show how emotion can be generated in situations which typically arouse justice concerns, and we have suggested that these justice-related emotions instigate deliberations about whether or not a particular situation is fair. From this perspective, the emotions typically associated with judgments of justice are seen as being a contributing cause of a decision that a transaction is unfair, rather than being a consequence of these judgments. This approach suggests that factors affecting emotion may be important elements in the process by which decisions to demand reparations, and to attempt to reorganize the means of distribution of resources are made.

We feel that the potential impact of emotion on this process is rather dramatic. This paper has been an attempt to highlight some of these impacts. The ultimate test, of course, is empirical, and we await further research to delineate the emotional bases of injustice.

ACKNOWLEDGMENT

Completion of this paper was facilitated by support from a National Institutes of Mental Health Training Grant in Measurement of Affect and Affective Processes (PHS T32 MH 15789-14). We would also like to thank A. George Alder, Guillermina Jasso, Barry Markovsky, Robert Sinclair, and one anonymous reviewer for helpful comments on an earlier version of this manuscript.

NOTES

1. Or, at least as fair as the best available alternative (see Thibaut & Kelley, 1959).

2. This form of the function is equivalent to Jasso's because $\theta \ln(R_A/R_B) = \log_{JI}(R_A/R_B)$, with $\theta = 1[1n(JI_{AB})]$.

3. Recently, several authors have begun to examine when people might use other justice principles (see Deutsch, 1985; Mikula & Schwinger, 1978; and Leventhal, Karuza, and Fry, 1980 for examples). To date, these theories have been less formal. We believe that what we say can be applied to these theories as well. However, the equity principle has been far and away the most

common basis for justice research, and therefore we will not focus on these so-called "multiprinciple' approaches in this paper.

4.   Sinclair & Mark (1992) argue that mood affects justice judgements because of the effects of moods on cognitive processing strategies. In particular, people in good moods tend to process information in a simpler, more heuristically based, nonsystematic way. In contrast, people in bad moods tend to process information more carefully and systematically (see Fiedler, 1991; Schwarz, 1990, and Sinclair & Mark, 1992 for reviews). Thus "if positive moods lead to nonsystematic processing, failure to discriminate among stimuli, and thus the use of broader categories, then people in good moods should perceive less variation in fairness as rewards become more or less equitable, relative to subjects in other moods." (Sinclair & Mark, 1992, p. 176).

There appears to be some inconsistency between these findings and other research on categorization and mood, however. Mark & Sinclair (1992) showed the most pronounced effects for depressed subjects. Happy subjects seem only slightly different from subjects in a neutral mood in their fairness jugements. However, previous research on the effects of mood on categorization seems to suggest that only *positive* moods lead to categorization differences, relative to no mood controls (e.g., Isen & Daubman, 1984; and Murray et al., 1990), and if negative moods have any effect on categorization, it is in the *same* direction as the effect of positive mood (Isen & Daubman, 1984). Sinclair, Mark, and their colleagues (Sinclair, 1988; Sinclair and Mark, 1991 and 1992; and Sinclair, et al., 1992) have begun to provide evidence consistent with their notion that negative moods do lead to narrower or more precise categorization, and therefore, consistent with their proposed explanation for the effects of mood on justice judgements. This issue deserves further attention. Nonetheless, it is clear that negative moods *do* increase the likelihood that inequitable distributions will be seen as unfair, relative to equitable distributions.

5.   This effect is asymmetrical. For subjects who had been called on a sunny day, calling attention to the weather had no effect on their reported well-being. However, since we are concerned predominantly with the negative emotions that arouse injustice deliberations, this asymmetry does not alter our conclusions.

6.   See Fazio and Cooper, 1983; Zanna and Cooper, 1976; and Zillman, 1978 and 1983 for further discussion of the theoretical and empirical background of this misattribution paradigm.

7.   This is not an exhaustive presentation of affect control theory. Interested readers can consult other sources (Heise, 1979, 1985; and Smith-Lovin and Heise, 1987) for a complete specification of the theory.

8.   The coefficients in these equations are unstandardized, in order to allow prediction of the *actual* values of the transient impressions. All EPA ratings were scored using a metric ranging from -4 to $+4$, with an assumed interval scale (see Smith-Lovin, 1987, p. 43 for details).

9.   All of the generative aspects of affect control theory—predictions of behavior, of emotions, and of identity and trait attributions—are obtained by manipulating empirically based equations mathematically under the constraints of theoretical assumptions. We refer the reader to Heise (1987 and 1989) for the actual derivations.

10.   Theoretically, affect control theory could be applied using individualized fundamental sentiments. However, culturally shared meaning would be a major component of these sentiments. In applications of the theory, the initial sentiments are always taken to be the culturally shared, out-of-context sentiments (i.e., the mean EPA ratings).

REFERENCES

Adams, J.S. 1965. "Inequity in Social Exchange." In *Advances in Experimental Social Psychology,* Vol. 2, edited by L. Berkowitz. New York: Academic Press.

Averett, C. P., and D. R. Heise. 1987. "Modified Social Identities: Amalgamations, Attributions, and Emotions." *Journal of Mathematical Sociology* 13: 103-132.

Austin, W., N.C. McGinn, and C. Susmilch. 1980. "Internal Standards Revisited: Effects of Social Comparisons and Expectancies on Judgments of Fairness and Satisfaction." *Journal of Experimental Social Psychology* 16: 426-441.

Berger, J., M.H. Fisek, F.Z. Norman, and M. Zelditch, Jr. 1977. *Status Characteristics and Social Interaction: An Expectation States Approach.* New York: Elsevier.

Berger, J., M. Zelditch, Jr., B. Anderson, and B.P. Cohen. 1972. "Structural Aspects of Distributive Justice: A Status Value Formulation." Pp. 119-146 in *Sociological Theories in Progress*, edited by J. Berger, M. Zelditch, Jr., and B. Anderson. Boston: Houghton Mifflin.

Berger, J., M.H. Fisek, R.Z. Norman, and D.G. Wagner. 1983. "The Formation of Reward Expectations in Status Situations." Pp. 127-168 in *Equity Theory: Psychological and Sociological Perspectives*, edited by D.M. Messick & K.S. Cook. New York: Praeger.

Blalock, H.M., Jr., and P.H. Wilken. 1979. *Intergroup processes: A Micro-Macro Perspective.* New York: Free Press.

Deutsch, M. 1979. "A Critical Review of 'Equity Theory': An Alternative Perspective on the Social Psychology of Justice." *International Journal of Group Tensions* 1: 20-49.

_____. 1983. "Current Social Psychological Perspectives on Justice." *European Journal of Social Psychology* 13: 305-319.

_____. 1985. *Distributive Justice: A Social-Psychological Perspective.* New Haven, CT: Yale University Press.

Fazio, R. and J. Cooper. 1983. "Arousal in the Dissonance Process." Pp. 122-152 in *Social Psychophysiology: A Sourcebook*, edited by J.T. Cacioppo and R.E. Petty. New York: Guilford.

Fiedler, K. 1991. "Emotional Mood, Cognitive Style, and Behavior Regulation." Pp. 100-119 in *Affect, Cognition and Social Behavior: New Evidence and Integrative Attempts*, edited by K. Fiedler and J. Forgas. Toronto: C.J. Hogrefe.

Gallagher, D. and G.L. Clore. 1985. "Effects of Fear and Anger on Judgments of Risk and Evaluations of Blame." Paper presented at the annual meeting of the Midwestern Psychological Association, Chicago, Ill.

Goffman, E. 1963. *Stigma: Notes on the Management of Social Identity.* Englewood Cliffs, NJ: Prentice-Hall.

Gollob, H. F. 1968. "Impression Formation and Word Combination in Sentences." *Journal of Personality and Social Psychology* 10: 341-353.

Gollob, H.F., and B.B. Rossman. 1973. "Judgments of an Actor's 'Power and Ability to Influence Others.' "*Journal of Experimental Social Psychology* 9: 391-406.

Harris, R.J. 1976. "Handling Negative Inputs: On the Plausible Equity Formulae." *Journal of Experimental Social Psychology* 12: 194-209.

Heise, D. and E. Lewis. 1988. *Programs Interact and Attitude: Software and Documentation.* Dubuque, IA: Wm. C. Brown Publishers, Software.

Heise, D. R. 1966. Social Status, Attitudes and Word Connotations." *Sociological Inquiry* 36: 227-239.

_____. 1969. "Affective Dynamics in Simple Sentences." *Journal of Personality and Social Psychology* 11: 204-213.

_____. 1970. "Potency Dynamics in Simple Sentences." *Journal of Personality and Social Psychology* 16: 48-54.

_____. 1979. *Understanding Events: Affect and the Construction of Social Action.* New York: Cambridge University Press.

_____. 1985. "Affect Control Theory: Respecification, Estimation, and Tests of the Formal Model." *Journal of Mathematical Sociology* 11: 191-222.

_____. 1989. "Effects of Emotion Displays on Social Identification." *Social Psychology Quarterly* 52: 10-21.

Heise, D.R. and L. Smith-Lovin. 1981. "Impressions of Goodness, Powerfulness and Liveliness from Discerned Social Events." *Social Psychology Quarterly* 44: 93-106.

Heise, D. R., and L. Thomas. 1989. "Predicting Impressions Created by Combinations of Emotion and Social Identity." *Social Psychology Quarterly* 52: 141-148.

Higgins, E.T. 1987. "Self-discrepancy: A Theory Relating Self and Affect." *Psychological Review* 94: 319-340.

Homans, G.C. 1961. *Social Behavior: Its Elementary Forms.* New York: Harcourt, Brace, and World.

Humphreys, P. and J. Berger 1981. "Theoretical Consequences of the Status Characteristics Formulation." *American Journal of Sociology* 86: 953-983.

Isen, A. 1987. "Positive Affect, Cognitive Processes, and Social Behavior." Pp. 203-253 in *Advances in Experimental Social Psychology,* Vol. 20, edited by L. Berkowitz. New York: Academic.

Isen, A. and K.A. Daubman. 1984. "The Influence of Affect on Categorization." *Journal of Personality and Social Psychology* 47: 1206-1217.

Jasso, G. 1978. "On the Justice of Earnings: A New Specification of the Justice Evaluation Function." *American Journal of Sociology* 83: 1398-1419.

_____. 1980. "A New Theory of Distributive Justice." *American Sociological Review* 45: 3-32.

_____. 1983. "Social Consequences of the Sense of Distributive Justice: Small-group Applications." In *Equity Theory: Psychological and Sociological Perspectives,* edited by D.M. Messick & K.S. Cook. New York: Praeger.

_____. 1986. "A New Representation of the Just Term in Distributive-Justice Theory: Its Properties and Operation in Theoretical Derivation and Empirical Estimation." *Journal of Mathematical Sociology* 12: 251-274.

_____. 1990. "Methods for the Theoretical and Empirical Analysis of Comparison Processes." Pp. 369-419 in *Sociological Methodology 1990,* edited by C.C. Clogg. Washington, DC: American Sociological Association.

Jones, E.E., A Farina, A.H. Hastorf, H. Markus, D.T. Miller, and R.A. Scott. 1984. *Social Stigma: The Psychology of Marked Relationships.* New York: W.H. Freeman.

Kemper, T. D. 1990. "Social Relations and Emotions: A Structural Approach." In *Research Agendas in the Sociology of Emotions,* edited by T. D. Kemper. Albany, NY: State University of New York Press.

Lakoff, G. 1987. *Women, Fire, and Dangerous Things: What Categories Reveal about the Mind.* Chicago: The University of Chicago Press.

Leventhal, G.S., J. Karuza, Jr., and W.R. Fry. 1980. "Beyond Fairness: A Theory of Allocation Preferences." In *Justice and Social Interaction: Experimental and Theoretical Contributions from Psychological Research,* edited by G.M. Mikula. New York: Springer-Verlag.

Lerner, M.J. 1980. *The Belief in a Just World: A Fundamental Delusion.* New York: Plenum Press.

Lind, E.A. 1992. "The Fairness Heuristic: Rationality and 'Relationality' in Procedural Evaluations." Paper presented at the Fourth International Conference of the Society for the Advancement of Socio-Economics, Irvine, CA.

Lind, E.A. and P.C. Earley. 1992. "Procedural Justice and Culture." *International Journal of Psychology* 27: 227-242.

Lind, E.A. and T.R. Tyler. 1988. *The Social Psychology of Procedural Justice.* New York: Plenum.

MacKinnon, N. J. Forthcoming. *Analyzing Social Interaction: A Cross-Cultural Study in Affect Control Theory.* Albany, NY: State University of New York Press.

MacKinnon, N.J. and D.R. Heise. 1993. "Affect Control Theory: Delineation and Development." In *Theoretical Research Programs: Studies in the Growth of Theory,* edited by J. Berger and M. Zelditch, Jr. Stanford, CA: Stanford University Press.

Mark, M.M. and R.C. Sinclair. 1992. *Mood, Perceived Justice, and Categorization Breadth: A Processing Strategy Interpretation.* Unpublished manuscript, Pennsylvania State University, University Park, PA.

Markovsky, B. 1985. "Toward a Multilevel Distributive Justice Theory." *American Sociological Review* 50: 822-839.

Melton, R.J. and S. J. Scher. 1992. "The Role of Emotion in Judgments of Transgressors." Paper presented at the annual meetings of the Midwestern Psychological Association, Chicago, IL.

Messick, D.M. and K. P. Sentis. 1979. "Fairness and Preference." *Journal of Experimental Social Psychology* 15: 418-434.

_____. 1983. "Fairness, Preference, and Fairness Biases." In *Equity Theory: Psychological and Sociological Perspectives,* edited by D.M. Messick & K.S. Cook. New York: Praeger.

Mikula, G. and T. Schwinger. 1978. "Intermember Relations and Reward Allocation: Theoretical Considerations of Affects." In *Dynamics of Group Decisions,* edited by H. Brandstätter, J.H. Davis, and Heinz Schuler. Beverly Hills, CA: Sage.

Murray, N., H. Sujan, E.R. Hirt, and M. Sujan. 1990. "The Influence of Mood on Categorization." *Journal of Personality and Social Psychology* 59: 411-425.

O'Malley, M.N. and D.K. Davies. 1984. "Equity and Affect: The Effects of Relative Performance and Moods on Resource Allocation." *Basic and Applied Social Psychology* 5: 273-282.

Osgood, Charles E. 1962. "Studies of the Generality of Affective Meaning Systems." *American Psychologist* 17: 10-28.

Osgood, C. H., W. H. May, and M. S. Miron. 1975. *Cross-Cultural Universals of Affective Meaning.* Urbana: University of Illinois Press.

Russell, J. 1991. In Defense of a Prototype Approach to Emotion Concepts." *Journal of Personality and Social Psychology* 60: 37-47.

Schwarz, N. 1983. "Stimmung als Information: Zum einfluß von Stimmungen auf die Beurteilung des eigenen Lebens." (Mood as Information: The Influence of Moods on Judgment of My Life.) In *Bericht über den 33. Kongreß der Deutschen Gesellschaft für Psychologie in Mainz 1982* (Report of the 33rd Congress of the German Association of Psychology in Mainz, 1982), edited by G. Luers. Göttingen: Hogrefe [Cited in Schwarz 1990].

Schwarz, N. 1990. "Feelings as information: Informational and Motivational Functions of Affective States." Pp. 527-561 in *Handbook of Motivation and Cognition: Foundations of Social Behavior,* Vol. 2, edited by E.T. Higgins & R.M. Sorrentino. New York: Guilford.

Schwarz, N. and G. Clore. 1983. "Mood, Misattribution, and Judgments of Well-being: Informative and Directive Functions of Affective States." *Journal of Personality and Social Psychology* 45: 513-523.

_____. 1988. "How Do I Feel About It? The Information Function of Affective States." Pp. 44-62 in *Affect, Cognition, and Social Behavior,* edited by K. Fiedler and J. Forgas. Lewiston, NY: Hogrefe.

Schwarz, N., W. Servay, and M. Kumpf. 1985. "Attribution of Arousal as a Mediator of the Effectiveness of Fear-Arousing Communications." *Journal of Applied Social Psychology* 15: 74-78.

Sinclair, R.C. 1988. "Mood, Categorization Breadth, and Performance Appraisal: The Effects of Order of Information Acquisition and Affective State on Halo, Accuracy, Information Retrieval, and Evaluations." *Organizational Behavior and Human Decision Processes* 42: 22-46.

Sinclair, R.C. and Mark, M.M. 1991. "Mood and the Endorsement of Egalitarian Macrojustice Versus Equity-based Microjustice Principles." *Personality and Social Psychology Bulletin* 17: 369-375.

_____. 1992. "The Influence of Mood State on Judgment and Action: Effects on Persuasion, Categorization, Social Justice, Person Perception and Judgmental Accuracy." In *The Construction of Social Judgments,* edited by L. Martin & A. Tesser. Hillsdale, NJ: Erlbaum.

252   STEVEN J. SCHER and DAVID R. HEISE

Sinclair, R.C., Mark, M.M., Weisbrod, M.S., & Enzle, M.E. 1992.*The Effect of Mood on Categorization Breadth: Flexibility or Differential Processing Strategy?* Unpublished manuscript, University of Alberta, Edmonton, Alberta, Canada.

Smith-Lovin, L. 1987a. "Impressions from Events." *Journal of Mathematical Sociology* 13: 35-70.

_____. 1987b. "The Affective Control of Events Within Settings." *Journal of Mathematical Sociology* 13: 71-101.

_____. 1982. "A Structural Equation Model of Impression Formation." Pp. 195-222 in *Multivariate Applications in the Social Sciences,* edited by N. Hirschberg and L.G. Humphreys. Hillsdale, NJ: Lawrence Erlbaum.

_____. 1988. *Analyzing Social Interaction: Advances in Affect Control Theory.* New York: Gordon and Breach Science Publishers. (Reprint of a special issue of *The Journal of Mathematical Sociology,* Vol. 13.)

Smith-Lovin, L., and W. Douglas. Forthcoming. "An Affect Control Analysis of Two Religious Subcultures." In *Social Perspective in Emotions,* Vol. 1, edited by V. Gecas and D. Franks. Greenwich, CT: JAI Press.

Stouffer, S., E. Suchman, L. DeVinney, S. Star, and R. Williams. 1949. *The American Soldier: Adjustment During Army Life.* Princeton, NJ: Princeton Univeristy Press.

Thibaut, J.W. and H.H. Kelley. 1959. *The Social Psychology of Groups.* New York: Wiley.

Thibaut, J.W. & L. Walker. 1975. *Procedural Justice: A Psychological Analysis.* Hillsdale, NJ: Lawrence Erlbaum.

Thoits, P.A. 1990. "Emotional Deviance: Research Agendas." In *Research Agendas in the Sociology of Emotions,* edited by T. D. Kemper. Albany, NY: State University of New York Press.

Tyler, T.R. and E.A. Lind. 1992. "A Relational Model of Authority in Groups." Pp. 115-191 in *Advances in Experimental Social Psychology,* Vol. 25, edited by M. Zanna. New York: Academic.

Walster, E., E. Berscheid, and G.W. Walster. 1973. "New Directions in Equity Research." *Journal of Personality and Social Psychology* 25: 151-176.

Walster, E., G.W. Walster, and E. Berscheid. 1978. *Equity: Theory and Research.* Boston: Allyn & Bacon.

Zanna, M. P. and J. Cooper. 1976. "Dissonance and the Attribution Process." Pp. 199-217 in *New Directions in Attribution Research,* Vol. 1, edited by J.H. Harvey, W. Ickes, & R.F. Kidd. Hillsdale, NJ: Lawrence Erlbaum.

Zillman, D. 1978. "Attribution and Misattribution of Excitatory Reactions." Pp. 335-368 in *New Directions in Attribution Research,* Vol. 2, edited by J.H. Harvey, W. Ickes, & R.F. Kidd. Hillsdale, NJ: Lawrence Erlbaum.

_____. 1983. "Transfer of Excitation in Emotional Behavior." Pp. 215-240 in *Social Psychophysiology: A Sourcebook,* edited by J.T. Cacioppo and R.E. Petty. New York: Guilford.

WHAT IS A GROUP?
A MULTILEVEL ANALYSIS

Per Månson

ABSTRACT

Small group research has in later years developed into many specialized areas, but it has not revived its former position within either social psychology or social science in general. One of the reasons for this could be ambivalence about the basic concept, the group, and its ontological and epistemological foundations. Without any ambitions to solve this basic problem, this article discusses the concept of group, and relates the group level of analysis in social science to subgroup and supergroup levels of analysis of social reality. Problems emanating from the use of laboratory groups as a prototype of an abstract group are related to the differences between ontological abstract and concrete groups and epistemological abstract groups.

INTRODUCTION

Everywhere I turn the small group is being rediscovered. In the psychiatric hospital ward, a patient's relations with other patients have been found to influence the course of this

Advances in Group Processes, Volume 10, pages 253-281.
Copyright © 1993 by JAI Press Inc.
All rights of reproduction in any form reserved.
ISBN: 1-55938-280-5

rediscovery. In the classroom, the extent to which the instructor is "teacher-centered" or "learner-centered" appears to affect the learning process. In industry, "brainstorming" in "group-think" sessions has been introduced as a creative problem-solving technique. Psychologists, who used to be content to describe the subject's response to the color wheel or the rat's response to the maze, now study the influence of group norms on individual judgement or the ways in which groups of animals influence the behavior of their fellows. Sociologists, who might once have studied whole socities or institutions, now record the behavior of small groups either in the laboratory or in the field. (Hare 1962, p. v.)

In the 1950s and 1960s, there was a flourishing optimism in the literature about the status and future progress of small group research. In an article from 1954, it is shown that the bibliographic references to small group studies had increased enormously in the 1940s and the beginning of the 1950s (Strodtbeck & Hare, 1954, p. 110). In the following years, numerous books and studies on small group research continued to be published, and in the second edition of *The Handbook of Social Psychology* from 1968, one whole volume out of five was dedicated to "Group Psychology and Phenomena of Interaction."

But then something happened. In 1973, small group theory was characterizes as "the light that failed" (Mullins, 1973). One year later, Steiner meant that small group research had lost much of its influence and vitality after its rapid growth in the 1940s and 1950s (Steiner, 1974). Steiner associated this loss of influence with the increasing individualism of American social psychology in general in the 1960s. His explanation to this was that the American society in the 1950s and beginning of 1960s was a serene society, where "only a few wrong-headed deviants disturb(ed) our tranquillity... The Eisenhower years were a tranquil interlude during which American society must have resembled a swarm of single-minded ants" (Steiner, 1973, p. 105).

In such a societal situation, social psychologists, according to Steiner, are more inclined to study either individuals or large organizations and institutions, which also could be an explanation for the growing difference between the sociological and psychological branches of social psychology in general. But because American society experienced a great turmoil in the latter half of the 1960s and in the early years of the 1970s, Steiner predicts that studies on small group processes once more will grow, but "with a lag about eight or ten years behind" (1973, p. 105).

Accordingly, in the beginning of the 1980s, there should have been an increasing research activity on group processes. Did this happened? In another article Steiner answers the question:

[I] am now forced to acknowledge that my foresight was somewhat faulty. Although there has been a detectable resurgence of groupy research, the increase has been small by comparison with the upsurge of the late 1940s and early 1950s. For the most part social psychology has remained a discipline that examines the activities, cognitions, and feelings of single individuals. The dynamics of human behavior are still generally concieved to be lodged deep inside the actor himself. (Steiner, 1983, p. 541.)

So, although there was a modest revival of group research in the late 1970s, Steiner's basic hypothesis of a direct influence of a societal turmoil on the magnitude of social psychological group research has to be modified. As is usual in science, he introduces new counteracting variables. This is not the place to analyze these variables, most of which are concerned with the *differences* between the turmoil of the 1930s and the turmoil of the late 1960s and early 1970s. There is also one "psycho-economic" variable introduced, which is concerned with how researchers behave when their jobs are scarce. All this makes the relations between the state of the society and the amount of social psychological group research so complicated that Steiner is very cautious in his future prediction: "Eventually the group will rise again, but eventually may be a long time. Soothsayers ought not to specify precise dates" (p. 547).

Besides his rather poor prediction of 1974, and the cautious prediction of 1983, Steiner's method of analyzing small group research is very interesting. It is a sociology-of-knowledge analysis, but what makes it interesting is that Steiner does not use any argument or results from group research himself. He could have done so, treating the social psychologists as a group, and scrutinized, for example, how different networks develop in different areas of social psychology, how different status organizing processes work inside and outside the different groups in social psychology, and how results of previous group research could be used, at least partly, to explain the dependent variable, the amount of social psychological research on small groups. Of course, this would have been a very difficult task, but it is a telling fact that he did not try or even discuss it.[1]

In this way, Steiner is an example of what he himself is exploring; the vanishing use of group variables to explain people's behavior. Instead, on one hand he introduces an individual psychological variable (how people, including academicians, behave in time of scarcity of jobs) and, on the other hand, he turns to "macro variables," the culture mood, the role of mass media, the decline of legitimicy of political institutions etc. He treats his research subjects, social psychologists, as free individuals in a macro societal context, without any intermediatory group level. His argument on the use of group research suggests that this is exactly where small group theory could be strong: groups as intermediary links between individuals and society.

What has happened to small group research after the middle of the 1980s is very difficult to say in general. Without being an expert in the field, it is my impression that small group research has regained some of its development potentials, in line with Steiner's more cautious prediction from 1983. The present series, *Advances in Group Processes*, is one example. The research program on "status organization processes" at Stanford University (Berger, Rosenholtz & Zelditch, Jr., 1980; Wagner & Berger, 1990; and Berger & Zelditch, Jr., 1991), the structural small group theories in Iowa (Ridgeway, 1983; Markovsky, 1987; Markovsky & Willer, 1988; and Lawler, Ridgeway,

& Markovsky, 1993), are others. Some of this group research is centered around the problems of power, status, and just distribution in groups, and especially the latter phenomenon of distributive justice has been an active field of small group social psychology (Törnblom, 1992). In Europe, Henri Tajfel has developed "the 'minimal group paradigm', [which] has greatly stimulated theoretical thinking in the field of sociopsychological studies of intergroup relations...[and] about a dozen or more volumes (...) have since appeared" (Doise, 1988, p. 100). Here, too, studies on social identity and social categorization in group formation have been an important area of research (Turner, 1987).

Also, the more psychologically oriented small group research has developed, but has, according to one overview, "especially in recent years, been relatively *problem oriented*" (Davies and Stasson, 1988, p. 248). Group problem solving, risky shift, attitude polarization, and social impact are some examples of the scattered subjects of small group research in the 1980s. But our general impression is that small group research, both in its psychological and sociological variants, is still missing as a coherent and fascinating area of research inside social science, at least as compared to "the golden age" of the 1940s and 1950s. Small group research is now scattered over many different areas, which study different subjects from different theoretical points of view. The area has developed many specialized fields, where different research programs and research traditions develop without much reference to each other, and where a basic research program of studying and finding out how groups work in general more or less has been abandoned. There are, of course, both societal and scientific reasons for this, but it could also be due to a more basic problem within the study of small groups. This basic problem could be seen as a combination of two closely related questions; "*What is a group?*" and "*How it is possible to study a group?*" These ontological and methodological questions have, in our opinion, still not been satisfactory analyzed in the tradition of small group research.

In this article, I discuss some of the basic ontological and methodological problems connected with the scientific study of small groups. I shall do this with the help of a multilevel analysis, and try to find the general types of relations that exist between different levels of social reality. The article ends by a short discussion of what these difficulties could mean for the future status and nature of group research. First, the location of small group studies inside the field of social science is discussed.

GROUP THEORY, SOCIAL PSYCHOLOGY AND THE LEVELS OF SOCIAL SCIENCE

Human beings live as social beings by their very nature always together with other human beings. Most of this living together takes place in different kinds of groups, and in this respect *all* social sciences can be regarded as a group science. There are, of course, a few examples of persons, who for religious or other reasons have chosen to isolate themselves, but one can argue that they carry other persons and different reference groups mentally inside themselves in their physical loneliness. But this fact only says that humans are *social* creatures, and it does not us tell anything about either how groups work or how different groups influence individuals. The study of group and influence processes has developed into a specialized field and research tradition in one of the subdisciplines of the social sciences, social psychology.

There are many different definitions of the field of social psychology, and it is common to differentiate between two or more kinds of social psychology. Most writers on the subject differentiate between a Psychological Social Psychology (PSP), and a Sociological Social Psychology (SSP), but there are writers who also differentiate between three och four kinds of social psychology (Boutilier et al., 1980; House, 1977; McMahon, 1984; and Stryker, 1977, 1989). Why there are so many different definitions of social psychology is due to the basic theme of the field: the relation between mental and social factors in human life. In other words, social psychology is, by definition, a part of the social sciences which tries to relate the "inner" mental world of human beings to their "outer" social world. As many have pointed out, social psychology is thus a "border crossing" discipline, which, amongst other things, means that it tries to combine different levels of human reality.

In social science it is customary to differentiate between levels of reality and between levels of analysis. Without pretention to cover all that is possible in a science of society, one can distinguish between the following levels:

1) Intraindividual (or mental);
2) Individual;
3) Interindividual (or relational);
4) Group;
5) Organizational;
6) Institutional; and
7) Social system or macro social structure.

As can be seen, the levels go from micro to macro. The group level is placed in the middle, and, together with the organizational level, the group level is sometimes referred to as the "meso level." Of course, there will be different points of views of the exact definitions of these levels and of the exact borderlines between them, but, in general, the following definitions could work:

(1) *Intraindividual level.* Here, the scientist studies how subindividual entities, such as biological, chemical, or psychological factors, influence each

other in the individual. Most psychiatric theories, and some psychological theories, including the "classical" Freudian theory, are situated at this level. The basic model here is to analyze how different "elements" of an individual are structured, how they influence each other, and how they affect the individual's behavior, and in some cases also how they influence higher levels.

(2) *Individual level.* Here, the individual is the basic entity of study. The individual is seen as a self-containing system, which influences other individuals, and these influences explain what is happening in the social life. Most traditional psychology is situated at this level, as is much of social psychological research from Tripplet's experiment and so on. The basic assumption here is that "only individuals exist," and that all super levels must be reduced to this individual level.[2]

(3) *Interindividual level.* Here, the *relation* between two or more humans is the basic object of study. The individual in itself is seen as "an empty abstraction," and significant parts of what is happening on levels 1 or 2 are explained by what is happening on level 3. Also, higher levels of society could be explained by phenomena of this level. Most of symbolic interactionist theory and parts of modern psychoanalytic theory (object relation school) work on this level.

(4) *Group level.* Here, *structured* relations between humans are the focus of study. This means that on this level, the researcher studies individuals and their relations, though in a very special way. The group must, in one way or another, be a structured system, which has well defined boundaries to other relations or other parts of social life. To talk about a group is, in essence, to talk about these boundaries between the group and the rest of the world, or between the group and other groups.

(5) *Organizational level.* An organization is an intended social system, with goals and means, organzational plans and patterns, with different positions and roles. An organization is differentiated from other levels exactly in that it is consciously created in order to fullfill some explicit tasks. As everyone knows, it is a totally different question if these tasks are fullfilled, and if the chosen means or organizational plan is the best for the fullfillment of the tasks.

(6) *Institutional level.* The concept of social institution is one of the most important concepts in (Durkheim-inspired) sociology. It indicates a superindividual, and, from the point of view of the single individual, *unconsciously created* entity of social life. Here, the researcher studies large clusters of roles, norms, values, and prescriptions in society that has come to life in the societal development. As such, social institutions, of course, need human beings, but the basic idea of the concept is that social institutions are "external" and "constraining" in relation to the individual, and that there are no individual explanation to social institutions.

(7) *Social system or macro social structure level.* At this level, "whole" societies, often some kind of national state or even systems of national states,

are the object of study. A social system is often conceived of as being a system of social institutions, which, especially in structural functionalist analysis, are regarded to fullfill some "functional requisita" for the equilibrium of the social system. The concept of social system or social structure is a concept of "the whole," in relation to "the parts" of society. It is not logically necessary, but often "the system" is seen to have priority over "the parts."

It is of course, possible to talk about other levels of analysis in social science, for example, class, culture or phenomenological levels. But these levels could probably be defined using the above-mentioned levels. Because of this, it appears that the level model above represents a rather common way of looking at the different levels of social life. These levels do not necessarily go from "small" to "big" entities, because there are, for example, big groups and small organizations, but, in general, they are ordered from micro to macro phenomena in the order above.

With the help of this model it is possible to discuss what group theory in social science is or could be. A simple answer would be that group theory is the part of social science that studies phenomena on the fourth level. But that would be an insufficient answer, because it only says that group theory *does not* study phenomena on intraindividual, individual, interindividual, organizational, institutional or social system levels, is not even true. It is rare in the social sciences that a researcher only studies *one* level, as a self-containing entity. On the contrary, most social scientists study *how different levels affect each other,* for example, how intraindividual factors affect the behavior of an individual, how groups affect individuals, or how social systems affect social institutions (or vice versa). That is one of the reasons why social psychology is such a problematic field, since it is a field which by its very definition tries to relate different levels of reality to each other.

In social science, there is also a division of labor between different disciplines. Where psychology mostly studies phenomena on levels 1 and 2 (and, to a small degree, also level 3), sociology spans from level 3 to 7 (or, as in the cases of Homans or modern rational choice theory, from 2 to 7). Political or economic science can be defined, in sociological concepts, as disciplines which study particular social institutions or social systems, the political or administrative system, and the economic system, respectively. This does not mean that they only study these phenomena as institutions or systems. On the contrary, they also study political/administrative or economic phenomena on lower levels, which, for example, can be seen in the division between micro- and macroeconomics. Social psychology is probably the most complicated field in social science, because its object of study ranges from level 1 to 7, and, at the same time, it does not have a special social institution to study like political or economic science. In short, social psychology studies everything, from intraindividual to macrosystem processes. And, as social psychology, by its very definition, always relates (or should try to relate) one or more of the levels to each other, it is no wonder that social psychology, in August Comte's words, is the queen of the social sciences.

Different parts of the field of social psychology can, in many respects, be differentiated by which levels they incorporate in their studies. Another way of differentiating them is through their various meanings of the key concepts in the word social psychology: social and psychological, and the way they relates the areas of reality to each other.

While most of PSP regards the mental factors as situated *inside* the individual and has developed concepts for this (personality, cognition, motivation, emotions, behavior, and so on), and often treats the social factors as undifferentiated "other people" (out there), SSP has less developed concepts of the mental factors, but much more developed concepts of the social factors (race, class, gender, organization, social structure, culture, and so on). This is not surprising. Because PSP comes from and is a part of psychology, psychological concepts are much more elaborated here than in SSP. And, of course, the reverse is true of SSP. A genuine social psychology would be a science that conceptually has developed both sides of the relation. The need for these double elaborations of concepts is the background for the building of interdisciplinary research centers in the 1940s, and also why so much emphasis was placed on interdisciplinary research projects in the so-called "crisis debate" in social psychology in the 1970s in order to overcome the supposed crisis of the field.

Group theory can be defined as the part of social psychology where factors on the group level are the object of study, either as independent or dependent variables. A theory of group structure or group dynamics which have group variables as both dependent and independent variables is also possible, but most group theories relate the group level to one or more other levels of social reality. Like social psychology in general, "current theory and research on small groups can be traced to both psychological and sociological predecessors" (Collins and Raven, 1968, p. 102). I think that the psychological group theories are mostly concerned with relating levels 1 and 2 to the group level, while sociological group theories either treat the group level as an independent level, or relate the group level to higher levels. Lewin's Field Theory, considered as a motivational theory, can be seen as a "paradigmatic example" of a relation downwards from the group level, and the following statement can be regarded as typical for a sociological group theory:

> Studies of groups, I feel, need to be placed in a perspective of larger organizational structures. Ours is a society of large and complex organizations. The most difficult human relations problems are found within these organizations. We cannot afford to lose sight of the small group, but neither can we afford to study it in isolation. (Whyte, 1951, p. 297.)

Thus, psychological group theory (PGT) uses the group as "the social factor," which often constitutes the independent variable. The mental or individual factors are often regarded as dependent variables (most cognitive or individual behavior). The general problem of PGT is, therefore, not separated from the general approach to the study of social behavior in psychological social psychology. The best and most often quoted definition of PSP is Gordon Allport's: "With few exceptions, social psychologists regard their discipline as an attempt to understand and explain how the thought, feeling, and behavior of individuals are influenced by the actual, imagined, or implied presence of other human beings" (Allport, 1968, p. 5). In PGT, the expression "presence of other human beings" refers to a group of people, and using Allport's definition one can say that PGT studies "how a group influences thoughts, feelings and behavior of an individual."

In sociological group theory (SGT), the problem is a bit different, because in the sociological context the small group is viewed as a part of society, and not only as something which influences individuals. Although Homans many times has been critized for his "individualistic fallacy," he has expressed this point very clearly:

The study of the human group is a part of sociology, but a neglected part. As the science of society, sociology has examined the characteristics and problems of communities, cities, regions, big organizations like factories, and even smaller social units that make up these giants. In doing so, it has followed the order of human experience, for the first and most immediate social experience of mankind is small group experience. From infancy onward we are members of families, childhood gangs, shool and college cliques, clubs and teams— all small groups. When, as grownups, we get jobs, we still find ourselves working with a few persons and not with the whole firm, association, or goverment department. We are members of these larger social organizations, but the people we deal with regulary are always few. They mediate between us and the levithans. The group is the commonest, as it is the most familiar, of social units, and on both counts it is at least worth study as any of the others. Sociology might have begun here. (Homans, 1950, p. 1-2).

In sum, small group theory can be seen as a part of social psychology in general, which shows the same cleavage between a psychological oriented research tradition and a sociologically oriented research tradition as the field of social psychology in general.[3] Thus, the basic problem of social psychology, how the relation between the mental and the social processes works in human life, is also a basic problem in small group theory. Because of its embeddedness in the field of social psychology, small group theory has a weak position in nonsociological psychology and nonpsychological sociology. One of the main reasons for this, I think, is the difficulties with the basic concept itself, the group, and its place in social science in general. So, what is a group, and how is it related to other levels of the social reality?

THE REALITY OF GROUPS

But what, one might ask, are small groups? More familiar to us are concepts of the individual, the organization, the institution, and the large society. Yet between the single person and organization are units composed of two or more persons who come in contact for a purpose and who consider the contact meaningful. Some, like families, are relatively seperate, while others, like boards of directors, are parts of larger units. Still others include a construction gang, a hunting party, a town committee, a ceremonial dance team, a bomber crew, and an athletic team. There are many more. In fact, with some 3.2 billion individuals in the world and with each one on the average belonging to five or six groups, and allowing for overlap, an estimate of the total number of small groups existing now would be as high as from four to five billion. When we add past and future groups to the present groups, we see that the universe of small groups runs into many billions—many more than societies, even more than seperate individuals (Mills, 1968, p. 45).

Throughout history people have joined together in groups to accomplish a wide range of purposes. Men and women form personal relationships to procreate, to raise families, and/ or to make the business of the day-to-day living more interesting and meaningful. These family groups are probably among the oldest and most basic types of groups, but many other types come readily to mind: decision making groups, discussion groups, political groups, athletic teams, committees, fraternities...The list could be extended ad nauseam. It is apparent that a large proportion of human behavior occurs in groups. Goverments conduct much of their business in groups, whether they be the Central Committee of a socialistic society, the President's Cabinet in a republic, or a group of select advisers to the head of a dictatorship. Industrial organizations employ committees to improve quality of decisions or to reduce the probability of defective decisions or products; educators employ group discussions to facilitate learning; college students form groups to facilitate social interaction, to promote common causes, or to protest injustices. It has been estimated that there are four or five million groups in existence at any given time (Shaw, 1981, p. 1).

The two quotations describe the small group very much in the same way, and many other quotations from the literature on small groups could be added. The quotations, and most of the literature on small group, say that small groups can be found everywhere there are human beings, and that there are different kinds of groups, families, friendship groups, working groups, military groups, political groups, sport groups, and so forth. The groups are purposefully formed by individuals who want to gain something out of it. In summary, groups can be regarded as something *universal,* which means that they are transhistoric and transcultural, they are frequently *purposefully* formed, and there are *many kinds* of groups. If one adds to this that the group is often seen as consisting of two or more persons, one can define a group as something that comes into existence whenever two or more people meet and interact.

This view says that the basic substance of a group is the social interaction that takes place in it. In this psychological view of groups, the concepts of interaction and group are very close. This can be seen, for example, in *The Handbook of Social Psychology,* whose fourth volume is titled "Group Psychology and Phenomena of Interaction." The group is something that is a part of "the social," that is, an entity "out there" from the perspective of the individual. In this tradition, thus, group processes are closely associated with influence processes, which is obvious in the famous experiments of, for example, Sherif and Asch.

Since this way of defining social groups does not seperate *different kinds* of social influences, it cannot either differentiate between different types of groups or between different parts of the social structure that influence the individual. For example, in a family, which undoubtedly consists of a number of persons, who (at least the parents) have formed their family for some purpose, the influence processes are based on the *institutional character* of the group. That is, it is based on the superindividual norm system which determine the family as a social institution. In a friendship group, on the other hand, there could also exist superindividual norms. But these norms are rules how to behave toward friends, not how to behave toward husband, wife, and children. Therefore, the influence processes are likely to work differently in a family and a friendship group respectively. To abstract this difference between these two kinds of groups is to regard them just as groups, not determined by anything else than being groups.

To reach the level of abstraction where all social relations are considered group activity, one has to lose all concrete determinations which make one group a family, another a board meeting, a third an athletic team, and so on. Maybe there are certain general social laws operating when people meet (Goffman, for example, has tried to show general patterns when people meet and interact in his dramaturgial social psychology), but they are general social interaction laws. The level of abstraction is here very important, because if a social group is regarded equivalent to social life in general, then the conclusion must be *that there is no special group level,* and that groups are formed on every level of the society. This means that group theorists do not study social processes on the group level, but instead they study what can be called *patterned sociability.*

The concept of patterned—or structured—sociability, comes from the symbolic interactionist tradition, where the Swedish social psychologist Johan Asplund has coined the concept of social responsivness. This concept refers to a basic human inclination to be social in oneself. That is, not being social as a means to an end, which the psychological group theory sometimes presupposes is the reason why individuals form groups, but socially for its own sake. An unsocial individual is, therefore, in the symbolic interaction tradition, not a human being at all. From the point of view of this concept, all the above mentioned examples in the quotations are, therefore, not only examples of different group activity, they are also, and more, examples of the basic social nature of man.

The important difference here is that the psychological social psychology takes its point of departure in the individual, who is regarded the carrier of the mental aspect, and the group is something "out there." The individual chooses to "go out there" and form a group or join an already existing group. This type of group theory, therefore, studies what happens with the individual when s/he chooses to be social and enter social relations. The individual comes first, both in reality and in theory. The social world is secondary, and the scientific problem is what happens to the individual when s/he enters the social world. Because of this, the basic research model of psychological group theory is that the group processes are independent variables, and the individual or subindividual processes are dependent variables. Or, to put it more simply: What happens with the individual when s/he decides to be social?

In symbolic interactionism, it is the other way around. Here, the social is primary, and the mental is secondary. One could even say that the social life is the cause of the individual life. The social is not just something "out there" for the individual, but the base of his/her individual existence. When people marry, work, meet friends, engage in athletics, and so forth, it is not primarily due to individually based instrumental reason, but to the basic fact that to be and stay human, we need other human beings. That is why we engage in relations all the time, that is why we influence each other, and that is why we also form more stable forms of social life.

The concept of patterned sociability refers to the fact that when we engage in relations, we do not do it in a haphazard way, but in a structured way. For the individual, the social world "out there," to which s/he belongs through her/his inner mental world, is always structured in different patterns. These patterns could be situated on different levels, such as the group, organizational, institutional or system levels. To relate oneself to one of these levels is not the same thing as to relate to just another individual. If, for example, a woman plans to marry, she does not only think of the future husband (if she is not madly in love), she also thinks about the wedding, maybe the church, the wedding party, her relatives, and so on. This means that she is not only relating herself to the husband, but also to the institutionalized rituals of the wedding, and to the whole social institution of marriage.

If, on the other hand, a boy joins an athletic club, he does it because the club is an organization with the explicit goal of practicing athletics. The purposive element in group forming or group joining is not only related to other concrete people, but also to elements of the social structure. These elements are always situated at different levels of the social reality, and the individual reacts to this structured social reality. This concept of patterned sociability is, I think, a parallel to Mead's concept of the generalized other.

Thus, in symbolic interactionist tradition, all the levels, from intraindividual to macro system level, are related to each other in the concrete act of the individual. The individual level is a point of intersection between the intraindividual level on one hand, and the relational to the social system levels on the other. The social is, therefore, both "outside" and "inside" the individual, and the social influences do not only work when an otherwise solitary individual decides to meet other otherwise solitary individuals. The social influence processes operate all the time and on every level of social reality. But, the mechanisms behind this influence upon individuals are quite different from each other.

First, the way intraindividual factors influence the individual is through thinking, planning, reflecting, feeling, and unconciously steering by repressed feelings. Second, the way relations with other people influence the individual is basically through the process of role-taking. Third, the way an organization influences the individual is through the goals and means of the organization, organizational plans and patterns, and, in reality, also the informal patterns that always develop in every formal organization because of basic human social responsiveness. These informal patterns can been regarded as a part of the relational level, and they work therefore through the same role-taking processes.

Fourth, the way a social institution influences the individual is as a *conscience collective,* that is, through a general knowledge of "how to do things." Here, we find cultural goals and values, norms, and so forth, which are generally accepted in the society. And, finally, the macro social system influences the individual through the generalized character of the interdependence of all the social relations, organizations and social institutions in a certain society influencing the individual through social relations, organizations, or institutions. Any change in the social system will always have the character of changes in relations, organizations, and/or institutions. When the social system an individual lives in is related to *another* social system, for example, when "America" is related to "Sweden," this relation is mediated through personal relations, organizations, or institutions. Thus, the social system seldom affects the individual directly from its own level, but the influence is mediated through other levels.

How does a group in abstracto influence the individual? It is impossible to answer this question as long as the concept of group exists on such an abstract level that makes it work on all social levels. If to form or join a group is equivalent to entering into any kind of interaction with other human beings in any social circumstances, then the group influences can work in many different ways, through role-taking, organizational plans, superindividual norms and value systems and the inner interdepedence of the social system. A group is always embedded in some social circumstances, and dependent on the concrete circumstances, the group will influence the individual in different ways.

Therefore, a group in abstracto is a group that is just a group, not a family, not a friendship group, not a president's cabinet or anything else, just a group. If we abstract all concrete determination, we have some people, two or more, who interact with each other. What is it? It is nothing else than two or more persons who interact with each other. But is this a group? In this question is buried one of the greatest problems in the whole small group theory tradition.

LEVELS OF GROUP REALITY

Imagine three friends who happens to meet in the street by accident. They decide to go to a café and drink a cup of coffee and talk. Do they form a group? If we use the common definition of a group, we must say yes, because there are two or more persons who interact with each other, and in this interaction they influence each other. After their talking they part and go home. The group did not exist before they met, and it ceases to exist when their meeting is over. This type of group is a result of the friends' decisions to spend some time together, whatever their different reasons could have been.

This type of group has an empirical existence only for a short time. Before the meeting it is not a group, and after the friends have left it only exists as a *memory* of the meeting. The group was formed in the everyday ongoing social life, and not for any other reason than to meet. This group can, therefore, be called a *spontaneously formed* group, or a *relation seeking* group. The group level is determined by level 3, the interindividual or relational level. It is, of course, also possible to argue that the group was formed because of the personalities of the friends, and, thus to use phenomena on level 1 and 2 to study it. Also, level 6, the social institution level, could be of interest if one, for example, tries to understand how this group works and how it affects the individuals in the group through institutionalized rules of behavior among friends.

But imagine that this group meets regularly on a certain day each week. Perhaps the members meet each other at a pub at the same time every Friday afternoon, in order to spend some time together, drink beer, talk, and relax after their work. In this example, the friends do not form the group by chance or impulse. Their regular meetings enter into an *institutionalizing process.* This implies that the group does not exist empirically only once, but that it also exists between the meetings in the heads of the group members and influences their behavior. The group has no obvious existence outside their meetings, but it has become a social fact through its influence on the social behavior of its members (and maybe some of their relatives, too).

Accordingly, when social relations are being institutionalized, the group also gets another form of existence. What ontological form of existence such a pattern of institutionalized relations has is, of course, a matter of discussion. From a Durkheimian point of view, however, one can say that this kind of group exists as a social fact *sui generis,* that is, on its own. This means, that the group will also exist in the social structure, when it does not have an empirical existence. Perhaps it will be named "Friday club," and maybe one of the members is absent one week, and another friend joins the group instead. Some group norms could have developed about how to drink, what to talk about, and so forth, like in Sherif's experiment. Thus, when a newcomer joins this "Friday club," he will have to adjust to these group norms, and learn to behave according to them.

But the processes of institutionalization of the social relations could be of very different kinds, depending on whether the relations are characterized by personal interest, or if they are a part of an already existing organization or social institution. This means that there are at least two different kinds of groups. The first type of group is characterized by *personal relations,* and in the group process there does not exist any superindividual position network with social roles. The group norms, which develop in the friendship group, are connected with *interpersonal roles* (for example, the role of jester, philosopher, and so forth) which are tied to the particular persons, not to superindividual positions. Accordingly, this type of group and its group norms are determined by the lower levels of social reality.

Another type of group is determined by the superindividual social structure, in which the group is embedded. In these type of group, the influence processes and the behavior of the participants are mediated through an organization or social institution. The individuals are, in principle, exchangeable, but they influence the concrete working of the group with their personalities, ambitions, goals, and behavior. The difference between this B-type and the A-type group is their opposite determination by other levels of social reality. In summary, the group level can be determined from two entirely different directions, one from "below" and one from "above":

Group A: Level 1, 2, or 3 determine level 4

Group B: Level 7, 6, or 5 determine level 4

Or in one combined model:

Level 1, 2, or 3 ———> A   level 4   B <——— level 5, 6, or 7.

But does a separate group level exist, which is not determined either "from below" or "from above"? This question is a part of the larger question of the forms of existence of all levels of social reality, or to put the question more concretely: Do the intraindividual, individual, relational, group, organizational, institutional, and social system levels exist separately from each other in human life without all the other levels? There are many answers to this question, but, in this paper, we take the point of view that these levels can only exist together in a structured social totality. To speak of levels inside a totality implies, therefore, understanding the concept of level as an analytical concept, not as an isolated ontological concept.

Everything in the human world, from intraindividual phenomena to social system phenomena, is ontologically related. To speak of different levels of human social reality is to carry out a methodological process of abstraction. There are no social systems without mental processes, individuals, relations, groups, organizations, or institutions. It is possible to argue in the same way for every separate level. Accordingly, just as there does not exist any ontologically isolated level in the social world, there could not exist any isolated group level in reality. This means that groups are always related to and determined by other levels of social reality.

GROUPS IN THE EMPIRICAL REALITY AND IN THE SOCIAL STRUCTURE

To say that groups do not exist as isolated and undetermined phenomena in the social world is, however, not the same thing as to say that they do not exist at all. Obviously, there are groups in the real world, and the question is, what kinds of social phenomena are they? In the literature on group theories there are many definitions of the group, for example the following:

> Perhaps the easiest way to indicate what is meant by small groups is to list some examples. Friendship groups, committees, a corporate board, the president and his advisers, families, the lunch group at work, a dating couple, the city counsel, and a teenage gang are all examples of small groups. What have they in common? First of all, they are all small enough that the members can deal with one another on a *face-to-face* basis... Each meets on regular basis and shares an identity or sense of purpose. Thus, each small group has a character of its own and is a recognizable social entity to its members and to outsiders. This is what distinguishes a group from a simple collection of people. Six strangers in an elevator are not a small group (Ridgeway, 1983, p. 3).

> Groups are social aggregates in which members are interdependent and have at least the potential for mutual interaction. In most groups, members have regular face-to-face contact... The definition emphasizes that the essential feature of a group is that members influence one another in some way (Sears et al., 1988, p. 377).

These two quotations share the basic idea that there must be interdependence between the members in a group and that its members meet in face-to-face situations, but they differ in their concept of social structure. Ridgeway, who is a sociologist, says that the members of a group must share some kind of "identity or sense of purpose," and that the group must be a "recognizable social entity to its members and outsiders." This means that the group, besides consisting of some people who meet in the empirical reality, also must be regarded as a *social fact* in the social structure. Thus, some strangers in an elevator can not be regarded as a group. They miss the predetermined social structure which make a genuine group a social fact or a *socially structured entity*.

The second quotation is written by psychologists. They define the group as "a social aggregate" where the members influence *each other*. The *group* does not influence its members, since it does not exist as a fact of its own in the social structure. Even the strangers in the elevator could in this definition be seen as a group. They make "a social aggregate" who have "a potential for mutual interaction" (imagine what will happen if the elevator stops?) and they can definitely interact and influence each other. But, they do not do it in a *socially structured* way; that is, the people who happen to meet in the elevator do not exist as a social fact in the social structure. (It is, of course, possible to say that there exist rules explaining how *not* to behave in an elevator.) Thus, Sears et al. do not regard the group as a separate entity in the social structure. Their group is determined by the interaction and influence processes *between* the members, not by its existence as a set of shared rules, norms, and identities. These two quotations reflects very well the difference between the sociological and psychological way of regarding the group.

But it is also possible to argue in another way, if the expression "potential for mutual interaction" means that the group also exists as a social fact when the members of the group do not "have regular face-to-face contact." In this way, the group is defined as both the *empirical face-to-face interaction,* and as the *potential* meeting of the members. The concept of "potential meeting" here means not only that a meeting can take place by chance; it refers to the *prestructured* relations and interaction processes that will develop when the potential meeting becomes a factual meeting, in other words, the group as a social fact. This is the same difference made explicit with the two examples above. This signifies that this group has *two forms of existence,* one in the empirical world of concrete acts, and one in the world of social structure.

The difference between these two forms of existence, empirical and social, has been conceptualized in many ways in social science, for example as the difference between the world of actors and the world of social structure. In the Marxist tradition, it is comparable to the difference between essence and forms of appearance (concrete praxis). A similiar, but not identical, discussion of the problem of different levels of group existence can be found in Lawler, Ridgeway, and Markovsky (1993), where the problem is related to micro- and macrostructures, respectively:

Microstructures consist of (1) a set of people (2) their relationship, and (3) their mutually contingent behavior. Microstructures emerge in encounters and depend for their endurance on the repetitive actions of the individuals involved. They are created whenever people interact purposely for more than a few moments and are especially important when the same set of people meet in repeated encounters...

In contrast, a macrostructure is a network consisting of (1) a set of positions with (2) a set of relations among the the posititions, and (3) a set of activities attached to the positions. The positions may be occupied by individual or corporate actors, and they endure beyond their occupants' tenures. In microstructures, positions exist as structured expectations for particular individuals, whereas in macrostructures the expectations that come to be associated with the positions also become independent of the particular occupants...(Lawler, Ridgeway, and Markovsky, 1993, p. 273).

The difference between this way of regarding groups and the way presented in this paper is that Lawler, Ridgeway, and Markovsky do not conceptualize the actor on level 1 or 2, that is on the intraindividual and individual levels. In their "structural social psychology," they use the concept of *minimal persons* to define the actor in a "minimal" way, that is, as an actor who does not possess "a wide variety of capacities, dispositions, opinions, behavioral proclivities, experiences, and so on" (p. 4). When these "minimal persons" meet and repeat their encounters, a microstructure develops, but without any determination of the personalities of the concrete individuals who form the group. This means that phenomena on the relational and group level are determined by the interaction of abstract individuals, not by the empirical "maximal" persons who interact in the encounter. "Microstructures" are group structures, which are only determined by the interaction process of the group, and, accordingly, "macrostructures" are organizations, institutions and/or social systems.

In comparision to this, we regard the difference between the empirical encounter and the existence of a social structure on both the micro and macro levels. Of course, there are great differences between the microstructure that develops when individuals meet and interact and repeat their meeting on one hand, and the encounters which take place in an already defined pattern in the macrostructure, on the other. But this difference is a difference between level 3 and levels 5, 6, and 7. In the model of Lawler, Ridgeway, and Markovsky, levels 1 and 2 are absent, and the only determinations of the individuals are that they are "purposive and responsive," and that individuals "use encounters as a means to achieve individual and/or collective ends, and in doing so, they take account of each other and respond in ways that are contingent upon their perceptions of one another's actions and motives" (p. 5). That is, the microstructure is only determined by the *abstract sociability* of the individuals, not by their concrete *individual personalities*.

The model presented in this article tries to determine the difference between the group as an empirical phenomenon, and the group as a phenomenon in the social structure. An empirical group is defined as the empirically meeting of individuals, and the group as a social fact is defined as a group that has an existence in the social structure *sui generis*. The question is whether there could develop a group in the micro social structure, which is not at all determined by the macro social structure, for example, fraternal groups that are not determined by institutionalized norms of friendship behavior. An "oversocialized conception of man" (Wrong, 1961) would deny this. However, this question is a part of the ongoing discussion between psychologists, who regard personality as important, and sociologists, who regard social facts as vital for understanding the empirical behavior of the individual. Fortunately, this basic and unresolvable question does not have to be settled in this analysis.

Accordingly, the forms of existence of the group could be defined by its existence on the empirical and social structural levels, respectively. With different combinations we find four types of groups:

		Social Structural	
		Yes	*No*
		---	---
Empirical	Yes	*Group A*	*Group B*
	No	*Group C*	*Group D*

Group A has both an empirical and a social structural existence. This implies that it exists both as a phenomenon when concrete individuals meet, and as an abstract phenomenon in the social structure. Various types of groups are possible. One type is the above-mentioned institutionalized friendship group, which exists in the micro social structure of its members and probably their close relatives. Here, the limits of behavior are set by the personalities of the members and, eventually, by superindividual friendship norms. Another type of structural influence is the limits of behavior set by superindividual norms on a board meeting or a training group in an athletic club. They are both parts of an organization, with goals, division of labor, and different positions. Here, the limits of the behavior of the group members are determined by the prescriptions of the positions of the organization, and, to an extent, by the concrete interaction on the intraindividual level.

Group B is a temporary group, which does not develop any superindividual norm structure. The members of this group meet once, and when the encounter is finished the group dissolves, whatever the reason for the temporary group formation might have been. Many more or less spontaneous encounters in everyday social life could be classified this way; from a group talk in the street to party group formation. Here, people meet, interact, and in the encounter they influence each other. But, there has not developed any group feeling or any behavioral norms if one happens to meet again.

Group C has only a social structural existence, which means that it is only a potential group, without any empirical encounter of concrete individuals. A table of organization for a firm is an example of this, though when the positions are filled with people and they meet each other, this type of group goes to category A. One can also regard abstract patterned social relations in, for example, a family as being this type of group. This means that they are totally determined by their social structural existence, and not at all by the personalities of and the relations between the individuals who are potential position occupants.

Group D, finally, does not have either an empirical or a social structural existence. It is a group that does not exist at all. In this category, one can place all social interaction that takes place between two or more persons who do not form a group or think of themselves as belonging to a group. Much of the social interaction that takes place in everyday life, for example, in shops, at work, in buses, or in the street, are examples of this kind of interaction. Therefore, we suggest that it is wrong to define *all* social interaction that takes place as group interaction. The interaction can only be regarded as group interaction when there are definitive dividing lines between the interaction inside the group and the interaction with the rest of the social world.

In summary, group A has an existence both empirically and in the social structure, group B has only an empirical existence, group C has only a social structural existence, and group D has no existence at all. The problematic cases are B and C, not A or D, because they represent a fully existing group and a nonexisting group, respectively. But what types of groups are B and C, and how can they be used in an analysis of the study of groups?

METHODOLOGICAL ABSTRACTIONS
AND LABORATORY GROUPS

B and C represent two opposite types of groups, and they are determined by different social laws. Group C is a part of the social structure in society and can only be understood in relation to the organization or institution of which it is a part. Group B is an empirical group, which is determined by the members' personalities and by the forms of interaction they have. If one should abstract these groups from their determinations, C from the organization/institution by which the group is defined and B from the people involved and the circumstances in which they meet, none of these groups exist in reality. What remains is only an abstract group, a group of nobody and anybody, a group without any concrete determinations. Here, lies one of the basic problems for a researcher who wants to study groups.

There are a lot of different research methods in social science, and group researchers have used them all. But, as in social psychology in general, there has been a rather strong orientation towards the laboratory experiment. The reason for this is obvious: In the laboratory, the reseacher can control "disturbing influences," manipulate the independent variable, and measure the values of the dependent variable. The reason to use the laboratory experiment is, therefore, not the question, but the problem is what kind of group the researcher really uses in his study.

The basic design of a group experiment is to gather some people, who do not know each other before, and who are not a part of a specific organization or social institution. These people are made to form a group, which is not to be determined either by the individuals personalities, or by their belonging to an organization or a social institution. This means that the ideal laboratory group is not determined by anything else than by being a group, that is, some people who happen to meet without any prestructured relations. It is an abstract B-group, a group that has empirical but no social structural existence, and which is not influenced by the empirical individuals, the researcher wants to create in the laboratory.

But, in reality, instead of being an abstract B-group, this kind of group is, inevitably, also a concrete empirical group, namely, the group consisting of the experimental staff and those who have volunteered to be in the group experiment. They form a concrete empirical group in the above mentioned sense. It is impossible to create an *methodologically* abstract group which is supposed to represent an *ontologically* abstract group, since ontological abstract groups do not exist in reality.

The basic onthological assumption of many small group experiments is that the group level could be studied on its own, without any determinatons either from "below" or "above." This is necessary, if one wants to find general group laws, either of the influence of membership in a group on an individual, or group interaction patterns in itself. But the experiment always takes place in an already extant organization or social institution, which determines the structure of relations and the patterns of interaction in the experimental group, including, for example, who is the leader, who gives the orders and who obeys, who defines the situation, who has what responsibility, and who is taking the role of the other. If the common definition of the situation, for example, will be "an experiment which takes place inside scientific social psychology," the research subjects probably will behave as the research leader tells them to behave, and what will happen in this concrete group will be dependent on the design of the experiment (Miller, 1972; Orne, 1962). This is the problem of the vividly discussed artifact bias of the social psychological experiment (Rosnow and Aiken, 1973). Partly due to this basic problem, some social psychological experimental researchers have tried to work with deception as a methodological strategy to overcome these artifacts. But then the design opens itself to ethical problems (Gross and Flemin, 1982; Kelman, 1967; Schlenker and Forsyth, 1977; and Strickler, 1967), and it does not solve the basic problem of the definition of the situation.

The basic problem with the laboratory experimental study of small groups is that it rests on an ontological abstract conception of a group in general, that is, of a social phenomenon that only exists on level 4. This abstract group, which, in the laboratory, is a methodological based scientific abstraction, is formed by abstract individuals, who represent individuals in general. When these abstract individuals meet and form an abstract group (without history or future), they represent a pure group phenomenon, which is freed from all determination from either level 1, 2, and 3 or level 5, 6, and 7. In the experimental group, there are no concrete individuals, with hopes and dreams, feelings and reflections, no purpose among them to form the group as a group for themselves, no specific relations, and no determination of any supergroup social structure. It is only a group of people.

What has been created in this scientific endeavor is a group which is a double abstraction. This scientific concept of the group disregards the influence on the group from both subgroup and supergroup levels, which is why it is possible to talk about a *double* abstraction. But in reality, this doubly abstracted group can not exist. That is why, one can suspect, many group experiments actually study something that does not exist. Undeniably, a lot of social psychological group experiments have taken place, and if one carries the argument above to its logical end, one must come to the conclusion that they did not study what they thought they did.

What have small group experiments really studied, then? One can say that they have studied a concrete group of people who, for different reasons (out of curiosity, to get money, to learn something, and so forth) have decided to participate in the experiment. There, they form a group that someone else has decided to form. Thus, the research experimenter has priority in the process of defining the situation. Therefore, s/he can influence the interindividual relations that will arise. This advantage of the leader defining the situation also holds in the case of deception. If, on one hand, the deception works, then the group will be formed according to the will of the research leader. If, on the other hand, the deception does not work for one reason or another, then the group will be formed by what the research subjects *think* is the will of the researcher. In the latter case, perhaps the researcher will not learn what s/he intends to learn. What will actually happen in the group is formed by the leader's *social* will; that being the will as the co-actors in the situation have interpreted it.

This implies that the real, concrete B-type group working in the laborary consists *both* of the research subjects and the experimental staff. This concrete group is affected by phenomena on level 1, 2, or 3, not least by the personality and thinking of the research leader and his/her behavior toward the other group members. But, more important is the fact that this concrete group is also determined by supergroup phenomena, that is, by the organizational and institutional levels. An experiment always takes place inside the institution of science and in the organizational setting of the symbol of science, the university or some other research institute. So, this *ontological concrete group* in the laboratory will be determined both by subgroup and supergroup levels, while at the same time the *methodologically abstract group* is supposed to be the object of study.

The paradox of the laboratory experimental groups in social psychology is, thus, that the group is influenced by numerous determinants, which the experimental designer tries to control or eliminate. The basic problem is that it is impossible to create a group that will not at all be determined by any other social levels. The methodological abstract group, which is created in the laboratory, has no ontological equivalent, and it does not matter how much the researcher tries to abstract from concrete determinations. The objective situation of the laboratory experiment is already determined both from below and above.[4]

But if it is impossible to create an onthologically abstract group with the help of methodological abstractions, it would be better to accept this and consciously study the variables one tries to abstract or control. This means to bring the excluded variables back in to the experiment, and incorporate in the research design such levels as "personality," "sociability," and "communication capability," for example, comparing two groups from different organization or social institutions. This means that both the sub- and supergroup levels will be a part of the research design.

If, for example, one wishes to study the attitudes and communication patterns towards the other sex in problem-solving groups, it would be interesting to match four different groups against each others: married men, married women, unmarried men and unmarried women. The differences between these four groups is that they have different "values" on two social institutional variables: marriage and sex, and these differences can explain a great deal of the results of the experiment. Or, if one wanted to study status organization processes in group forming, it could be of interest to match different types of personalities against each other, or even to see how hunger affects the status processes (let one group eat very much just before the experiment, and the other starve for eight hours). In this way, the excluded variables in most of the laboratory experiments could be studied as a part of the determinating pattern.

Of course, this suggestion is not something new in the history of social science, and not even in social psychology. But inside the tradition of small group theory, this knowledge does not seem to have been very important. Therefore, on one hand a psychological small group theory ought not only study influence processes, either on the interindividual or individual level, but also how these subgroup levels influence the group. On the other hand, a sociological group theory ought not only study how group structure develops out of different interaction structures and different position values, but also how macro sociological levels affect different group variables. This implies a study of group processes as dependent variables, and to use processes on both the sub- and the supergroup levels as independent variables or, in other words, to bring both psychology and sociology back to the study of small group processes.[5]

CONCLUSIONS

As early as 1956, Warriner pointed out the basic insecurity with the concept of group. Even though that time was a very promising decade for small group research, he meant that

> The term "group" is an ancient one in social science, but despite its antiquity there is little agreement on the nature and reality to which it refers—or even if it refers to any reality at all. This problem has been fundamental to many of the arguments of the past. There have been times when we assured ourselves that the issues were resolved only to find that they have arisen again in somewhat different form. (Warriner, 1956, p. 549.)

Nearly forty years later, these words seem remarkably modern and fresh. In this article, we have used some of the discussion and development in social theory since 1956 to discuss the reality of groups and the problem with the division of group theory into psychological and sociological parts. One result of the discussion was that there is a need to abandon the concept of a "pure experimental group," that is, a group that is not determined by anything else than itself.

The desire to "open up" the study of small groups through bringing in more psychological, relational, organizational, and institutional variables into small group theory is, of course, very close to the desire to open up social psychology in general to both psychological and sociological factors, which was very often called for during the crisis debate in the 1970s. Another theme, which is also parallel to an ongoing discussion in social psychology, could be added: the desire to open up the study of groups to *history* (Gergen, 1973). A real existing group always has a concrete historical existence; it always exists at a certain place and in a certain time. If a group always is embedded in historical circumstances, and is determined by both the concrete people who are its members and the social structural setting, then it follows that it is impossible to generalize studies on particlar groups to *all* groups. Some group processes, common to all kind of structured social relations and its interaction processes, without any personal, organizational, or institutional concretization, perhaps exist, but they are more general human interaction patterns, not something special when people act in a *group*.

That does not mean, of course, that all results from the tradition of small group research are uninteresting. Quite to the contrary, both in prior history and and in contemporary small group studies there are a lot of fascinating discoveries about the nature of group life. But, it implies that contemporary small group researchers ought to concretize their theories, that is, to bring in the excluded sub- and supergroup variables, in order to make them available to a broader spectrum of social science. This means a conscious opening up of the small group research to both psychological and societal levels.

The revival of small group research, which was predicted by Steiner (1983), has to some extent come true during the last half of the 1980s. But from a broader social science perspective, this zrevival has up to now, not influenced the general knowledge of the concrete social world we are living in. In this article it is argued that one of the basic problems in the tradition of small group research, both in its psychological and sdc ociological orientations, is the attempt to find abstract group interaction and influences laws with the help of methodological abstract group concepts. Of course, there are many exceptions to this general trend, but as a scientific project, this view of modern group theory does not seem totally untrue. Some of the strongest and most viable group research has included studies on more or less concrete group phenomena, for example, studies on juries, which could incorporate other levels in their analyses. For the future development of small group studies, a more conscious incorporation of sub- and supergroup levels in the analyses could be of importance. It does not follow that laboratory experiments must be abandoned, but it implies that the objectively existing variables on sub- and supergroup levels ought to be incorporated in the research design.

The paradox is thus that for small group theory to be as viable and expanding a field as it was in the 1940s and 1950s, it must abolish itself as a special field, which only studies the group level or how the group influences individuals.

There are different phases in the development of a scientific tradition. One phase is characterized by the basic questions of the field, constituting the research object and developing of research methods. Another phase can be characterized as a "normal science-phase" in a Kuhnian sense, and still another by doubts and reorientations. In Kuhn's scheme, the last phase could lead to a scientific revolution, where new paradigms develop. In the history of group research, the first three phases have undoubtely existed, but the last revolutionary phase never came into being. After the crisis debate in the 1970s, both social psychology in general and small group research in particular continued to be more and more specialized, that is, continued its phase of normal science.

To make small group research as relevant to other fields of social science as it "objectively" is, due to the central position of the group level in social life, it is necessary to develop models of how different levels of social reality both affect and are affected by the group level. It is, of course, impossible in a short article to show how such a model could be developed, but a multilevel and multiperspective group analysis based on a historical awareness (groups are always formed and live in concrete historical circumstances), can be one of the important points of departure.

NOTES

1. In their analysis of the small group theory in American sociology, Mullins & Mullins use a network analysis of the rise and spread of small group theory in the US, that maybe could have been used in Steiner's analysis of the Group Theory School (Mullins and Mullins, 1973, chap. 5).

2. The classical example of this is Floyd Allpord's words from 1924: "Social psychology must not be placed in contradistinction to the psychology of the individual; it is a part of the psychology of the individual, whose behavior it studies in relation to that sector of his environment comprised by his fellows" (Allport, 1924, p. 4).

3. An investigation I made among psychologists and sociologists in Sweden on how they regard social psychology shows very clearly that there exists one psychological and one sociological social psychology in Sweden. For example, in their definition of the field of social psychology, psychologists much more often answered that the relations "intra-psychological/individual," "individual/individual," and "individual/group" are objects of study in social psychology, while sociologists more often answered "individual/group," "individual/social environment," or "individual/society." The relation "individual/group" was the only one that was among the three most mentioned relations by both psychologists and sociologists (Månson, 1990, pp. 14-18).

4. Asplund means that most of the social psychological group experiments have not studied group processes at all. Instead, he argues, they study phenomena on level 3, the relational level. For Asplund, "a group in proper meaning consists of a larger or smaller number of persons, who have a common history and a common future together with shared tasks and goals" (Asplund 1987, p. 23). Obviously, this is not the kind of group that is studied in the research laboratory. Instead of studying a group "in proper meaning," group theorists have studied something more basically, which he calls *elementary communion* (p. 24).

Asplund argues that it is possible to speak of a group only when it has both empirical and social structural existence ("a numer of persons who have a common history and a common future together with shared tasks and goals"). This is the category A above. He does not accept either group B or C as genuine groups. This implies that a group at the same time is defined *both* by the concrete people who are its members and by the social structure which exists in their collective identity. A real group is always determined by both lower and higher levels of social life. One can, perhaps, even say that a group is precisely that social phenomenon which exists at the point of intersection between the intraindividual and individual levels on one hand, and the organizational and institutional levels on the other. Therefore, the group could be regarded, as so many group theorists have done, as a point in social life where macro sociological phenomena interact with micro sociological, psychological, and mental phenomena.

5. One conscious attempt to relate the group level to subgroup levels is made by Turner in his "Self-Categorization Theory" (Turner, 1987). Here, he refers to the "good agreement about the major features of group formation..., 'identity', 'interdependence' and 'social structure' " (p. 19), and build parts of his theory on the relation between group level and individual and interactional level. However, he does not discuss any supergroup levels, but conscious about that, Turner calls his theory "a theory...of the social psychological aspects of the group, of the psychological processes underlying and making possible the transformation of human individuals into social groups" (p. 2). Discussion about linking the group level to supergroup levels can be found in Taifel (1981), and a try to open social psychology to the social institution of ideology is made by Billing (1982). See also the discussion of "the individualistic fallancy" in group theory and social psychology in Turner and Oakes (1986).

REFERENCES

Allport, F. H. 1924. *Social Psychology.* Boston: Houghton Mifflin.

Allport, G. W. 1968. "The Historical Background of Modern Social Psychology." Pp. 1-80 in *The Handbook of Social Psychology,* edited by G. Lindzey & E. Aronson.

Asplund, J. 1987. *Det social livets elementära former* (The Elementary Forms of Social Life). Göteborg: Korpen.

Berger, J. and M. Zelditch, Jr. 1991. "A Revised Bibliography of Expectation State Research." Working paper No. 100-1. Stanford University.

Billing, Michael. 1982. *Ideology and Social Psychology: Extremism, Moderation and Contradiction.* Oxford: Basil Blackwell.

Boutilier, R. G., J. C. Roed, and A. C. Svendsen. 1980. "Crisis in the Two Social Psychologies: A Critical Comparison." *Social Psychology Quarterly* 43:5-17.

Cartwright, D. & A. Zander. 1960. *Group Dynamics. Research and Theory,* 2nd ed. Evanstone, IL: Row, Peterson.

Collins, B. E. and B. H. Raven. 1968. "Group Structure: Attraction, Coalitions, Communication, and Power." Pp. 102-104 in *The Handbook of Social Psychology,* 2nd ed., Vol. 4, edited by G. Lindzey & E. Aronson. New York: Random House.

Davis, J. H. and M. F. Stasson. 1988. "Small Group Performance: Past and Present Research Trend." In *Advances in Group Processes,* Vol. 5, edited by E. Lawler. Greenwich, CT: JAI Press.

Doise, W. 1978. *Groups and Individuals: Explanations in Social Psychology.* Cambridge: Cambridge University Press.

————. 1988. "Individuals and Social Identies in Intergroup Relations." *European Journal of Social Psychology* 18:99-111.

Gergen, K. J. 1973. "Social Psychology as History." *Journal of Personality and Social Psychology* 26:309-320.

Gross, A. E. and I. Fleming. 1982. "Twenty Years of Deception in Social Psychology." *Personality and Social Psychology Bulletin* 8:402-408.

Hare, A. P. 1962. *Handbook of Small Group Research.* New York: Free Press.

Hare, A. P., E. F. Borgata, and R. F. Bales. 1967. *Small Groups. Studies in Social Interaction.* Revised ed. New York: Alfred A. Knopf.

Hewstone, M., W. Stroebe, J. P. Codol and G. M. Stephenson. 1988. *Introduction to Social Psychology. A European Perspective.* Oxford: Basil Blackwell.

Hogg, M. A. and D. Abrams. 1988. *Social Identifications. A Social Psychology of Intergroup Relations and Group Processes.* London: Routledge.

Homans, G. C. 1950. *The Human Group.* New York: Harcourt, Brace & World.

House, J. S. 1977. "The Three Faces of Social Psychology." *Sociometry* 40: 161-177.

Jackson, J. M. 1988. *Social Psychology, Past and Present. An Integrative Orientation.* Hillsdale, N. J.: Lawrence Erlbaum.

Jones, E. E. 1985. "Major Development in Social Psychology During the Past Five Decades." Pp. 47-107 in *The Handbook of Social Psychology,* 3rd ed., Vol. 1, edited by G. Lindzey & E. Aronson. New York: Random House.

Kelman, H. C. 1967. "Human Use of Human Subjects: The Problem of Deception in Social Psychological Experiments." *Psychological Bulletin* 67:1-11.

Lawler, E. J., C. Ridgeway, and B. Markovsky. 1993. "Structural Social Psychology and the Micro-Macro Problem." *Sociological Theory* 11:268-290.

McMahon, A. M. 1984. "The Two Social Psychologies: Postcrisis Directions." *Annual Review of Sociology* 10:121-140.

Markovsky, B. 1987. "Toward Multilevel Sociological Theories: Simulations of Actor and Network Effects." *Sociological Theory* 5:101-117.

————, D. Miller and T. Patton. 1988. "Power Relations in Exchange Networks." *American Sociological Review* 53: 220-236.

Miller, A. G. (ed.). 1972. *The Social Psychology of Psychological Research.* New York: Collier-MacMillan.

Mills, T. M. 1968. "On the Sociology of Small Groups." In *American Sociology. Perspectives, Problems, Methods,* edited by T. Parsons. New York: Basic Books.

Moreno, J. L. "Old and New Trends in Sociometry: Turning Points in Small Group Research." *Sociometry* 17: 179-193.

Mullins, N. C. 1973. *Theory and Theory Groups in Contemporary American Sociology.* New York: Harper & Row.

Månson, P. 1990. "Socialpsykologin i Sverige—resultatet av en enkät" (Social Psychology in Sweden: The Result of a Questionnaire). *Sociologisk forskning* (Sociological research) 4:3-21.

———— 1993. "The History and Existence of Symbolic Interactionism in Sweden." *Symbolic Interaction.* Forthcoming.

Orne, M. T. 1962. "On the Social Psychology of the Psychological Experiment." *The American Psychologists* 17:777-783.

Patnoe, S. 1988. *A Narrative History of Experimental Social Psychology. The Lewin Tradition.* New York: Springer Verlag.

Rabbie, Jacob M. and Murray Horowitz. 1988. "Categories versus groups as explanatory concepts in intergroup relation." *European Journal of Social Psychology* 18:117-123.

Ridgeway, C. L. 1983. *The Dynamics of Small Groups.* New York: St. Martin's Press.

Rosnow, R. L. and L. S. Aiken. 1973. "Mediation of Artifacts in Behavioral Research." *Journal of Experimental Psychology* 9:181-201.

Shaw, M. E. 1981. *Group Dynamics. The Psychology of Small Group Behavior.* 3rd edition. New York: McGraw-Hill.

Schlenker, B. R. and D. R. Forsyth. 1977. "On the Ethics of Psychological Research." *Journal of Experimental Social Psychology* 13:369-396.

Sherif, M. and C. W. Sherif. 1969. *Social Psychology.* New York: Harper International.

Sears, D. O., L. A. Peplau, J. L. Freedman & S. E. Taylor. 1988. *Social Psychology.* 6th edition. Englewood Cliffs. N.J.: Prentice Hall.

Shibutani, T. 1961. *Society and Personality. An Interactionist Approach to Social Psychology.* Englewood Cliffs, N.J.: Prentice-Hall.

Steiner, I. D. 1974. "Whatever Happened to The Group in Social Psychology?" *Journal of Experimental Social Psychology* 10:94-108.

———— 1983. "Whatever Happened to the Touted Revival of the Group?." In *Small Groups and Social Interaction,* Vol. 2, edited by H. H. Blumberg, A. P. Hare, V. Kent, and M. Davies. Chichester, England: Wiley and Sons.

Stricker, L. J. 1967. "The True Deceiver." *Psychological Bulletin* 68:13-20.

Strodtbeck, F. L. and A. P. Hare. 1954. "Bibliography of Small Group Research." *Sociometry* 17: 107-177.

Stryker, S. 1977 "Development in 'Two Social Psychologies'. Toward an Appreciation of Mutual Relevance." *Sociometry* 40:145-160.

———— 1989. "The Two Social Psychologies: Additional Thoughts." *Social Forces* 68:45-54.

Tajfel, Henri. 1979. "Individuals and Groups in Social Psychology." *British Journal of Social and Clinic Psychology* 18:183-190

———— 1981. *Human Groups and Social Categories: Studies in Social Psychology.* Cambridge: Cambridge University Press.

Tibaut, J. W and H. H. Kelly. 1967. *The Social Psychology of Groups.* New York: Wiley.

Törnblom, K. Y. 1992. "The social psychology of distributive justice." In *Justice: Interdisciplinary Perspectives.* edited by K. Scherer. Cambridge: Cambridge University Press.

Turner, J. C. 1987. *Rediscovering the Social Group. A Self-Categorization Theory.* New York: Basil Blackwell.

———. 1988. "Comments on Doise's Individual and Social Identies in Intergroup Relations." *European Journal of Social Psychology* 18:113-116.

Turner, J. C., and Penelope J. Oakes. 1986. "The Significance of the Social Identity Concept for Social Psychology With Reference to Individualism, Interactionism and Social Influence." *British Journal of Social Psychology* 25:237-252.

Wagner, D. D. and J. Berger. 1990. "Status Characteristics Theory: The Growth of a Program." A Working Paper from The Center for Sociological Research. Dept. of Sociology.

Warrener, Ch. 1956. "Are Groups Real? A Reaffirmation." In *Am. Soc. Review* 21:549-554.

Whyte, W. F. 1951 "Small Groups and Large Organizations." Pp. 297-312 in *Social Psychology at the Crossroads,* edited by J. H. Rohrer and M. Sherif. New York: Harper.

Wrong, D. H. 1961. "The Oversocialized Conception of Man in Modern Sociology." *Am. Soc. Review* 26:183-193.

PRIOR SOCIAL TIES AND MOVEMENT INTO NEW SOCIAL RELATIONSHIPS*

Richard T. Serpe and Sheldon Stryker

ABSTRACT

This paper investigates the question of whether prior ties to family impede or facilitate the development of new social ties. This question is addressed from the perspective of identity theory and sociology of the family. Using a sample of 320 first semester university freshmen measured at three time points throughout the fall semester, the analysis focuses on the effect of family commitment at the time the subjects moved to campus on the development of new friendships, dating patterns, and joining organizations. The results clearly demonstrate that the higher the level of prior family commitment, the more subjects developed new friendships, the stronger their dating patterns, and the more organizations they joined.

Does the existence of strong prior ties to a given set of role partners abet or inhibit the development of new relationships with new role partners? Whatever the answer to this question, does the answer hold in general, or is the answer

* This paper is the product of a fully collaborative relationship.

Advances in Group Processes, Volume 10, pages 283-304.
Copyright © 1993 by JAI Press Inc.
All rights of reproduction in any form reserved.
ISBN: 1-55938-280-5

conditional on particular kinds of new relationships? And, if the answer to this second question is that the answer to the first cannot be made without taking into account particularities, what might those particularities be? These and other related questions are pursued in this paper.

Translating these questions into the specifics of the research reported below, the empirical question raised in the research is: Do prior ties to family in the form of involvement in activities with other family members relate either positively or negatively to movement into new relationships—new friendships, new dating partners, new organizational memberships and activities—when persons move to a university community at some distance from their home communities?

According to one possible reading of Identity Theory (Stryker, 1968; Stryker, 1980; Stryker, 1987; Stryker, 1992; Stryker and Serpe, 1983; and Serpe and Stryker, 1987), extant ties to family ought make it more difficult for persons leaving their home communities to involve themselves in new relationships of whatever kind. In the language of the theory, existing ties to particular social networks represent commitments. According to the logic of the theory, commitment to a given network of social relationships increases the salience of the identity—the internalized role designation carrying behavioral expectations which, when enacted, represent playing out the role—through which one enters that set of role relationships. The high commitment to a given network of relationships and the consequent high salience of the associated identity, argues the theory, lead to greater sensitivity to stimuli cuing behavior in line with the identity, greater likelihood of interpreting events as calling for identity-relevant behavior, and so forth, and, in general, active efforts to seek out opportunities to enact the identity in the context of the given networks of relationships that have underwritten the identity. Such active efforts may well include spending time, money, and energy to return to one's home community in order to pursue home-based relationships, thereby lessening resources that could be put to use in the pursuit of new relationships elsewhere. If one makes the reasonable assumption that resources of various kinds—time, money, energy—have at least some limits, not necessarily closely specified, family commitment and highly salient family-related identities should lead persons to seek out and to create opportunities to enact the family identities in existing family network contexts, and thereby should constrain or suppress entering new relationships.

However, identity theory also recognizes (Stryker and Serpe, 1982; Serpe, 1987;Serpe and Stryker, 1987) that the social psychological process just elaborated is not static and that it occurs in a social structural context. The larger frame from which the theory develops, a so-called structural symbolic interactionism (Stryker, 1980), finds its dynamic in the person's changing locations in social structure, and envisages social structure (or, better, social structures) as boundaries enlarging or diminishing the probabilities that particular kinds of social relationships will develop, be maintained, or disappear. Such structures include race, ethnicity, class, gender, occupation, community, ecology, and so forth, all of which constrain in one way or another the opportunities for entering into particular kinds of social relationships and, therefore, for the creation, maintenance, and enactment of particular kinds of identities.

Given the empirical questions being raised in this paper, structural constraints are embodied in the specification that we are studying persons who leave home communities to attend a residential college. The distance between home community and university town implies, minimally and whatever the level of resources available to the person, decreased opportunity to interact with other members of one's family who remain in the home community.

Under this circumstance, a further reading of identity theory underwrites the argument that persons will try to construct new social relationships in order to enable them to act out salient prior identities, based on high levels of prior commitments, in appropriate new social networks. By way of example, if structural circumstances prevent persons with high familial commitment and highly salient familial identities from interacting with (narrowly defined) family members, thus preventing them from enacting familial roles and reinforcing underlying levels of commitment and identity salience, the theory leads us to expect that they will seek to establish quasifamilial or familylike relationships with new sets of others—roommates, friends, coworkers, etc. To make the example more concrete, the theory suggests that if a person with strong family commitments and an identity of family member that reflects a "little brother" content cannot enact that identity with "real" family members, that person will seek friends in relationship to whom the identity can be enacted.

Identity theory further argues that if such construction of social relationships cannot be managed, the salience of the identities that can no longer be enacted will diminish with the consequence that the underlying commitment will then also diminish (Serpe and Stryker, 1987). Extending this line of argumentation, the theory implies that even should new relationships allow for the expression of preexisting highly salient identities, commitment to the older social networks that once underwrote those identities will decrease. To put these matters in the language of the empirical events studied, identity theory leads us to expect that prior close ties to family will lead to attempts to maintain those ties; if those ties cannot be maintained, attempts will be made to substitute equivalent relationships; and, to the extent that these substitutions take on higher levels of commitment and higher levels of salience in identities associated with these substitute relationships, commitments to family will weaken.

An additional theoretical argument can be made with respect to the expected effects of prior ties to family and perhaps more broadly. While not an argument that derives from identity theory, it has implications for that theory and its future development. The early childhood development literature (e.g.,

Ainsworth, 1985; Ainsworth, 1989; Bretherton, 1985; and Bretherton, 1990) strongly suggests that it is the presence of an infant's mother that provides the infant with sufficient security to explore "dangerous" segments of its environment. There is an equivalent suggestion in Corsaro's (1985) work on nursery school children, his observation being that secure friendship ties give a child the freedom to explore novel aspects of the nursery-school world. More specific in its implications for the present research, however, is the argument thoroughly embedded in the literature of family sociology and social psychology (e.g., Merton, 1968; Parsons, 1964; and Parsons, 1965), namely, that a major task of the contemporary family is to launch its progeny into the world as independent adults. Relatedly and in particular, this literature asserts that an important part of the ideology of the contemporary middle class family (Walters et al., 1964; and Zunich, 1961), is the obligation of parents to let go of their children, to grant children, and even require of them, the independence enabling them to emerge from the protected family cocoon. Correlatively, there is presumably an obligation on the part of the children to accept and exercise this independence. This idea suggests the hypothesis that the greater the commitment to the family of origin, the more readily and completely will the offspring enter new social relationships characteristic of university life. This hypothesis is based on the premise that parents in families of origin will seek to "launch" their progeny into the world outside the family itself, and on the further premise that high commitment of offspring to their families of origin and the consequent high salience of their identities as children will lead to behavior consistent with parental launching efforts.

With respect to the development of identity theory, the validity of this hypothesis would argue strongly the need to incorporate directly into the empirical examination of the theory considerations of culture in the form of institutionalized norms that may be important parts of the meaning-content of roles and therefore of identities. Institutionalized norms in the form of role expectations have entered the theory through the concept of identity, which is defined as a set of expectations or meanings attached to an internalized role designation. To this point, however, by and large the theory as well as tests of it have focused on more or less formal, structural issues and have not given much attention to the cultural content of identities. They have concentrated on such matters as the effect of strength of commitment to networks of role relationships on the location of a related identity in a hierarchy of salience of all the identities making up a person's self, and the subsequent effect of identity salience on aspects of role performance, *without* taking centrally into account the substance, content, or meaning of the roles and identities involved. However, it is necessary to have some sense of those meanings in order to predict just what behavioral consequences will follow from commitment and identity salience. Stated more generally, to understand the potential consequences of persons holding multiple identities and playing out multiple roles in multiple networks of social relationships, theorists and researchers can no more concentrate on social or self structures to the exclusion of culture and the manifestations of culture in the context of personal systems of meaning than they can concentrate on the latter to the exclusion of the former. Evidence that this is the case, in the form of tracing out the variable consequences of employment in the lives of married men and women, can be found in the work of Simon (1992).

In the context of the present research, cultural norms that lead parents to expect their children to become independent of themselves, emotionally as well as financially, and that lead offspring to be responsive to those efforts, define an important part of the meanings or the cultural content of parental and offspring roles and identities. That implies, once again, that evidence supporting the hypothesis that links high prior family commitment to subsequent greater entry into new social relationships will in turn call for the inclusion of the cultural content of roles and identities in tests of identity theory.

It may be useful to restate in summary terms the hypotheses guiding the research being report as a prelude to a description of the design of the research and a report of its findings. Drawing directly on Identity Theory, we hypothesize that:

(1) When structural conditions permit, persons with high levels of commitment to family prior to entering a university setting will continue to manifest high levels of interaction with other family members.
(2) Given structural conditions that impose some level of deterrence to maintaining high levels of interaction with other family members, persons with high levels of commitment to family prior to entering a university setting will form new social relationships that are "quasifamilial" or familylike (i.e., close, intense, etc.) And,
(3) The formation of new social relationships will over time decrease persons' levels of commitment to family.

Not drawn directly from Identity Theory but relevant for the development and test of the theory is the hypothesis that:

(4) Persons with high levels of family commitment will form more new relationships in general when they move into a university setting. This hypothesis assumes that parents are concerned with launching their offspring as independent adults and that offspring share this concern.

Research Design and Data

New freshman students entering a residential university located in a small midwestern university town in the Fall of 1982 were the subjects for the research being presented. The university is populated largely by students drawn from middle-class backgrounds. Almost all of these students have homes located at a distance from the town, on average just under 190 miles; thus, they had to leave home, family, and friends to go to the university. However, the homes of most are not so far from school that ties to family and friends in home communities cannot be maintained. In all, then, such students had available to them both opportunities to maintain old ties to others, and to establish new ties. Because of this, reasonable variation in measures of commitment could be expected. As a matter of convenience, subjects were recruited through introductory psychology classes; they met course obligations by participating in this research.

A sample of three hundred and twenty (320) students was recruited. Two-thirds ($n = 215$; 67 percent) of the students were female, one-third ($n = 105$; 33 percent) were male. The sample appeared to be generally representative of freshman college male and female students (see Serpe, 1987, for relevant details), and only minimal differences existed between male and female subsamples.[1]

Subjects completed a self-administered questionnaire in September, 1982, shortly after moving from home to campus. They completed these questionnaires again in October and in December of 1982. The questionnaires called for subjects to provide information about themselves in relation to five roles and identities, all of which were aspects of their positions as students: academic, athletic/recreational, extracurricular (nonorganizational), friendship or personal involvement, and dating. They also provided information about themselves in relation to their familial roles and identities. This report focuses on familial, friendship, dating, and organizational activities.

Variables and Measures

The analyses to be presented make use of a number of variables, and measures thereof, that are either explicitly or implicitly contained in the introductory paragraphs of this paper.

Family Commitment

Identity theory defines commitment as the costs entailed in relationships that would be foregone if the individual were no longer to have an identity and play a role; the theory recognizes two types of commitment: interactive and affective. For reasons inherent to the larger project from which this report derives, only the former was measured with respect to the family. The measure asked how often subjects engaged in various activities with other family members: see them, have meals with them, have a conversation with them at home, have a conversation with them outside the home, have a telephone conversation with them, and work on household projects with them. Subjects responded to these stimulus items on a scale of 1 = never, 2 = seldom, 3 = once a month, 4 = once a week, 5 = several times a week, and 6 = daily. Family commitment was measured at two points in time, September (T1) and December (T3). Within the analysis presented here, family commitment is measured by a scaled item for each time point. The scale represents the averaged score over the six items. The T1 measure had an alpha reliability of .84; the T3 measure had an alpha reliability of .87.[2] Ranges, means, and standard deviations on these and other measures used in the analyses are given in Table 1. The difference in T1 to T3 family commitment means reported there has an obvious source in the fact that most students were at some distance from home at T3.[3]

Friendship

The analyses reported use two measures of friendship patterns: number of *friends*, defined as "people you know and do things with," and number of *close friends,* defined as "people you can count on for help." At T1, the friends measure refers to persons in subjects' home towns; at T2 and T3, the measure refers to new friends or new close friends made at school. The covariance analysis reported below uses a latent construct labelled "Friends" which is made up of the "new friends" and "new close friends" items as measured at T2.

Dating

Dating activity was measured through *dating frequency* and *dating intensity.* The former is a single-item measure generated by asking how often subjects dated, and scoring the responses from 0 (never) through 1 (once a month), 2 (twice a month), 3 (once a week), 4 (several times a week), to 5 (daily). The latter is a variable measuring the nature of the dating relationship, scored from 1 (not currently dating) through 2 (occasionally dating, but not regularly), 3 (regularly dating several people), to 4 (regularly dating one person). The latter reflects "going steady," and is presumed to imply greater intensity in the dating relationship. Dating frequency and intensity were measured at all three time points. The latent construct *dating* used in the reported covariance analysis is made up of the T2 dating frequency and dating intensity variables.

Number of organizations

Subjects were asked at T2 and T3 whether they had joined an organization associated with one or another of four student-related roles and identities that were of concern in another part of our research. Number of organizations is simply the sum (1 to 4) of their yes/no responses.

Distance from home

This is a measure, in estimated miles, of how far subjects' homes were from the campus. The descriptive statistics given in Table 1 for this measure suggest that a few subjects' homes were in or not far from the town in which the university is located, while a very few subjects' homes were in countries other than the United States. Important for understanding and interpreting the findings presented later in this paper is the fact that, on average, the university is 187.17 miles from subjects' home towns.

Table 1.   Means, Standard Deviations, and Ranges

Variable	Mean	S.D.	Min	Max
Family Commitment				
Time 1	4.60	.86	1	6
Time 3	2.90	.67	1	6
Number of Friends				
Time 1	15.86	15.97	1	80
Time 2	12.42	11.97	1	90
Time 3	14.49	14.03	1	85
Number of Close Friends				
Time 1	3.87	4.71	0	24
Time 2	3.98	3.99	0	30
Time 3	4.24	3.61	0	21
Dating Frequency				
Time 1	2.27	1.24	0	5
Time 2	2.37	1.34	0	5
Time 3	2.31	1.31	0	5
Dating Intensity				
Time 1	2.61	1.81	1	4
Time 2	2.67	1.14	1	4
Time 3	2.79	1.17	1	4
Number of Organizations[1]				
Time 2	1.67	.92	0	4
Time 3	1.54	.87	0	4
Distance	187.17	273.80	1	4000
Sex	.67	.47	0	1

Note: [1] We do not have a measure of the number of organizations the repondents belonged to before coming to the university.

Results

Table 2 reports the zero-order correlations among the variables used in the covariance structure analysis of the model given in Figure 1.[4] The model, estimated using LISREL VII, inquires into the impact of commitment to family prior to moving off to school (i.e., at T1) on the development of new social relationships, in the form of new friendships, organizational memberships, and dating activities at T2. It inquires further into the impact of these new T2 social relationships on subsequent (i.e., T3) family commitment. Since female subjects—by prior expectations based on considerable sociological and social psychological research as well as by the evidence of data in this study—are more committed to family than are males, gender is entered into the model as a control. So, too, for obvious reasons, is the distance of school from home.

The fit of the data to the model is clearly very good. The chi-square of 18.17 with 18 degrees of freedom, the probability level of .445, and the fit index of .967 all support that assertion, as does the relatively high explained variance (.53) of the ultimate dependent variable, T3 family commitment. Only statistically significant coefficients are shown in Figure 1. Gender is indeed correlated with T1 family commitment, with females having greater commitment than males, and there is a significant effect from gender to T3 family commitment, indicating that being female contributes more to family commitment than does being male. As is reasonable given extant literature, females make fewer friends and fewer close friends, at least in the roughly six-week period after they leave home for the university that is, between T1 and T2. And as expected, distance from home diminishes T3 family commitment.

More directly related to identity theory arguments, as well as to the "launching" hypothesis presented earlier in the paper, are the coefficients linking T1 family commitment to T3 family commitment, T1 family commitment to T2 social relationships, and T2 social relationships to T3 family commitment. The stability coefficient of .559 that links T1 and T3 family commitment suggests that family commitment is not an ephemeral phenomenon, but that persons tend to retain their levels of commitment to their families over time. Recalling that commitment scores reflect levels of interaction with other family members, this finding supports the hypothesis (Hypothesis 1) that persons with initial high levels of interaction will continue to manifest high levels of interaction with other family members.

At the same time, the size of the stability coefficient for commitment suggests that a fair amount of change occurs in commitment to family when offspring move out of their family homes to go off to college. Clearly, higher commitment to family prior to movement to school commitment leads persons to make more, and more close, friends when at school. It also leads to more, and more intense, dating and more joining of organizations. Recalling that the situation is one in which efforts to maintain one's commitment to family are constrained

Table 2. Zero-Order Correlations

	Num New Friends	Num Close Friends	Dating Frequency	Dating Intensty	Numorg	Family Com T1	Family Com T3	Miles	Sex
NUMFRD	1.000								
NUMCF	.282	1.000							
DAT FRE	.022	-.022	1.000						
DATEINT	-.004	.042	.558	1.000					
NUMORG	.195	-.127	-.070	-.110	1.000				
FAMCOM3	-.116	-.134	-.207	-.176	-.121	1.000			
FAMCOM1	.086	.100	.141	.118	.129	.344	1.000		
MILES	.045	-.081	-.023	-.138	.018	-.234	-.002	1.000	
SEX	-.158	-.091	.029	.093	-.031	.183	.136	-.024	1.000

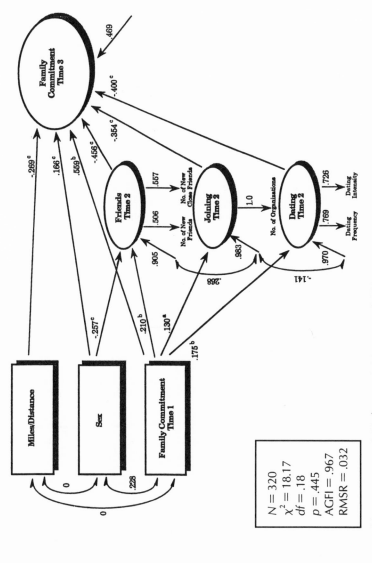

Notes: $a = p. < .05$; $b = p. < .01$; $c = p. < .001$.

Figure 1. Path Model for Effects of Family Commitment on Social Relationships Over Time

N = 320
χ^2 = 18.17
df = .18
p = .445
AGFI = .967
RMSR = .032

by the distance between the university and the location of one's family, and therefore any effort to reconstruct in a literal sense one's family relationships is almost necessarily doomed, these data suggest that persons closely tied to their families at home seek out new relationships of all kinds, as the "launching" hypothesis (Hypothesis 4) predicts. These data also are consistent with the "substitution" hypothesis (Hypothesis 2), which asserts that persons will seek satisfying substitutes for vacated family ties. That is, one way of understanding these findings is to see the establishment of particular new social relationships at school as a way of reestablishing or reconstructing conditions, disrupted by the movement from home, for identities whose content rests on warm, secure, intimate relationships. Another, but not competing, way of understanding these findings is to suggest that middle class families in which parent-child ties are strong, and in which parents are presumably sensitive to their obligation to launch their child, are more successful in doing so than are families in which ties are less strong.

Assuming the correctness of these identity theory-related understandings, we can observe that the significant negative coefficients linking the T2 social relationships to T3 family commitment imply that the very acts of making more friends and more close friends, joining more organizations, and dating more frequently and intensely begin to diminish subsequent commitment to family, thus supporting Hypothesis 3. In a sense, the very behaviors that are the product of reconstruction efforts at the same time contribute to a process by which new relationships are constructed and (by implication from identity theory) old identities are changed and new identities are created. Assuming the correctness of the "launching" argument, we can observe that the strategy is apparently successful, in that it does indeed diminish commitment to family and in that sense enlarge independence from family.

Exploratory Analyses

We can turn to a closer inspection of the data available to see whether these interpretations are reasonable, and whether they can be refined. Our first step in this closer inspection examines the linkages between scoring either very high or very low on the T1 family commitment measure and entering social relationships through making friends and close friends, dating, and joining organizations. T1 family commitment scores were categorized as very low, low, high, and very high; 54 (16.9 percent) of our subjects fell into the very low category, 95 (29.7 percent) into the low category, 114 (35.6 percent) into the high category, and 57 (17.8 percent) into the very high category.[5]

Using this four-category family commitment variable, we then examined via one-way analysis of variance, the relationship between levels of family commitment and mean scores on the number of friends, number of close friends, dating frequency, dating intensity, and number of organizations measures for each of three points (T1, T2, and T3) at which information on these variables was gathered. These mean scores are provided in Table 3, as are the levels of statistical significance obtained in the analyses. The number of friends at T1 and the numbers of new friends at T2 and T3 are linked in an orderly, incremental way to the categorical family commitment variable, with those subjects scoring highest on family commitment having the largest number of friends and new friends at the three time points, and those subjects scoring lowest on family commitment having the fewest. Only the family commitment-new friends relationships at T2 and T3 achieve statistical significance; however, it is precisely these relationships that are most important to the theoretical issues of concern in this paper, since T1 friendships must be a function of long-term accumulation reflecting a great variety of unknown sources and influences.

In general, the number of close friends by family commitment analysis provides similar findings. While only the family commitment-new close friends at T3 relationship is statistically significant, just one reversal mars the orderly progression of the mean number of close friends at T1 or new close friends at T2 and T3 over the four categories of family commitment: it is the subjects with low family commitment at T1 who have the fewest new close friends at T2, rather than the subjects with very low T1 family commitment.

With respect to the analyses relating family commitment and dating behaviors, none of those that deal with sheer number of dates achieve conventional levels of statistical significance. Nor do we find in these analyses the same orderly progression of mean number of dates as we move from the very low to the very high family commitment categories. Unlike dating frequency, the data for dating intensity appears to reinforce the theoretical interpretation offered earlier. That is, we find in the dating intensity analyses orderly movement from lowest mean scores attached to those subjects with very low family commitment at T1 to highest mean scores attached to those subjects with very high family commitment at T1. Despite the fact that the findings for T2 dating intensity fall short of being statistically significant, these analyses as a whole argue that it is those whose T1 family commitment is highest who turn more often and more quickly to more exclusive, and therefore presumably more emotionally intense and intimate, forms of dating. The means presented in the dating behavior tables suggest that, at least generally, as dating intensity goes up, dating frequency goes down. Persons who "go steady," that is, practice exclusive dating, appear to be distributed erratically through the distribution of dating frequency; some date often and some date less often. This is even more the case among those with the highest T1 family commitment, since they are most likely to be "going steady." All of this suggests that casual dating does not serve as well as more intense dating to "replace" the warmth and intimacy of family relationships.

Table 3. Means and One-Way ANOVAs
By Level of Family Commitment: Time 1

	Level of Family Commitment: Time 1				
	Very Low	Low	High	Very High	Sig.
Number of Friends					
Time 1	8.15	9.42	11.02	11.47	ns
Time 2	10.92	11.12	13.56	14.81	.05
Time 3	9.54	11.25	12.97	14.42	.05
Number of Close Friends					
Time 1	3.17	3.59	3.97	4.12	ns
Time 2	3.66	3.09	4.59	4.70	ns
Time 3	3.42	3.65	4.32	4.85	.05
Dating Frequency					
Time 1	2.01	2.21	2.43	2.27	ns
Time 2	2.43	2.53	2.35	2.33	ns
Time 3	2.24	2.71	2.54	2.15	ns
Dating Intensity					
Time 1	2.15	2.75	3.15	3.45	.05
Time 2	2.55	2.57	2.68	2.90	ns
Time 3	2.35	2.80	3.05	3.61	.05
Number of Organizations					
Time 1	Not Available				
Time 2	1.02	1.31	1.39	1.59	.05
Time 3	1.15	1.37	1.42	1.77	.05
Family Commitment					
Time 3	2.56	2.75	2.99	3.28	.01

We did not ask subjects about organizational memberships they may have held prior to attending college, so our measure of number of organizations joined refers only to T2 and T3. The orderly, monotonic increases in mean number of organizations joined from those with very low family commitment at T1 to those with very high family commitment, which are statistically significant at both points in time, repeats a now familiar pattern.

Finally, with respect to our exploration of the data through one-way analysis of variance, it is those whose initial family commitment is very low whose family commitment at a later time, T3, is also low; and it is those whose initial family commitment is very high whose later family commitment is also high. Again, the progression through the categories of family commitment is monotonic, and the finding is statistically significant.

Explorations of the data available were continued through a number of analyses, most taking the form of 2 × 2 analyses of variance, asking whether predeparture data for school friendships or close friendships interacted with predeparture family commitment to affect the development of relationships after moving to the university community. We found no interaction effect in any of the 2 × 2 analyses. (The complete analyses, for the sake of economy, are not presented here.) We did find a significant main effect of number of friends at home, T1, on the number of new friends at T2 ($F = 6.627; p < .01$), with those subjects having more friends at the earlier time making more new friends at the later time. This finding is clearly consistent with the identity theory claim that persons seek to recreate interactional conditions supportive of existing identities when they are removed from the prior relationships from which existing identities derive.

The analysis of a possible interaction effect of T1 family commitment and T1 number of friends on number of new close friends at T2 produces both a significant main effect of T1 family commitment ($F = 8.998; p < .01$) and a significant main effect of T1 number of friends ($F = 4.468; p < .05$). It is those whose earlier family commitment is high who make more close friends later; and, again, it is those with larger numbers of friends at home who have more new close friends later. The latter finding reinforces the meaning attributed to the finding for new friends as a whole; the former finding reinforces the argument that those with close ties to family tend to seek out familylike relationships (in terms of warmth, intimacy, etc.) when immediate links to family are diminished.

In this search for T1 family commitment and T1 friends interaction effects, family commitment at T1 has a significant effect on the number of organizations joined ($F = 5.547; p < .05$), a finding that replicates one presented earlier.

Comparable analyses inquiring into the possibility of T1 family commitment and T1 number of close friends interaction effects on subsequent social relationships produced comparable results. The data exhibit a significant main effect of earlier close friends on T2 number of new friends ($F = 12.693; p < .001$), significant main effects of T1 number of close friends ($F = 8.108; p < .01$), and T1 family commitment on T2 number of new close friends ($F = 14.676; p < .001$), as well as a significant main effect of prior family commitment on later organizational joining ($F = 7.879; p < .01$). The direction of these effects are all the same as indicated earlier for the analyses involving T1 number of friends.

A final set of analyses examined whether distance from home interacted with family commitment at T1 in impacting school-based social relationships. In these analyses, distance from home was dichotomized, creating a low category consisting of persons for whom school is 150 or fewer miles from home ($n = 169$, 52.8%), a high category of persons for whom school is more than 150 miles from home ($n = 151$, 47.2%). In general, the answer to the interaction question is negative, with the one interesting exception. First, however, it should be noted that these analyses produced two statistically significant main effects. One emerged in a 2×4 analysis (the two distance from home categories by the four T1 family commitment categories used earlier); in this analysis, distance from home is related to number of new close friends such that the greater the distance, the larger the number of new close friends ($F = 5.642$; $p < .05$). This finding, as was true of other, earlier findings, is open to an interpretation stressing attempts on the part of persons whose preexisting relationships are cut to recreate such relationships.

The single exception to the lack of interaction effects in the series of analyses of variance completed occurred in an analysis involving dating intensity, in which the second significant main effect also appeared ($F = 9.104$; $p < .01$). The main effect indicates that the greater the distance from home, the higher the level of dating intensity. The interaction effect indicates that the combination of very high family commitment and greater distance from home is particularly potent in producing persons who develop the most intense dating pattern ($F = 13.452$; $p < .001$). Once again, this set of findings fits well a basic identity theory theme argued in this paper, namely, that persons seek to recreate in new social relationships conditions permitting the maintenance of identities created in prior social relationships.

SUMMARY AND DISCUSSION

We began this paper with an interest in the question of whether prior ties to family, under circumstances that made maintenance of those ties difficult, impeded or facilitated the development of new social ties; whether the answer to this question varied by type of new social ties; and whether, if it did vary by type of social ties, we could discern any reason for such variation. In the language of identity theory, our questions concerned the consequences of prior commitments for movement into new social relationships. In particular, our concern was with how prior commitment to family related to the development of new social relationships in a university context, when attendance at the university entailed movement to a location at some distance from home community. Our concern was as well as with the question of how these new social relationships impact subsequent family commitment.

This more focused concern stems from two sources: the sociology of the family and social psychology. The former contains the idea that the contemporary American family, especially the middle-class family, seeks to launch its offspring as independent adults. The latter, particularized in identity theory, lays theoretical ground for understanding the processes by which social relationships are initiated, maintained, renewed, and changed. More specifically, identity theory argues that the salience of role-related identities reflects commitments to social networks in which those identities are played out, and that the salience of role-related identities can be expected to lead to behavior that reinforces existing commitments. The theory further argues, however, that when structural circumstances impede the reinforcement of extant commitments, salient identities will lead to attempts to create new relationships in which those identities can be enacted, and that these new relationships can function as substitutes in this sense for the relationships disrupted by structural circumstances. Finally, the theory argues that "investing" in these new relationships will be followed by decreased commitment to the old relationships.

These ideas were examined in the context of a data set whose subjects, *n* = 320, were students leaving their home communities to attend a residential college. Leaving home communities for school means that existing commitments to family, friends, and dating partners are in some degree disrupted—the degree being, in part, a result of the distance between home and school—and means, further, that opportunities to enter new social relationships are created.

Questionnaire-based measures of commitment to family, friendship patterns, and dating relationships prior to departure for school were taken, with these measures repeated either once or twice in a subsequent four-month period. In addition, the study has available a measure of organizations joined at school and estimates of the distance between home and campus for each subject. Analyses, both in the form of a structural equation model and analyses of variance, relate family commitment, number of friends, and number of close friends, and dating frequency and intensity at a first point in time to number of friends and number of close friends, dating frequency and intensity, and number of organizations joined at a second point in time, and related the T2 measures to family commitment at the third point in time. Gender and distance from home entered the structural equation model as controls. The analyses of variance included one-way analyses of changes in our variables by T1 levels of family commitment, and 2 × 2 analyses intended to discover possible interaction effects on development of new relationships of interactions between T1 levels of family commitment and other T1 social relationships.

The structural covariance analysis clearly indicates that, with gender and distance between home community and the university controlled, the higher the level of *commitment to family* prior to departure for school, the more subjects make new friends and new close friends at school, the more school-based organizations they join, the more they date, and the more intense the dating relationships they enter. Conversely, of course, the less initial commitment to family, the less likely are subjects to form new social relationships. These effects occur in the context of an important degree of stability in family commitment (in the time period studied); that is, there is evidence that higher earlier commitment to family is associated with greater later interaction with other family members, a result predicted in Hypothesis 1, drawn from identity theory.

There is no evidence in this analysis that strong preschool family commitments impede the development of new social relationships; rather, the evidence is consistent with the idea that families whose ties to their offspring are close manage to successfully launch their children into extra-familial relationships, while families whose ties to their offspring are less close are less successful in that task. Thus, the evidence is consistent with Hypothesis 4, the "launching" hypothesis. Further, this analysis also indicates that the more new relationships of any kind are entered, the lower subsequent commitment to family. That is, at the same time that close family ties work to abet the formation of new social relationships, they work to increase children's independence from their family, a finding predicted by Hypothesis 3, also drawn from identity theory. Finally, the structural covariance analysis provides evidence that, along with new relationships generally, persons with an early strong commitment to family seek to enter close or intense social relationships when they leave home for school. They form more close friendships and engage in more exclusive dating relationships than do persons with lower levels of family commitment.

The analyses of variance serve largely to reinforce the conclusions reached through the structural covariance analysis, but also permit some refinement of those conclusions. In particular, that it is the intensity of the dating relationships entered rather than the frequency of dating at school that most closely relates to levels of prior family commitment requires consideration. First, this finding suggests that subjects seek to maintain identities reflecting the very warmth and intimacy of their family relationships by finding new relationships also based on warmth and intimacy. Second, this suggestion implies that high levels of early family commitment do not translate indiscriminately into high levels of other forms of social relationships; rather, the consequences of high commitment to family appear to be at least somewhat particular to relationships that reproduce the affective quality of close familial ties.

The analyses intended to seek out possible interaction effects did not, in general, reveal any. The single exception occurred in the impact of distance from home and family commitment on dating intensity and indicated that the greater distance-greater family commitment combination is especially strong in producing new social relationships that can approximate the warmth and intimacy of family ties. These interaction-effect analyses also make a point relevant to identity theory: the claim that persons seek to reproduce, in new settings, relationships that allow for the enactment of previously established identities receives support from the findings that relate numbers of new friends and new close friends to numbers of friends and close friends in home communities.

It must be remembered that the processes under discussion occur under particular structural conditions which make it at least somewhat difficult to maintain prior closeness to family and which also open up possibilities for new relationships. Thus, the basic identity theory model of the relationships between commitment (and, by implication, identity) and subsequent role-related behavioral choices is supported by the evidence deriving from the structural covariance analysis. There are two important provisos that need be asserted, however. The first is that structural location is a strong conditioner of the nature of those relationships: movement some distance from home represents a structural impediment to the maintenance of "old" relationships and creates opportunities for the development of new relationships. The second proviso is that the culture-based content of the norms of social units of which persons are members must be taken into account in understanding the possibilities inherent in the commitment-salience-role choice model. The latter proviso reflects the success of the "launching" hypothesis and, consequently, the import that can be attached to the reasoning underwriting that hypothesis. That reasoning incorporates the presumption that cultural norms leads (especially) middle-class parents to roles and identities stressing obligations to move offspring out of the family nest to live as independent adults, and leads offspring to roles and identities incorporating the reciprocal obligation to so move.

The "substitution" hypothesis presented and examined in this paper has its origin in identity theory. The theory asserts the motivational character of identities. It argues that persons will seek to produce or to reproduce relational contexts which permit behavioral expressions of salient identities rooted in high levels of commitment to social networks incorporating roles related to the identities. (The theory argues reproduction under permitting structural circumstances; it argues production of "substitute" relations under structural circumstances impeding reproduction.) It may be, at least in some part, that the movement of persons into new relationships is a product of socialization. In the present context, it may be that families in which offspring develop high levels of commitment are families which socialize their offspring in strategies and tactics easing entry into new social relationships. It may also be, again at least in part, that offspring in families to which they are highly committed develop strong "dependency needs" satisfied only in equivalent (to family) relationships, and these needs have little to do with the imperatives of identities. The manner in which such possible accounts of our findings relate to, as well as the degree to which they constitute alternatives to, an identity theory account are the stuff of future theoretical consideration and empirical research. Pending that future consideration and research, we believe the identity theory account that has been provided warrants the credence we have given it.

Accepting this last assertion, there are, it would appear, three major conclusions that can be reached on the basis of the theoretical and empirical analyses presented in this paper. First, identity theory-based predictions regarding the formation of new social relationships under the structural conditions described stand up well, thus lending support for the underlying identity theory arguments themselves. Second, the data are consistent with the longstanding argument, from the sociology of the family, that families play a major role in successfully launching their offspring as independent actors. Third, and juxtaposing these first two conclusions, identity theory arguments based on purely formal relations of commitment, salience, and role performance must incorporate the culture-based content or meanings of social roles and identities premised on these roles to enable sensible predictions of social outcomes.

NOTES

1.   The predominantly female composition of our sample may reflect gender-related biases in recruitment into beginning psychology classes. As noted, differences between male and female subsamples in a variety of social characteristics were minimal; and, in any event, the reported analyses control for gender. We did not obtain information on families' socioeconomic status (SES), and so could introduce no controls for SES into analyses. However, the undergraduate population in the university from which the sample was drawn is relatively homogeneous with respect to SES as well as other characteristics, for example, ethnicity: that population is largely middle-class and white. Too, a sample drawn as we have drawn ours can be expected to be even more homogeneous in these respects than a general sample drawn from the university at large. While on theoretical grounds we would not expect major deviations in the fit of the models estimated across samples of students from diverse class or ethnic backgrounds, it is certainly possible that such deviations could occur. In particular, it may be that the "launching"-based hypothesis will hold better in the white, middle-class segment of the American population than elsewhere.

2.   The data reported on various measures are for males and females combined. We initially estimated a groups model assuming "process" differences between males and females. The groups model indicated that there were two significant differences in the structural coefficients out of a possible seven between males and females. However, the magnitude of the differences was quite small and the effect on the level of significance of the overall model was trivial. All other structural coefficients could be constrained to be equal. The fit of model using sex as a control was equal to the fit of the groups model.

3.   Family commitment could have been treated as a latent construct rather than a scale in this analysis. After an examination of the empirical information, the decision was made to present the most parsimonious model. First, the reliability coefficients for the scales were high and the correlation between family commitment at T1 and T3 was only .344. The concern here lies in the problem of either an overestimate or underestimate of the stability coefficient in the covariance model. The model which estimates family commitment as a latent construct presents no additional information and the difference in the stability coefficients is trivial, with the full model estimating the stability coefficient as .567, as opposed to .559 reported here. Consequently, we do not believe that failing to correct for measurement error in the family commitment variable has attenuated the stability coefficient.

4.   Figure 1 presents all of the estimated coefficients in the model except for those estimated in the measurement matrix for the observed variables at T2. These include the diagonal and one correlated error term. The following are the estimates of the diagonal: number of new friends = .743; number of new close friends = .691; number of organizations = 0 (as a count of

organizations, this variable is assumed to be measured with no error); dating frequency = .409; dating intensity = .473, and family commitment at T3 = 0. The only correlated error estimated in this model is between number of new friends and number of organization. This coefficient = .322.

5. There were no clear breaks in the distribution of T1 family commitment scores to guide the categorization process. Our aim in categorizing subjects was to permit a focus on reasonably homogeneous extreme categories while at the same time retaining a sufficient number of cases in every category to permit sound statistical analysis. Subjects, obviously, were not assigned randomly to family commitment categories. It is conceivable that location in given categories reflects a social desirability effect that is also present in reports of dating, friendships, and organizational memberships. We deem this highly unlikely given, for example, variation in relationships between family commitment and close versus all friendships, as well as variation in relationships between family commitment and dating patterns.

REFERENCES

Ainsworth, M. D. S. 1985. "Patterns of Infant-Mother Attachments: Antecedents and Effects of Development." *Bulletin of New York Academy of Medicine* 61: 771-791.

———. 1989. "Attachments Beyond Infancy." *American Psychologist* 44: 709-716.

Bretherton, I. 1985. "Attachment Theory: Retrospect and Prospect." Pp. 3-35 in *Growing Points of Attachment Theory and Research,* edited by I. Bretherton and E. Everett. Monographs of the Society for Research in Child Development. New York.

———. 1990. "Open Communication and Internal Working Models: Their Role in the Development of Attachment Relationships." Pp. 57-113 in *Nebraska Symposium on Motivation, 1988: Socioemotional Development,* Vol. 36, edited by R. Dienstbier and R. A. Thompson. Lincoln, NE: University of Nebraska Press.

Corsaro, William. 1985. *Friendship and Peer Culture in the Early Years.* Norwood, NJ: Ablex.

Merton, Robert K. 1968. *Social Theory and Social Structure,* 3rd Ed. New York: Free Press.

Parsons, Talcott. 1964. *The Social System.* New York: Free Press.

———. 1965. "The Normal American Family." Pp. 31-50 in *Man and Civilization: The Family's Search for Survival,* edited by S. M. Farber, P. Mustacchi, and R. H. L. Wilson. New York: McGraw-Hill.

Serpe, Richard T. 1987. "Stability and Change in Self: A Structural Symbolic Interactionist Explanation." *Social Psychology Quarterly* 50: 44-55.

Serpe, Richard T. and Sheldon Stryker. 1987. "The Construction of Self and the Reconstruction of Social Relationships." Pp. 41-66 in *Advances in Group Processes,* Vol. 4, edited by E. J. Lawler and B. Markovsky. Greenwich, CT: JAI Press.

Simon, Robin W. 1992. *Spouse, Parent, and Worker: Gender, Multiple Role Involvement, Role Meaning, and Mental Health.* Unpublished doctoral dissertation. Indiana University.

Stryker, Sheldon. 1968. "Identity Salience and Role Performance: The Relevance of Symbolic Interaction Theory for Family Research." *Journal of Marriage and Family* 30: 558-564.

———. 1980. *Symbolic Interactionism: A Social Structural Version.* Menlo Park, CA: Benjamin-Cummings.

———. 1987. "Identity Theory: Developments and Extensions." Pp. 89-104 in *Self and Identity: Psychosocial Perspectives,* edited by K. Yardley and T. Honess. London: Wiley.

———. 1992. "Identity Theory." Pp. 871-876 in *Encyclopedia of Sociology,* edited by E.F. Borgatta and Marie L. Borgatta. New York: Macmillan.

Stryker, Sheldon and Richard T. Serpe. 1982. "Commitment, Identity Salience, and Role Behavior." Pp. 199-218 in *Personality, Roles,and Social Behavior,* edited by W. Ickes and E. Knowles. New York: Springer-Verlag.

_____. 1983. "Toward the Theory of Family Influence in the Socialization of Children." Pp. 47-74 in *Research in the Sociology of Education and Socialization,* Vol. 4, edited by A. Kerckhoff. Greenwich, CT: JAI Press.

Walters, J., R. Connor, and M. Zunich. 1964. "Interaction of Mothers and Children from Lower-Class Families." *Child Development* 35: 433-440.

Zunich, M. 1961. "Lower-Class Mothers' Behavior and Attitudes Toward Child Rearing." *Psychology Reports* 29: 1051-1058.

Advances in Group Processes

Edited by **Edward J. Lawler, Barry Markovsky, Cecilia Ridgeway,** and **Henry A. Walker,** *Department of Sociology, University of Iowa*

REVIEWS: "A major impression one gets from this volume is that far from being dormant, the social psychology of groups and interpersonal relations is quite vibrant, and very much involved with compelling problems."

"Concerns about the imminent demise of group processes as an area of study in social psychology are clearly exaggerated. But should doubts remain, they ought to be allayed by the range and quality of the offerings in this volume."

". . . should be of interest both to specialists in group processes and to sociologists who are interested in theory, particularly in theoretical linkages between micro and macro analysis. Because many of the papers offer thorough reviews and analyses of existing theoretical work, as well as new theoretical ideas, they are also useful readings for graduate students."

— *Contemporary Sociology*

Volume 9, 1992, 280 pp. $73.25
ISBN 1-55938-516-2

Also Available:
Volumes 1-8 (1984-1991) $73.25 each

Advances in Research and Theories of School Management and Educational Policy

Edited by **Samuel B. Bacharach,** *Department of Organizational Behavior, Cornell University* and **Rodney T. Ogawa,** *School of Education, University of California, Riverside*

This new annual review concerns itself with organizational behavior in schools and school districts. Specifically, the review will attempt to expand the field of educational administration by integrating the perspectives of management theory, organizational sociology, organizational psychology, industrial psychology, organizational behavior, human resources management and industrial relations. Theoretical, research and policy papers are invited. Summary or review papers are especially encouraged. Among the early papers received, are papers dealing with such topics as: organizational theory, school management, collective bargaining in education, peer evaluation, teacher career development, etc.

Volume 2, 1993, 271 pp. $73.25
ISBN 1-55938-253-8

CONTENTS: Editor's Introduction. Building Commuinty: The Influence of School Organization and Managment, *Clarie Smrekar.* **Black Boxes, Contingencies and Quasi—Markets: A Theoretical Analysis of Cooperation Among Educational Organizations,** *Patrick F. Galvin.* **Adoption Revisited: Decision Making and School District Policy,** *Kyla L. Wahlstrom and Karen Seashore Louis.* **Consequences of Reform: Retention Rate Fluctuations in Dade County,** *Don R. Morris and Majorie K. Hanson.* **Patterns of Constraint and Opportunity: The Policy Environment of Three Coalition Schools,** *Mary Ann Raywid and Terry L. Baker.* **Building School Capacity for Effective Teacher Empowerment: Application to Elementary Schools with at—Risk Students,** *Henry M. Levin.* **Methodological Issues in the Multi-Level Analysis of School Environment,** *Jean Stockard.* **Toward a Multilevel Conception of Policy Implmentation Processes Based on Commitment Strategies,** *Kenneth A. Leithwood, Doris Jantzi, and Byron Dart.*

Also Available:
Volume 1 (1990) $73.25

55 Old Post Road # 2 - P.O. Box 1678
Greenwich, Connecticut 06836-1678
Tel: (203) 661-7602 Fax:(203) 661-0792

Research in the Sociology of Organizations

Edited by **Samuel B. Bacharach,** *New York State School of Industrial and Labor Relations, Cornell University*

Associate Editors: **Peter Bamberger,** *Bar Ilan University,* **Pamela Tolbert,** *Cornell University* and **David Torres,** *University of Illinois, Chicago Circle*

REVIEW: *"Research in the Sociology of Organizations* is designed as an annual review of current work related to organizations. The editor apparently identifies individuals with major scholarly programs related to organizations and invites them to summarize their work. All six chapters in this, the second volume, clearly reflect several years of work by the authors. All are intellectually independent; other than emphasizing something about organizations, they have little in common."

— *Contemporary Sociology*

Volume 11, 1993, 297 pp. $73.25
ISBN 1-55938-462-X

CONTENTS: Preface, *Samuel B. Bacharach.* **Organizational Theories of the Labor-Managed Firm: Arguments and Evidence,** *Raymond Russell.* **Organizational Control and Informal Relations: The Paradoxical Position of Middle-Status Workers,** *Marcia Ghidina.* **Group Standards and the Organization of Work: The Effort Bargain Reconsidered,** *Randy Hodson.* **Loosely Coupled Organizations Revisited,** *Richard Ingersoll.* **Societal Coniderations in the Global Technological Development of Economic Institutions: The Role of Strategic Alliances,** *Richard N. Osborn and C. Christopher Baughn.* **Mental Piecework: Explaining Occupational Variation in Work Attitudes in a High-End Serice Sector Firm,** *Toby L. Parcel and Robert L. Kaufman.* **Flexibility in Work and Employment: The Impact on Women,** *Vicki Smith.* **Caucasians and Asians in Engineering: A Study in Occupational Mobility and Departure,** *Joyce Tang.* **Organizational Transitions: Form Changes by Health Maintenance Organizations,** *Douglas R. Wholey and Lawton R. Burns.* **Biographical Sketches of the Contributors.**

Also Available:
Volumes 1-11 (1982-1993) $73.25 each

Research on Negotiation in Organizations

Edited by **Roy J. Lewicki,** *College of Administrative Science, The Ohio State University,*
Blair H. Sheppard, *Fuqua School of Business, Duke University* and **Max H. Bazerman,** *Graduate School of Management, Northwestern University*

Volume 4, In preparation, Spring 1994
ISBN 1-55938-555-3   Approx. $73.25

CONTENTS: Preface, *Roy J. Lewicki, Blair H. Sheppard, and Robert Bies.* **PART I. GENDER. Getting Into the Conversations: Feminist Thought and The Transformation of Knowledge,** *Jean O'Barr.* **Preschool Negotiators: Linguistic Differences in How Girls and Boys Regulate the Expression of Dissent in Same-Sex Groups,** *Amy Sheldon and Diane Johnson.* **The Gender-Based Foundations of Negotiation Theory,** *Barbara Gray.* **PART 2. CONTEXT. Team Work: Linguistic Models of Negotiating Differences,** *Anne Donnellon.* **Negotiation in Context: Nobody Trusts Anybody Nowadays,** *Philip H. Mirvis.* **Engineering Consent in Formal Organizations,** *William G. Scott.* **Toward an Improved Framework for Conceptualizing the Coflict Process,** *Walter R. Nord and Elizabeth M. Doherty.*

Also Available:
Volumes 1-3 (1986-1991) $73.25 each

FACULTY/PROFESSIONAL *discounts are available in the U.S. and Canada at a rate of 40% off the list price when prepaid by personal check or credit card and ordered directly from the publisher.*

JAI PRESS INC.
55 Old Post Road # 2 - P.O. Box 1678
Greenwich, Connecticut 06836-1678
Tel: (203) 661-7602   Fax:(203) 661-0792